5

Beyond Belief: Randomness, Prediction and Explanation in Science

Editors

John L. Casti

Institute of Econometrics, Operations Research, and System Theory
Technical University of Vienna
Vienna, Austria

Anders Karlqvist

Royal Swedish Academy of Sciences
Stockholm, Sweden

CRC Press
Boca Raton   Ann Arbor   Boston

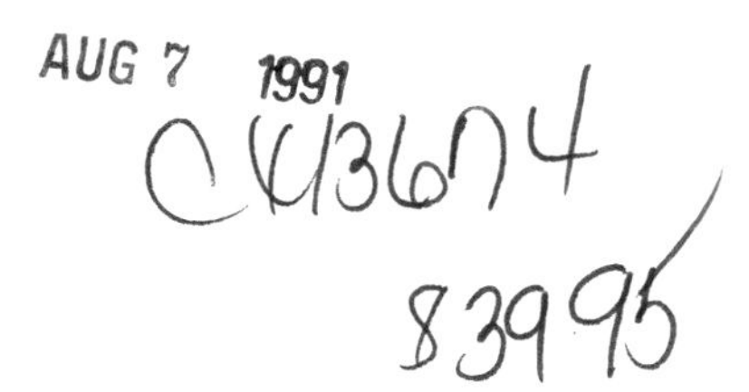

Library of Congress Cataloging-in-Publication Data

Beyond belief : randomness, prediction, and explanation in science / editors, John L. Casti, Anders Karlqvist.
p. cm.
Papers of participants at an annual workshop, held at the scientific research station of the Royal Swedish Academy of Sciences in Abisko, in May 1989, and organized by the Swedish Council for Planning and and Coordination of Research.
Includes bibliographical references and index.
ISBN 0-8493-4291-0
1. Science--Philosophy--Congresses. 2. Science--Methodology--Congresses. 3. Random variables. 4. Forecasting. 5. Explanation. I. Casti, J. L. II. Karlqvist, Anders. III. Sweden. Forskningsrådsnämnden.
Q174.B48 1990
502.8--dc20 90-15005
CIP

This book represents information obtained from authentic and highly regarded sources. Reprinted material is quoted with permission, and sources are indicated. A wide variety of references are listed. Every reasonable effort has been made to give reliable data and information, but the author and the publisher cannot assume responsibility for the validity of all materials or for the consequences of their use.

Direct all inquiries to CRC Press, Inc., 2000 Corporate Blvd., N. W., Boca Raton, Florida 33431.

International Standard Book Number 0-8493-4291-0

Library of Congress Card Number 90-15005
Printed in the United States

Preface

Beginning in 1983, the Swedish Council for Planning and Coordination of Research has organized an annual workshop devoted to some aspect of the behavior and modeling of complex systems. These workshops are held in the scientific research station of the Royal Swedish Academy of Sciences in Abisko, a rather remote location far above the Arctic Circle in northern Sweden. In May 1989, during the period of the midnight sun, this exotic venue served as the gathering place for a small group of researchers from across a wide disciplinary spectrum to ponder the twin problems of explanation and prediction—scientific style. The focus of the workshop was to examine the degree to which the science of today is in a good position to give convincing answers to the question: "Can we predict/explain the phenomena of nature and everyday life?"

In light of the extremely stimulating presentations and discussions at the meeting itself, each participant was asked to prepare a formal written version of his view of the meeting's theme. The book you now hold contains those views, and can thus be seen as the distilled essence of the meeting itself. Regrettably, one of the meeting participants, Itamar Procaccia, was unable to prepare a written contribution of his very provocative views due to the pressure of other commitments. Nevertheless, we feel that the eleven chapters presented here, spanning as they do fields as diverse as philosophy, physics, biology, and economics, give an excellent overview of how science stands today in regard to the foundational issues of prediction and explanation.

It is a pleasure for us to acknowledge the generous support, both intellectual and financial, from the Swedish Council for Planning and Coordination of Research (FRN). In particular, the firmly-held belief in the value of such theoretical speculations on the part of FRN Secretary General, Professor Hans Landberg, has been a continuing source of encouragement. Moreover, the ideas, guidance, and unflagging enthusiasm of Uno Svedin, also of the FRN, has contributed mightily to the success of these workshops. Finally, special thanks are due to Mats-Olof Olsson of the Center for Regional Science Research (CERUM) at the University of Umeå for his skills in attending to the myriad administrative and organizational details that such meetings necessarily generate.

August 1990

John Casti, Vienna
Anders Karlqvist, Stockholm

Contributors

William A. Brock—Department of Economics, University of Wisconsin, Madison, WI 53706, USA

John L. Casti—Institute of Econometrics, Operations Research, and System Theory, Technical University of Vienna, A–1040 Vienna, Austria

Brian Goodwin—Department of Biology, The Open University, Milton Keynes, MK7 6AA, UK

Peter T. Landsberg—Faculty of Mathematical Studies, University of Southampton, Southampton SO9 5NH, UK

Kristian Lindgren—Physical Resource Theory Group, Chalmers University of Technology, S–412 96 Göteborg, Sweden

Erik Mosekilde—Physics Laboratory III, The Technical University of Denmark, DK–2800 Lyngby, Denmark

Bertram G. Murray, Jr.—Department of Biological Sciences, Rutgers University, New Brunswick, NJ 08903, USA

R. T. Pierrehumbert—Department of Geophysical Sciences, University of Chicago, Chicago, IL 60637, USA

Otto Rössler—Institute for Physical and Theoretical Chemistry, University of Tübingen, D–7400 Tübingen 1, FRG

Robert Rosen—Department of Physiology and Biophysics, Dalhousie University, Halifax, Nova Scotia B3H 4H7, Canada

Rainer E. Zimmermann—Friedenauer Salon, Baumeisterstrasse 4, D–1000 Berlin 41, FRG

Contents

CHAPTER 10

CHAPTER 11

Introduction

John L. Casti and Anders Karlqvist

1. Prediction, Explanation, and the Rule of Law

Prediction and explanation are the twin pillars upon which the scientific enterprise rests. The whole point of science, along with religion, mysticism, and all of science's other competitors in the reality-generation game, is to make sense somehow of the worldly events we observe in everyday life. Thus, the whole point of the practice of science as science, is to offer *convincing* explanations for these events, as well as to back up those explanations with accurate predictions about what will be seen next. But what do we mean by a "scientific" prediction and/or explanation? In what way does it differ from predictions and explanations generated by employing the tools and techniques of religion, for example?

While the long answer to the foregoing query would take us too far afield in a brief essay of this sort, the short answer is rather easy to give. Scientific explanation and prediction is explanation and prediction *by rule,* or as it's sometimes put, *by law.* To take a simple example, the observed fact that copper is a good conductor of electricity and silicon is not, is explained by invoking the rule that chemical elements or compounds having a lot of easily detached electrons in their outer shell can use those electrons to conduct heat and electricity. On the other hand, the rule states that those elements like silicon that do not have an abundance of such "free" electrons are not good conductors. The situation is similar with regard to prediction. For instance, we can accurately predict that there will be a full moon on October 30, 1993 by appealing to the laws of celestial mechanics, essentially Newton's laws of gravitation and motion. And, in fact, when a rule like Newton's proves useful enough over a long enough period of time, it's often enshrined in the lexicon of science by being upgraded to a "law," viz., Newton's *law* of gravitation.

At this juncture a neutral skeptic might well object, saying that the other reality-generation schemes noted above are also based on rules. For example, many brands of mysticism involve various sorts of meditation exercises, exercises which their adherents claim will result in a kind of self-realization and enlightenment. It is then claimed that this enlightenment, in turn, enables one to at least explain, if not predict, the unfolding of everyday events. And just about everyone is familiar with Sunday-supplement seers, stock market gurus, and other types of visionaries who predict events of the coming year on the basis of crystal-ball gazings, tarot card readings, the interpretation of astrological configurations, the outcome of the Super Bowl, divine inspirations, and/or other procedures that might charitably be thought of as being based on the following of a set of rules. So in what way *exactly* do these rule-following schemes part company from what is customarily thought of as being a "scientific" procedure?

Basically, there are two properties tending to distinguish scientific rule-based schemes for prediction and explanation from their competitors. The first is that scientific schemes are *explicit,* i.e., the rules and the way they are to be applied are spelled out in sufficient clarity and detail that they can be used by anyone. In short, it doesn't require any special insight or inspiration to make use of scientific laws or rules. Of course, it may take years of training and a laboratory full of equipment to actually follow the rule. But what it doesn't take is any sort of private interpretation of what the rule means or how it is to be followed. Roughly speaking, such a set of rules can be encoded in a computer program, and the prediction and/or explanation encapsulated in the rules can be obtained by running this program on a suitable computer.

The second distinguishing characteristic of scientific rules is that they are *public.* Unlike many religions and other belief systems, there are no private truths in science. The laws of science are publicly available to all interested investigators. And, in fact, it's a necessary part of the scientific enterprise that the rules underlying any scientific claim for prediction or explanation be made available so that anyone who wishes to may test the claim using the same set of rules. This is the essential content of the testability of claims, one of the fundamental tenets of the ideology of science.

While there is a certain degree of explicitness and public availability of the rules underlying many of the belief systems noted above, it is in science that these features are most clearly seen. And it is these key features of science-based procedures for explaining and predicting that the reader should keep uppermost in mind as we proceed through the chapters in this book.

2. A Synopsis of the Book

Robert Rosen opens our quest for what science can know for sure by looking at the validity of Church's Thesis outside the restricted realm of theoretical computing. In loose language, Church's Thesis states that anything that can be obtained by following a set of rules can be obtained as the result of running a computer program on a universal Turing machine. The claim that it is possible to extend Church's Thesis to the world of nature and man, at large, constitutes the theoretical underpinning for the what's sometimes termed "strong-AI," i.e., the possibility to build a computer to think like a man. Rosen considers the plausibility of this claim, using Gödel's Theorem as a guide, concluding that the likelihood of Church's Thesis being true outside computer science is very low.

Still treading the narrow path between philosophy and natural science, in Chapter 2 Rainer Zimmermann examines some of the consequences of the anthropic cosmological principle. Advocates of this principle have used it to argue that many of the results of modern cosmology, like the expanding universe, can be deduced from the assumption that *Homo sapiens* occupy a privileged position when it comes to the kinds of phenomena that we can observe. Opponents claim that the principle is tautologous, inconsistent or, usually, both. In his contribution, Zimmermann shows that most of the inconsistencies in anthropic reasoning can be removed, once an integrative view is taken that abolishes the separation of man as subject and nature as object.

Moving from philosophy into more mainline physics, Chapter 3 by Peter Landsberg studies the problem of explanation in physics by means of the investigation of several important experiments (e.g., the falling of a body) and physical theories. From consideration of these experimental and theoretical results, Landsberg concludes that what is to count as an explanation in physics is intimately tied up with such things as the background knowledge of the investigator, the invention of new general concepts, and the mutation of reality itself.

A detailed look at the relativity principle of the 18th-century physicist Roger Boscovich constitutes the primary thrust of Otto Rössler's arguments given in Chapter 5. In this study Rössler shows how Boscovich's results prove that whether the observer is subject to internal motions or whether the world is are completely equivalent propositions. Thus, Boscovich's principle of relativity forms a more general theory of relativity than the far more celebrated arguments advanced by Einstein more than a century later. Rössler indicates how these Boscovichian claims pose serious questions as to the degree to which humans will ever be able to see the world as it "really is."

A central role in the problem of predictability of any dynamical process is played by the distinction between mere uncertainty or complication and genuine randomness. Chapter 5 by Kristian Lindgren gives a detailed account of some of the recent fascinating work on complexity and randomness in the behavior of discrete dynamical systems. After a review of the basic formalism, Lindgren shows the equivalence between n-dimensional cellular automata and $(n+1)$-dimensional models in statistical mechanics. Taken as a whole, this chapter provides a wealth of information about what it really means to be "random," as well as giving a number of ideas about how to use these various notions of randomness to tease out the deep structure of many physical processes.

Theoretical arguments and philosophical speculation are certainly two avenues for exploring the limits to predictability and explainability; following the tortuous path of real application is another. Chapter 6 by R. T. Pierrehumbert goes down this road in the exploration of that most challenging of applications—weather and climate forecasting. Focusing on the problem of computing the correlation dimension of the atmospheric attractor from actual observations, Pierrehumbert ultimately concludes that while there is considerable evidence for the existence of exploitable structure in the atmospheric equations, the rate at which predictability is lost itself turns out to be an unpredictable quantity!

In Chapter 7 we leave the realm of the so-called "hard sciences," turning to the life sciences with a thoughtful consideration by Bertram G. Murray, Jr. of the employment of the hypothetico-deductive method in biological modeling. Specifically, Murray explores the reasons why biologists have not been as successful as physicists in developing unifying, predictive, explanatory theory. Adopting the dichotomy of scientists as being either "diversifiers" or "unifiers," Murray claims that biologists are primarily the former and physicists the latter. This claim leads to what Murray sees as the differences in explanatory and model-building style between the physicists and the biologists.

The great *terra incognita* of modern biology is the problem of development. In Chapter 8 Brian Goodwin takes up the question of whether an organism's form can be understood as the result of a generative process with generic properties. The dominant view in biology today is that biology is a historical science, species being individuals and not types. If so, Goodwin argues, then biological form is not intelligible in dynamical terms and biology cannot be regarded as a rational science. The results of this chapter suggest that biological forms are indeed generated by dynamical attractors in morphogenetic space, so that biological forms may well be amenable to prediction and explanation of a scientific type.

Turning from the biological to the psychological sciences, Chapter 9 by Erik Mosekilde, Erik Larsen, and John Sterman describes an experiment with human decisionmaking behavior in a simulated microeconomic environment. The experiment involves asking participants to operate a simplified production-distribution chain to minimize costs. The authors show that performance is systematically suboptimal, and in many cases the decisionmakers are unable to secure stable operation of the system. As a consequence, large-scale oscillations and even chaotic behavior ensues. These results provide direct experimental evidence that chaos can be produced by the decisions of real people in simple managerial situations.

A problem of continuing interest is the question of whether it's possible to employ the techniques and tools of modern dynamical system theory and statistics to outperform other investors in speculative markets. This query is the theme song of Chapter 10 by William Brock. The chapter surveys some of the new methods of time-series analysis that are good at detecting low-dimensional chaos. After the development of these and related nonlinear correlation techniques, Brock considers the statistical issues raised by evaluating superior predictive performance of stock market technicians. One of his main conclusions is that too many of the methods used to model financial time series are context free, and that knowledge of where the series being modelled came from will have to used to do any better than the current procedures.

The final chapter by John Casti again takes up the problem of chaos in dynamical processes, but from the novel perspective of its relevance to the problem of mathematical truth. Folk wisdom has it that the presence of chaos in a dynamical process ensures that that process can never be completely predicted on the basis of observational data alone. Using arguments based on the equivalence between formal logical systems, Turing machines, and dynamical processes, in Chapter 11 Casti shows that while it may indeed be the case that chaos is a barrier to ever reaching the ultimate mathematical truth, it's also true that a world *without* chaotic phenomena would be one very impoverished in truths to begin with.

Taken as a whole, the contributions in this volume give what we think is a pretty good picture of the degree to which the science of today is in a position to predict and/or explain natural and human phenomena. But this picture is more like an Impressionist painting than a high-resolution photograph, and we can expect and hope for considerable sharpening of the details as time goes on. Hopefully, the material presented here will serve as a stimulus to that on-going effort.

CHAPTER 1

What Can We Know?

ROBERT ROSEN

Abstract

Through a series of transmutations, we argue that the question "What can we know?" is contingent on the answer we give to another question: "Is Church's Thesis true or false?". We then suggest a way to approach the latter question, using Gödel's Theorem as a guide.

1. Introduction

My assignment is to deal with the question "What can we know?", with the boundary condition that I preferably not exceed 45–60 minutes in doing so. However, I enjoy little challenges like this, and I will try. I will begin by imparting, at the very outset, one piece of absolutely certain knowledge, even more certain than the Cartesian Cogito: I shall not succeed in answering this question.

At this point, then, I could justly conclude my talk, by having imparted this piece of absolutely certain knowledge, and within the allotted time. However, I also know, though with much less certainty, that something more is expected of me; perhaps building a philosophy on my little grain of certainty, as Descartes did on his; or perhaps explaining why the paradoxical nature of this particular certainty precludes our knowing anything else. I will do neither; instead, I will try to persuade you that the question I have been posed is not really the right question.

Before undertaking this task, I should probably confess that I have no real credentials for talking about these matters at all. I am not by profession a philosopher; not an epistemologist or a metaphysician (although many of my colleagues may disagree). I am rather a theoretical biologist, hence a theoretician. My activities make me a consumer of philosophical concepts; a utilizer. At the same time, to the extent that I am successful in what I do, I may provide raw material for the true philosophers themselves to contemplate. These facts give me a somewhat different perspective than that of real philosophers, and such a shift in perspective can sometimes be illuminating.

Why, then, do I say that the question "What can we know?" is not the right question? I say this because an answer to this question would have to take the form of a list; a set. I have already given you a specimen from this set. I do not have to tell you how I came by this set of knowable things; why something is on the list or not; I merely have to produce the list itself. "What can we know?" "We can know these things"; "these things can be information" (cf. Rosen, 1988).

The fact is that no one would want such a list even if it were correct; no one would believe the list, or could believe it, except possibly on mystical or theological grounds. Indeed, it is clear at the outset that the very assertion of such a list runs right into a version of Russell's Paradox. Thus, insofar as the question "What can we know?" is properly answered by "We can know this, and this, and this," it is the wrong question.

Furthermore, what is this feature of "knowability" that characterizes the putative elements of such a set? Does it pertain to "we"; what can **we** (as opposed to someone or something else) know? Or does it inhere in a "knowable" thing itself; **what** can we know? Or both? Does it express a passive property or an active relation?

In what follows, I am going to replace the original question, "What can we know?" with another, more limited perhaps, but also more congenial; namely, "If we know something, what else do we know thereby?" We thus shift the ground completely; we are not talking any more about knowledge or knowability, but about entailment.

Now entailment is something I do think I know a little about, and indeed something I have already discussed at some length (cf. Rosen, 1985a). In the form of causal entailment, it is manifested in the external world of phenomena. In the form of inferential entailment, it underlies the world of formalisms, of mathematics. What I have called Natural Law (cf. Rosen, 1985a) allows us to compare systems of entailment; to attempt to bring them into some kind of congruence; i.e., to make models.

I am going to argue that "what we can know" about a system of entailment is embodied, in a precise sense which I hope to make clear, in the totality of its models. Some of "what we can know" about such a system pertains to one or another of its specific models; other things pertain to relations between such models; still others to the properties of the set (or better, the category) of all of its models.

Let me call attention at the outset to some features of the discussion to follow. First, as I have already said, the ground has been shifted away from some abstract principle of "knowability" to something more concrete; to what we can know about something; in this case, about a system of entailment. This kind of shifting is something I have had to do repeatedly; I have had to replace an abstract concept of adaptation with the prior concrete specification of the individual whose survival is favored thereby (Rosen, 1975); likewise, I have had to replace the abstract concept of cause with a prior specification of the effect which is to be entailed thereby (Rosen, 1985b). So it is here, too.

Another thing to notice is that I will, on the other hand, be talking about entailment as a thing in itself, divorced from any specific modifying adjectives like "causal" and "inferential." To talk about entailment in the abstract in this fashion is a somewhat radical thing to do. For example, it involves a higher level of abstraction than does mathematics itself. For any particular mathematical system, however formal it may appear, merely instantiates entailment, just as any particular set, say, instantiates the concept or abstract idea of "setness." However, as I hope to indicate through the course of the present discussion, despite its radical character, it seems a useful and necessary thing to do.

In fact, we hope to learn about entailment in the abstract precisely by considering its specific embodiments in mathematics and in the external world. To resuscitate an old and disreputable word, we can say that entailment itself is *immanent* in these embodiments, and hence we can learn something about it by studying them. Why else should a biologist (I am one) believe he can learn about life by studying organisms, or a chemist (I am not one) believe he can learn about chemistry by studying this or that molecule or reaction? Or, to come back to my subject, why should we believe that we can learn about knowing or knowability through particular knowables? The relation embodied in these beliefs is not that of general to particular, but rather is precisely an assertion of immanence.

Although this word "immanence" is coming to interest me more and more, and will in fact permeate the discussion below, I will not use it in what follows. I will merely point out that, for instance,

$$\text{SUBSET} : \text{SET} \neq \text{SET} : \text{“SETNESS”}$$

The relation on the left itself instantiates that of particular to general; the relation on the right instantiates the concept of immanence. They are obviously not the same, and the one on the right is much more interesting than the other one. But it would take us too far afield to pursue this here, and would certainly violate my imposed 45–60-minutes boundary condition.

2. Numbers and Number Theory

Since the time of Pythagoras at least, numbers have been considered the most absolutely certain things we can know, and entailment in the context of numbers is the most reliable kind of entailment there is. So let us look here first, using numbers and their entailments as a kind of parable or allegory for the questions with which we are concerned. That is, we will consider first the manifestation of entailment in this particular, venerable context, and then, via the modelling relation, export what we learn to other realms.

One of the reasons Number Theory is so reliable is that we can, like gods, literally create the numbers from nothing. A mathematical Descartes could force himself to doubt the existence of anything in mathematics, just as the real Descartes could doubt every fact of experience. But whereas the real Descartes concluded from this only that he existed, the mathematical Descartes could conclude something else; namely, that the empty set $\emptyset$ exists. It exists unconditionally, unfalsifiably, in defiance of anything and everything else, triumphantly. And this is enough.

If $\emptyset$ exists, as it must, then another set must exist; namely, the set $\{\emptyset\}$ whose only element is the empty set. But this set is not empty; it has an element in it. Give it a name; call it 1. If nothing exists, then 1 must exist; in Mathematics, if nowhere else, the very existence of nothing already entails the existence of something.

But now we have two things; nothing and something; $\emptyset$ and 1. Then a third thing must exist; namely,

$$\{\emptyset, \{\emptyset\}\} = \{\emptyset, 1\}$$

Give it a name; call it 2. Likewise a fourth thing must exist, and then a fifth, and then a sixth, etc. I believe it was Kronecker who said that the integers were given by God; all else is the work of Man. But as we see, God literally didn't have to do anything at all; Man could do it all by himself.

Number Theory is an attempt by Man to discover the properties of his creation; to know what further is entailed by these things, these integers that he has called into existence. For by calling the integers into existence, individually and collectively, we have also, as a kind of side-effect, created at the same time an enormous number of entailments that relate them. And even beyond that, we have created entailments that relate these entailments, and so on.

What these entailments may be, we do not know, at least not right away. We must discover them, and we must demonstrate that they are in fact entailments; we must embody them in theorems, and provide proofs of these theorems. A proof in this sense constitutes an explicit chain of entailment, carrying us from the creation itself to what we assert the creation has entailed.

3. Formalizations

The 1930s, however barren they may have been economically and politically, were excessively fertile years for science. Three things happened in those years that are pertinent for us. Let us review them briefly and then put them together.

First, Nicolas Rashevsky was in the process of constructing the first real theory of the brain, as part of his more general program of constructing a true "mathematical biophysics". He started by constructing a theory for peripheral nerve excitation; what he called an excitable element. Then he turned this into a theory of nerve conduction, by stringing excitable elements in series. Finally, he realized that a series of such elements was only the simplest kind of network that could be assembled from them, and that very simple networks could already do interesting, brainlike things. Thus, what started as a theory of peripheral nerve excitation culminated as a theory for the whole central nervous system. A decade later, Rashevsky's student Walter Pitts, together with Warren McCulloch, paraphrased these ideas in digital, Boolean form.

Second, that unique Englishman Alan Turing was independently developing the concept of a "mathematical machine," which now bears his name. In those days, the word "computer" meant a human being who was paid to do arithmetic. Turing, for various reasons which we shall discuss below, wanted to know what such a computer could in principle compute; his machines were in this sense "abstract computers," hence also to that extent abstract brains. But unlike Rashevsky's neural nets, they were simulations rather than models.

However, the most weighty event for our purposes had actually occurred in 1931. In that year, Kurt Gödel published his celebrated Incompleteness Theorem, which said something very important about Number Theory. In the rest of this paper we will be concerned with what he said about it, and with how, in conjunction with the work of Rashevsky and Turing, it bears not only on Mathematics but on the brain of the mathematician; on the creation and the creator simultaneously.

To put all these ideas into a proper context, we must realize that both Gödel and Turing were primarily motivated by a philosophy of mathematics itself, called Formalism, which had been developed earlier in the century by David Hilbert. Hilbert was (aside from Henri Poincaré) the most powerful mathematical mind of his time, and he was grappling with the notion of consistency. The almost unthinkable idea that Mathematics might be logically inconsistent had by 1910 become all too thinkable, partly because of the prior upheavals in geometry, and mostly because of the then newly-discovered "antinomies" in what is now called "naive" set theory.

Hilbert, of course, was convinced that Mathematics was logically consistent, and undertook to do whatever was necessary to make this manifest. This meant, of course, entailing consistency, and from something absolutely unassailable. His way of doing this is embodied in the concept of formalization, which we shall briefly describe.

Hilbert observed that a mathematical term, like the word "triangle," plays two different kinds of roles in traditional mathematics. On the one hand, it is a pattern of letters drawn from an alphabet of meaningless symbols, which appears in geometrical propositions, and is manipulated from one proposition to another according to certain rules. On the other hand, it also refers to something outside itself; namely, to a certain kind of geometric object. A mathematical proposition containing this word "triangle" thus can be said to refer to, or to specify, a quality or property of this external referent.

Thus, according to Hilbert, traditional Mathematics involves two different kinds of truth: one kind pertaining to form alone (e.g., to the word "triangle" as an array or pattern of letters, and how this array is to be manipulated from proposition to proposition), and another kind, which pertains to external referents. The former kind of truth we shall (for want of a better word) call syntactic; the latter kind we shall call semantic. (Hilbert himself used a somewhat different terminology, but I like mine better.)

Hilbert came to believe that all the troubles of Mathematics arose from the allowing of semantic truth into Mathematics at all. Accordingly, Mathematics would be made safe if and only if it could be transmuted into pure syntax, closed off from any and all external referents whatever. He thus came to believe that each and every semantic truth of Mathematics could in fact be faithfully captured by syntactic rules; that what appeared to require an external referent could be internalized by means of more syntax. Formalization embodied this philosophy, and dealt with the ways and means by which it could be accomplished.

In the Formalist view, then, Mathematics becomes identical with what is today called word processing, or symbol manipulation. It is a game of pattern generation, of producing new patterns (theorems) from given ones (axioms) according to definite rules (production rules). Mathematics thus becomes a kind of morphogenesis. What the ordinary mathematician calls Number Theory, according to this view, can be replaced by a completely equivalent formalization, but one which is no longer "about" numbers, or indeed "about" anything. Every entailment pertaining to numbers, every theorem of Number Theory, has a formalistic, purely syntactic counterpart. The virtue is that we can be sure, absolutely certain, that the formalization is consistent. We thus, according to Hilbert, must give up "doing Mathematics" in any ordinary sense, but at least we can be safe.

Hence the interest in "machines," especially the Turing machines. These machines in their behaviors are the ultimate in string processing and symbol manipulation. They thus embody the fruits of Formalization; they are purely syntactic engines. Hence also the excitement when it was realized that the neural networks of Rashevsky, at least in their Boolean version, themselves embodied "machines" or automata of precisely this type. In this picture, the brain itself was a syntactic engine. A fully consistent and satisfying picture, ranging from the biology of the brain to the inmost nature of mathematics itself, could be glimpsed; indeed, a highly seductive picture.

The only question was: Do things really work this way?

4. Gödel's Theorem

Gödel's Incompleteness Theorem devastated the entire Formalist program. In doing so, it said many interesting things, which we may state at the outset. It says that Mathematics is not, and cannot be made to be, identical with word processing or symbol manipulation. It says that Number Theory is already too rich in entailments (or equivalently, that machines are too poor in entailments) for there to be any kind of congruence between them. It says, in a nutshell, that one cannot forget that Number Theory is about numbers. It says that Mathematics, like natural language itself, cannot be cut off from all referents and remain Mathematics.

Basically, what Gödel showed is that, however one tries to syntacticize or formalize Number Theory, i.e., to express it in terms of a finite number of production rules for manipulating formulas or strings of symbols, there are entailments (theorems) in Number Theory having no counterpart in that formalization. In other words, no matter how we try to do it, we cannot capture all Number-Theoretic truth in terms of pure syntax; there is always a semantic residue that is not accommodated. This residue embodies those qualities or features of numbers that have failed to be syntactically internalized; i.e., expressed in terms of alphabetic patterns and rules for manipulation of these patterns. Hence, these are the features which we lose forever if we try to replace Number Theory by that formalization.

In a sense, Gödel's Theorem says that the entailments in Number Theory are of a higher cardinality than the entailments in a formalization (i.e., in a purely syntactic system). That is, no matter how one tries to count off the elements of the former against the elements of the latter, there will always be elements of the bigger set left over; uncounted. This is precisely what was meant when we said earlier that formalizations are too poor in entailment to be congruent to Number Theory.

So what we learn from Gödel's Theorem is that Number Theory, and hence Mathematics itself, is not just syntax. Syntax captures parts or fragments; we can capture different parts and different fragments with different syntax; but syntax alone, unaided, cannot capture all there is. Number theory transcends syntax.

On the other hand, Number Theory does not fail to be Mathematics simply on the grounds that it is not syntax alone. It does not abdicate its mathematical status because it cannot be formalized. It is still a system of entailment; what Gödel showed is that it is one of a different character from any purely syntactic system—just as Cantor showed that a transfinite set is of a different character than a finite one, but does not lose its mathematical status because no finite set can cover it.

5. *Complexity*

We will now take up the relations that exist between Number Theory and its formalizations. It is these relations that will finally let us make contact with the question with which we started, namely, "If I know something, what else is entailed thereby?"

The relation of Number Theory to any formalization of it is the relation of complex to simple. In these very explicit terms, a simple system (of entailments) is one whose entailments form a set of minimal cardinality; a complex system is any other system of entailment. Thus, Number Theory is complex; formalizations are simple. We see also that there are many ways for a system of entailment to be complex; only one way for it to be simple.

We shall argue that a formalization constitutes a (simple) model of that which it formalizes. If, as is the case with Number Theory, the formalization is simpler, then of course no one formalization will suffice; there will be many different simple models. We will look at the totality (the category) of all such models, and see how we may construct complex models from these simple ones. Because of the way we have tied "complexity" to the cardinality of something, we obviously need to invoke processes that go from lower cardinality to higher. In a word, we need to take limits in that category of simple models or formalizations.

For our purposes, the point of doing all this is that we are going to identify this category of simple models of a system of entailment with "what we can know" about it. We will soon transport the relation we have been discussing, between Number Theory and its formalizations, to the external world; to the relation between a material system and its models. When we do this, all of epistemology gets wrapped up in that category of models. "What we can know" about such a system thus becomes a question of how good we are at navigating in this category of models.

6. A Case Study: Church's Thesis

Church's Thesis, which also dates from the 1930's (in fact, from 1936, the same year that Turing published his first paper on machines), began as an attempt to syntacticize a certain property of (numerical) functions or predicates. That property is generally referred to as "effectiveness" or "effective calculability." To say that something is "effective" seems unbearably vague; it is very much like saying that something is "knowable." But Church thought he saw a way to replace that vague, semantic notion of "effectiveness" with an equivalent, precise, purely syntactic one; ironically, by in effect identifying it with syntax itself.

What Church asserted, roughly, was that something is "effective" if and only if it can be accomplished entirely by word processing. In more current language, something is "effective" if and only if we can program a universal Turing machine to do it. Thus, "effective" means "simulable" or "computable."

As I noted long ago (Rosen, 1962), through equivocation on the word "machine," Church's Thesis can be painlessly transmuted into an assertion about the material world itself; an assertion about what can be entailed in the causal world of material phenomena. That is, Church's Thesis can be interpreted as an assertion about the structure of the category of all models of any material system.

To see how easily this transmutation can take place, let me cite once again a quotation I have used before, taken from Martin Davis's influential 1958 book *Computability and Unsolvability:*

> For how can we ever exclude the possibility of our being presented some day (perhaps by some extraterrestrial visitors), with a (perhaps extremely complex) device or "oracle" that "computes" a noncomputable function? However, there are fairly convincing reasons for *believing that this will never happen.* (The emphasis is mine.)

Notice in particular the word "device," which means hardware. Davis is in effect asserting, on the grounds of Church's Thesis alone, the nonexistence of a certain kind of material system, in exactly the same way that, say, the Laws of Thermodynamics exclude perpetual motion machines. That is, Church's Thesis in this context becomes a profound assertion about chains of causal entailment in the material world.

What is that assertion? Explicitly, it's this: Every (mathematical) model of a material system must be simulable. That is to say, every model of a material system must be formalizable; expressible in terms of pure syntax, hence (as we have seen) freeable from all referents; expressible entirely as software (e.g., as a program).

Put in these terms, Church's Thesis assumes truly epic proportions. It entails, among other things, the following:

1. The category of all models of any material system has a unique largest element; a largest model, which already incorporates everything there is to know about that system.
2. That same category has a unique finite family of minimal or smallest elements ("atoms").
3. The largest model, and indeed every other model, is a direct sum of the smallest ones.
4. In that category, direct sums and direct products are naturally equivalent, implying that the objects in the category (the models themselves) are very linear.

Thus, Church's Thesis in the form we have given it entails, by itself, the corpus of contemporary physics; it does so without our having to know any physics. About the only thing I have not been able to show directly from Church's Thesis is that there is a finite, universal set of these "atoms" that uniformly suffices for all material systems.

The demonstration of these assertions cannot be contained in this narrow space. However, unlike Fermat, I have written them down elsewhere, in the manuscript of a monograph now being readied for publication.

The upshot of this discussion is that if we know Church's Thesis, many other things are entailed. Most of them are in fact limitations on entailment itself.

In Mathematics, Church's Thesis does no such damage, because whatever is not simulable is thereby merely relegated to the category of "ineffective." As we have seen, Gödel's Theorem already tells us that, in these terms, almost all of Number Theory is thereby rendered "ineffective." But the material world is different; whatever happens, or can happen in it, must thereby be "effective"; at least, it must be so in any normal usage of that term. Hence the equation "effective" = "simulable," which is the essence of Church's Thesis, boxes us in from the outset to a world of simple systems; a world of mechanisms.

As we have seen, such a world is actually extremely impoverished in entailment (in this case, impoverished in causal entailment). There is not much to know in such a world; not much to be known, even in principle.

7. *The Alternatives to Church's Thesis*

The upshot of our discussion so far is that, as far as the material world is concerned, the question "what can we know?" is intimately bound up with the truth or falsity of Church's Thesis.

If Church's Thesis is true, then every material system is a mechanism; everything is physics; the reductionists are right and science, in effect, becomes identified with a search for syntax.

If, on the other hand, Church's Thesis is false, then there are material systems that are complex; physics as we know it then becomes only the science of the simple; the reductionists are wrong; science itself acquires an ineluctable semantic aspect.

So the basic question becomes: Can we in fact know whether Church's Thesis, as an assertion about causal entailments in the material world, is true or not? And if so, how?

It's hard to answer such a question on the basis of evidence, in view of the elastic nature of syntax. For instance, in Mathematics, if someone comes along with a computer program that generates all presently known theorems of Number Theory, and asserts (in defiance of Gödel's Theorem) that his program syntacticizes all of Number Theory, I cannot refute him by proving a theorem which his program cannot. For he can always enlarge his program; in effect, add another epicycle, and repeat his assertion.

Thus, although I have plenty of evidence that, say, organisms are in fact complex, my evidence cannot compel or entail the falsity of

Church's Thesis. The problem here is very much akin to trying to prove to someone the existence of an infinite set, using only evidence obtained from finite ones, or the existence of an irrational number on the basis of evidence drawn from rationals. Indeed, the problems here are very closely related, because the complexity of a system means precisely that there are more entailments in that system than in any syntactic model of it.

On the other hand, there is likewise no way to establish the truth of Church's Thesis on the basis of evidence either. This fact is completely overlooked in contemporary physics, which is predicated on the belief that simple systems exist.

But all is not lost on this account. Evidence is not all there is. Gödel's Theorem was not proved on the basis of evidence. The Incompleteness Theorem in itself does not directly help us, since it pertains only to certain systems of inferential entailment. On the other hand, the way it was proved is of significance in this regard.

How was it proved? Roughly, by systematically representing assertions about numbers by numbers. In other words, it was proved by establishing numerical referents for such propositions; and conversely, by creating for some numbers (Gödel numbers) a corresponding propositional referent. Such referents obviously involve, from the outset, a semantic feature, and it is precisely this semantic feature that motors the argument.

This provides a hint as to where to look for complexity in the material world. Namely, a proposition about numbers constitutes a kind of model, or part of a model, of numbers themselves. What Gödel did was then like pulling the model inside the system being modelled. He showed, in effect, that there was always something "left over" in the system; that such a model could not be all of the system.

A system containing a model of itself, on the other hand, is precisely the kind of system I have called anticipatory (cf. Rosen, 1985a). And it is precisely in the context of such systems that the Gödelian argument can be reformulated in terms of causal entailment. I would then assert: If anticipatory systems exist, then Church's Thesis is false.

I stated at the outset that I would not answer the question "What can we know?" But I have perhaps provided a hint to how it might be approached. It seems to me that this hint is worth pursuing. In an inherently Gödelian way, it serves to pull epistemology within science, and hence to actually resolve epistemological questions on scientific grounds.

References

1. Rosen, R. 1962. "Church's Thesis and its relation to the concept of realizability in Biology and Physics," *Bull. Math. Biophysics,* 24, 375–393.

2. Rosen, R. 1975. "Biological systems as paradigms for adaptation," in *Adaptive Economic Models,* R. H. Day and T. Groves, eds., Academic Press, New York. pp. 39–72.

3. Rosen, R. 1985a. *Anticipatory Systems,* Pergamon, New York.

4. Rosen, R. 1985b. "Information and cause," in *Information Processing in Biological Systems,* S. L. Mintz and A. Perlmutter, eds., Proceedings of the 20th annual Orbis Scientiae, January 17-21, 1983, Miami, Florida, Plenum, New York, pp. 31–54.

5. Rosen, R. 1988. "Complexity and information," *J. Comput. Applied Mathematics 22,* 211-218.

6. Davis, M. *Computability and Unsolvability,* McGraw-Hill, New York, 1958.

CHAPTER 2

The Anthropic Cosmological Principle: Philosophical Implications of Self-Reference

R. E. ZIMMERMANN

Abstract

Some of the consequences of the anthropic cosmological principle are discussed. The basic evidence that originally gave rise to the anthropic concept is examined, and the example of the existence of extraterrrestrial intelligence is used to illustrate certain inconsistencies in the standard arguments. The chapter then attempts to develop a sound philosophical basis for a theory that integrates man and nature into the totality of the world. It is shown that most of the inconsistencies in the anthropic arguments can be removed, once an integrative view is taken that abolishes the separation of man as subject and nature as object, respectively. The theories of Bloch, Sartre, and Schelling are reviewed within this framework. The more explicit example of Penrose's twistor theory is also considered.

1. Introduction

The problem of explanation is of fundamental importance for man. And it is not only relevant for scientific purposes: The need to explain the perceived phenomena within the daily environment, or in nature, generates the continuing requirement of classifying, of systematic ordering, and of finding correlations and coherences. At the heart of these typically human activities is the need for orientation which arises mainly on account of man's possessing conscious self-reference (of knowing that he knows). In a wide range of fields of human activity, the process of orientation is carried out with a variety of methods and techniques. In fact, human life—as it transcends the need for mere reproduction—is chiefly characterized by permanent aesthetical production in a generalized sense: In the procedure of constituting orientation and establishing a consensus with others about the orientation thus achieved, humans are actually creating their own context (very much in the sense of a writer who creates an alternative world in one of his novels). Humans do this in a variety of ways: somewhat unsystematically and irrationally in daily life, depending mainly on tradition and individual (non-generalized) experiences and sudden intuition, or more systematically, but still in very subjective terms, as in the arts, or extremely systematically relying on very strict principles as laid down in rules and conventions based on logic and causality-based rationality, as in the sciences. Seen from this perspective, the interpretive activities of human life can be represented by an almost continuous spectrum of aesthetical production, ranging from the most formal, strict, and objective focusing of reality down to the least formal, most flexible, and highly subjective ways of description and explanation.

In physics, and in some parts of chemistry and biology, the recent appearance of the theories of self-organization and formation of structure (as they are associated with the names of Thom, Prigogine, Eigen, Mandelbrot, and many others), which are based on the results of nonequilibrium thermodynamics, stability theory, and the theories of nonlinear differential equations, has raised the hope that it will soon be possible to explain in more detail all those very complex processes in nature (and in the cultural domain) that up to now have resisted the efforts of physicists and mathematicians. Interestingly, one of the techniques usually applied in the process of explanation, the explicit prediction of phenomena, has been thrust into the background by these new theories. This is because it can be shown that at the critical points of nonlinear processes, spontaneous fluctuations and deterministic average behavior are of the same order of magnitude, contrary to what one would expect on the basis of classical physics. That is, the stochastic aspect of processes has become a generic property of macroscopic phenomena, not only of quantum phenomena or systems with a large number of components. This renders most statistical predictions useless, other than in those cases in which the goals of the statistical inference are extremely modest.

Most of the traditional instrumentalizing tendencies of science (treating nature as an object that is at man's hand and can be exploited without taking anything else into consideration, other than man's explicit needs) are based directly on notions of predictability, which can be used synonomously with availability. The challenging aspect of the new theories is that they seem to indicate a shift of perspective with respect to this instrumentalization of nature. In fact, it appears reasonable to assume that there is a deep fundamental relationship between the methodology of nonlinear processes and the attitude of research based on this new and reinterpreted epistemological background. This suggests that a new kind of dialogue between man and nature (Prigogine) is made possible when these aspects are explicitly taken into account. In this chapter we will concentrate on investigating the conditions that precede the perception and interpretation of phenomena, both according to the classical and the modern lines of argument. As a starting point for this discussion, we look at the anthropic concept in physics.

What we will do is first is have a look at the anthropic concept itself, as well as some of its immediate consequences, which are then discussed further in Section 2. In Section 3 we examine the basic evidence originally giving rise to the anthropic concept. The example of extraterrestrial intelligence is then used in Section 4 to illustrate certain inconsistencies in the standard arguments. The thesis is put forward that applying the Principle of Mediocrity rather than the Strong Anthropic Principle offers a smoother route to the modeling of the universe. The crucial aspect here is that traditionally man is separated from nature, due to the different paradigmatic domains of science that have been developed over the last two or three centuries. It is shown, however, that the apparent separation of man and nature in physics, as well as the apparent absence of problems associated with the notion of self-reference within this domain, are illusions. Hence, the development of a theory integrating man and nature into the totality of the world, transcending the limits of the language gap between formalized mathematics and substantial hermeneutics, is of fundamental importance in light of the results of the new theories mentioned earlier.

In Section 5 we try to develop a sound philosophical basis for such a theory, starting with the philosophy of nature originally introduced by Ernst Bloch. It can be shown that most of the inconsistencies in the anthropic argument can be removed, once an integrative view is taken that abolishes the separation of man as subject on the one hand, and nature as object on the other. An equivalent for subjectivity must be found that is adequate for nature. Section 6 gives a short review of the first philosophical model that tried to achieve such a unified view: Schelling's philosophy of nature. The parallels to the theories of self-organization are discussed, showing the deep insight and contemporary relevance of Schelling's ideas. The central role of aesthetical production within this context, especially with a view to science and philosophy, is then illustrated. We list a number of criteria for what a future "philosophy of totality" should look like. Finally, Section 7 gives a more explicit example of these ideas as seen in the twistor geometry of Roger Penrose.

2. Formulations: A Critique

In its original form, the anthropic concept was introduced by Brandon

Carter in 1974 when he stated that, "Our location in the universe is necessarily privileged to the extent of being compatible with our existence as observers."[1] This formulation, however, turns out to be not quite satisfactory, insofar as there is apparently no convincing reason for using the term "privileged." This is because it is impossible for us to say that a *nonprivileged* location in the universe (i.e., a *common* location) is **not** compatible with our existence; on the contrary, *a priori* we would not expect that to be the case since we are here.

There is a weaker formulation of the principle, however, which is not at all trivial: "We are a carbon-based intelligent life form around a star of spectral type G2, and any observation we make is necessarily self-selected by this fact."[2] But this does not tell us as much about the initial conditions of the universe—up to any degree of objectivity, provided we can consistently define such a concept—as it tells us about the human means of *perceiving* the universe. This means that in a sense humans create factuality in accordance with some initial projection by *reconstructing* it from its relative future—which is our present situation (and the present stage of the universe)—*into the past.* Though this method is common for daily life strategies, as is well known in the social sciences,[3] it is usually not employed in the physical sciences[4] (except perhaps in the works of Thomas Kuhn),[5] the reason being that "objectivity" is understood differently in physics, where the use of mathematical formalisms seems to offer a fixed (and observer-independent) view of how the world *really is* (at least in the macroscopic domain). On the other hand, the traditional separation of science and philosophy (a gap that's even worse than that separating the so-called *hard* and *soft* sciences), which originated in the 18th-century and is running into serious difficulties today, prevents any possible *rapprochement* of viewpoints.[6]

If applied to individual biographies or any local historical development of social groups, the *regressive* analysis of the successive stages of this reconstructing of reality may reveal the actual situation and choices of acting persons in the past. This can be achieved by gradu-

[1] Carter, 1974, p. 291

[2] Barrow and Tipler, 1988, p. 16

[3] Zimmermann, 1986a, 1988a

[4] Zimmermann, 1986b

[5] Kuhn, 1970

[6] I have discussed the danger of putting formalisms onto the basis of ill-received philosophical concepts in Zimmermann, 1988a

ally undoing the covering devices of the respective reconstruction that originally serve to smooth over the individual's activity (in a *post hoc* manner) with a view to a past that turns out to be its very basis.[7] This founding of the present by means of reconstructing the past is a common social process that can be derived from deep anthropological principles.[8] Results on the actual uncovering of this process can be achieved by hermeneutical techniques, which have been absent up to now in the physical sciences. This absence is due mainly to the fact that it is virtually impossible to develop a unified version of the formal language of mathematics and the hermeneutical *gestalt* languages.[9]

Now note the following fact: In general, the regressive analysis of projects reveals an obvious discrepancy between the actual project undertaken and the factual development which led up to its planning (because of the smoothing effect of the reconstruction). In fact, this discrepancy, once understood, gives a kind of qualitative measure for the indvidual freedom a person or a group has (and uses) when acting. Obviously, projects are undertaken by men themselves, but their meaning is relegated to earlier instances in a past that is more or less fictitious. There is no objective institution that defines a "first original project." Reconstruction by men alone is the necessary way for creating worlds. But it is more important to create the *meaning* of worlds than it is to create *real* worlds. Reconstruction is successful only because it is based on a general social *consensus* that ensures a consensus on the general procedures for producing meaning.[10]

The interesting point is that the production of *scientific* meaning is based on very similar processes. Basically, what is "true" in scientific terms is also decided by nothing more than a *common consensus* in the scientific community.[11] Certain standards that have been accepted according to this consensus constitute both acceptable evidence and method. In principle, this is more a social process than it is a scientific one, independent of the social context of scientists: In their *singular* role as humans, scientists close what Casti has called the "circle of self-reference" by indirectly investigating humanity when directly investigating nature.[12] Hence, the scientific process is neither egalitar-

---

[7]Cf. Zimmermann, 1990

[8]Sartre, 1985

[9]Jantsch and Waddington, 1976

[10]Habermas, 1981

[11]Casti, 1989, p. 10

[12]Ibid, p. ix

ian nor is it even necessarily logical in any absolute sense; but it is certainly meritocratic and ideological.[13] This is because verifiability of claims (independently by others) and peer review (especially the refereeing of papers) define the binding criteria for scientific work within the closed circuit of the scientific community (physicists tell people how the physical world **is**.).

On the other hand, scientists are members of social groups that pursue certain social activities that in turn define the social context of these groups, as well as their explicit cognitive structure. Therefore, group consensus (or partial group consensus within the scientific community, respectively) cannot be independent of external social (and, of course, anthropological) factors. Hence, to say the least, the results of physics and the view of the physical world derived from them, depend strongly on both a *global* and a *local* aspect: The global aspect is the anthropological basis of man's activities, giving a reference frame for the ways of perceiving, acting, and producing meaning. The local aspect is the explicit historical situation of the groups and individuals within which the production of meaning takes place. There is no reason to assume that for carbon-based intelligent life the anthropological details would not be structurally stable. Hence, seen globally, it's reasonable to restrict our discussion here to what we could call *equivalence classes* of the universe modulo this anthropological basis, which then indeed becomes a basis for self-selection. Seen locally, however, we cannot expect that the historical basis is anything other than an irreducible and contingent fact.

Note that this result correlates in a free and easy way with the fact that, technically speaking, all cosmological models represented by spacetime metrics as solutions of Einstein's equations can be determined uniquely only up to an isomorphism, and that we therefore deal with equivalence classes of metrics rather than with unique "individual" metrics.[14]

Comparing now the weaker formulation of the anthropic principle with the original (stronger) version, we find that there are basically two versions of the principle possible, according to whether one favors the "privileged" or the "non-privileged" view, respectively. One ver-

---

[13]Ibid, p. 14, 56ff

[14]Hawking and Ellis, 1973, p. 227ff. This is the basic reason for classifying the solutions of Einstein's equations according to the groups of motions or by algebraic properties (which means classifying according to the algebraic structure of the conformal curvature tensor) that they exhibit. See also Kramer, et al, 1980, in particular Parts II and III.

sion stresses the *particularity* of humans in the universe (introducing concepts of "privilege" from the outset), while the other stresses the *mediocrity,* or nonparticularity, of humans. From the discussion so far, we can conclude the following: If we relate the existence of carbon-based intelligent life to the existence of stars of G2 spectral type, then the empirical evidence about these stars should give us some idea about the distribution of carbon-based life forms throughout the universe. As we limit the discussion to these forms of life (by anthropological reasoning), we should assume that for any such distribution of life forms it's possible to deduce cognitive means by which we perceive the world, act in the world, and produce meaning. By the equivalence class argument, we can then also conclude that some consensus should be possible (about the question of how the world *is*) within the region of the distribution of carbon-based life forms (be they intelligent in some sense or not!).[15]

This conclusion speaks very much in favor of the mediocrity of human beings (provided the discussion is limited to carbon-based life forms and that empirical results establish the mediocrity of G2-type stars). We will show now that using principles of the strong form (i.e., assuming particularity of humans) means *assuming extra physical conditions* that are not empirically present when observing the universe **here and now.**

This can be seen in the following manner: At the basis of modern standard cosmology lies the *Copernican* cosmological principle, which implies that the universe is *isotropic* (mediocrity of direction) and *homogeneous* (mediocrity of location). This principle amounts to saying that physics is basically the same everywhere—and that therefore we cannot occupy a privileged position in the universe. This principle is used (in explicit mathematical terms) when deriving the standard cosmological solutions of Einstein's equations (of Friedman-Robertson-Walker-type).[16] These solutions give the best approximation for a model of the universe as it is at present. A number of tests (such as those concerning the microwave background radiation and the rate of expansion) have not been seriously challenged up to now. Measurements of the microwave background radiation in the universe show that the universe appears to be isotropic to within 0.2%, which is a strong

[15] So the problem is not really to achieve a discourse, once a contact has been established, but rather to find means of being recognized as a candidate for establishing such a contact in the first place. Cf. Casti, 1989, p. 409ff.

[16] Ryan, Shepley, 1975. See also Adler, Bazin, and Schiffer, 1965, p. 338ff, for an introduction.

result in favor of the Copernican principle.[17] Since according to the standard theory the universe is expanding, isotropy and homogeneity are properties that the universe has attained comparatively recently. Most likely the early universe was highly anisotropic and inhomogeneous,[18] but by an entropic process that is not yet clear[19] the universe has been "damped down" to its present state. Hence, standard cosmology uses the Copernican principle and *confirms it empirically;* thus, it also implies the *mediocrity of humans* in the universe. Any principle challenging this view should provide alternatives for the standard cosmology.[20] We should bear this line of argument in mind when now listing the basic anthropic principles.[21]

First of all, we distinguish between a weak and a strong version of the anthropic principle:

WEAK ANTHROPIC PRINCIPLE (WAP). *"The observed values of all physical ... quantities are not equally probable, but these quantities take on values restricted by the requirement that there exist sites where carbon-based life can evolve and ... that the universe be old enough for it to have already done so."*[22]

Note that the time required for stars to produce carbon is roughly equal to the lifetime of a star on the main sequence of its evolution, which is of the order

$$t_* \sim \left(\frac{Gm_p^2}{hc}\right)^{-1} \frac{h}{m_p c} \sim 10^{10} \text{ years},$$

where G is the gravitational constant, while m_p, h, and c refer to the mass of the proton, Planck's constant, and the speed of light, respectively. Hence, the universe must be at least as old as t_*, and at least ct_* in spatial extent.

We see that the WAP is not quite satisfactory, either. It can easily be restated in terms of a trivial fact, namely, that there should

[17] Hawking and Ellis, 1973, p. 348.

[18] Cf. Ryan and Shepley, 1975, p. 57 and, in particular, Zel'dovich and Novikov, 1983.

[19] For a general overview, see Ryan and Shepley, 1975 and Hawking and Israel, 1979.

[20] Peebles, 1971, p. 41. To my knowledge there is no alternative currently available, even taking into acount the criticisms of Peebles.

[21] Here we follow Barrow and Tipler, 1988, pp. 15–23.

[22] Ibid, p. 16.

be someone who can observe (otherwise there is no problem, at all), and that there should be something that can observed (otherwise we would face even more serious problems).

The only firm fact is that we know already that there are observers; hence, the relevant conditions have clearly been satisfied. What we do not know is whether this is so by chance or by systematic design. That is, the existence of carbon-based life and the fact of its having evolved give a *frame of reference* for checking on the consistency of data rather than being a requirement that restricts these data beforehand (unless, of course, one wishes to pursue one of the ancient "design" arguments). The strong version of the anthropic principle is very much formulated with this idea in mind:

STRONG ANTHROPIC PRINCIPLE (SAP). *"The universe must have those properties which allow life to develop within it at some stage of its history."*

This principle is obviously of a teleological nature, and it is comparable with both the *anthropomorphic* approach to nature (personalizing nature by some abstract human-like entity, e.g., a god) and with the *constructivist* approach in the strict sense (man constitutes nature by his activities). Variants of this SAP stress the design argument behind the scenes, this being clearly seen in their actual wording:

Variants of the SAP

(A) PURE DESIGN ARGUMENT. *"There is a universe designed* **with the goal** *of generating observers."*

(B) PARTICIPATORY ANTHROPIC PRINCIPLE (PAP). [23] *"Observers are* **necessary** *to bring the universe into being."*

(C) MANY-WORLDS CONCEPT. *"An ensemble of different universes* **is necessary** *for the existence of the universe."*

The *necessity* being expressed in the formulations of the variants of the SAP points simply toward the fact that the existence of humans (as a fact) is *assumed* to satisfy some goal in the past, this goal being to eventually produce observers. Obviously, these formulations do not address the following point: Certainly a number of physical conditions have been realized in the universe that did result in the development of humans. But there is neither any obvious necessity for our universe or

[23] Patton and Wheeler, 1975. As in cases A and C, the emphasis is mine.

for ourselves to develop, nor is there any hint as to whether a universe should really have the goal of producing observers. If, on the other hand, one wishes to simply state that a universe without any observer would have *no meaning* (as Wheeler has actually done), it would be possible to rephrase all of the SAPs according to our earlier discussion:

For ensuring the meaningfulness of his existence, man extrapolates the present model of the universe into the past in order to reconstruct the history of his becoming man. If not satisfied, he also defines a goal for the universe that later on coincides with his own, and places that goal at the beginning of the universe.

None of these variants has any physical necessity. But all of them have considerable *metaphysical* content. The most recent version of these principles has been formulated by Barrow and Tipler as a sort of *summa* of the aforementioned variants. In fact, this final version makes the design argument even more clear:

FINAL ANTHROPIC PRINCIPLE (FAP). *"Intelligent information processing must come into existence in the universe and, once it (does so), it will never die out."*

This version of the SAP has even a far-reaching conclusion (which is rather fatal, I believe): "Thus ultimately life exists in order to prevent the universe from destroying itself!"[24] This simply means that life is a sort of controlling device for ensuring the stability of the universe once it has been established. We can note the following:

(i) There is no qualitative difference between the FAP and the SAP or any of its variants. There is again a strong teleological structure in the argument.
(ii) It's difficult to understand why it should be necessary at all to ensure the stability of the universe (except for ourselves, of course).
(iii) Stability of the universe in the sense of Barrow and Tipler implies a cosmological model that is "open" (i.e., expanding forever). Neither is this assumption very satisfactory from the *Big Bang* point of view, nor has it yet been empirically verified (because the "mass problem" is not yet definitively settled).
(iv) According to the FAP, the development of the universe should be a direct function of the behavior of intelligent life (or its form of information processing). This is not a realistically tenable view (in fact, this is the view favored by astrologers).

[24]Barrow and Tipler, 1988, p. 674.

Before we come back to the design of principles, we first take a short look at the empirical evidence. We then give an example to clarify the caveats of argumentation before pursuing an alternate line of discussion.

3. Looking for Evidence

It is an established fact that the coarse-grained features of equilibrium states in the universe arise principally as a result of the actual numerical values of a small number of dimensionless constants.[25] Table 1 displays these *structure constants,* which determine a kind of fine tuning of the universe. In principle, these constants ultimately derive from the coupling parameters of the physical fields, i.e., they give a measure for the strengths of interaction among the elementary physical fields.[26] Consequently, we end up with the four basic coupling parameters (for the electromagnetic, gravitational, weak, and strong interactions, respectively), together with a cosmological term giving a measure of the evolution and structure of the large-scale geometry of the universe.

The unexpected feature of the set of structure constants is that they must remain structurally stable in order for the universe to be globally stable. That is, a small perturbation (less than, say, 10%) in the numerical magnitude of **any one** of these constants would destroy the overall balance of interactions in the universe. In fact, it can be shown explicitly that most of the processes leading up to the formation of matter as we observe it today could not even be initiated under these circumstances.[27]

A very general mathematical result provides a rigorous background for this conclusion. According to a 1971 argument by Ruelle and Takens, chaotic (though structured)[28] behavior of dynamical processes is more common than permanent stability.[29] If we visualize the universe as a dynamical system, say of the form

$$\dot{x} = F(x, \lambda); \quad x = (x_1, x_2, \ldots, x_n),$$

where x is a set of physical quantities (or, equivalently, a point in phase space), and λ is a set of structure constants,[30] then Ruelle and Takens

---

[25] Ibid, p. 367.

[26] For a general review see Gasiorowicz, 1966, p. 86ff or Bjorken and Drell, 1964, 1965, especially 1965, p. 90ff.

[27] Barrow and Tipler, 1988, chs. 5 and 6.

[28] Prigogine, 1986, p. 24, Nicolis and Prigogine, 1987, p. 57ff, p. 74.

[29] Ruelle and Takens, 1971.

[30] Barrow and Tipler, 1988, p. 254ff, p. 272ff.

Constant	**Quantity**
fine structure	$\alpha = e^2/\hbar c \sim 1/137$
electromagnetic coupling	$\beta = m_e/m_p \sim 1/1836$
gravitational coupling	$\alpha_G = Gm_p^2/\hbar c \sim 10^{-39}$
weak coupling	$\alpha_W = g_W^2/\hbar c \sim 10^{-13}$
strong coupling	$\alpha_S = G_S^2/\hbar c \sim 15$
cosmological structure	$\Omega_0 = \rho_0/\rho_c \sim G\rho_0/H_0^2 \sim q_0$

Table 1. Structure Constants and Fine Tuning. Here e is the charge of the electron, m_e is the electron mass, m_p the mass of the proton, G the gravitational constant, $\hbar$ is Planck's constant divided by 2π, c is the velocity of light, while g_W and g_S are the coupling parameters for the weak and strong interactions as determined by the Salam-Weinberg theory. Cosmological structure is defined by ρ_0, the universe's measured average density *at present,* and by ρ_c, the critical density which determines whether the universe is expanding or contracting. Alternatively, this can be expressed by q_0, which is the acceleration rate of the present expansion, or by an expression involving H_0, the Hubble parameter, which currently has the value $75 \pm 25 \text{km}^{-1}\text{Mpc}^{-1}$; $0 \leq q_0 \leq 2$. We have a flat, open, or closed geometry of the universe according to the values of these parameters:

flat geometry:	$k = 0$,	$q_0 = 0.5$,	$\Omega_0 = 1$
open geometry:	$k = -1$,	$q_0 < 0.5$,	$\Omega_0 < 1$
closed geometry:	$k = 1$,	$q_0 > 0.5$,	$\Omega_0 > 1$

Here k is the Gaussian curvature parameter from the Friedman-Robertson-Walker metric.

have shown that when $n \geq 3$, solution curves $x(t)$ will be unstable to changes in λ when these changes exceed some critical, *small* value. The conclusion is twofold:

1. The system is not chaotic at the outset. In this case, a small change in one of the components λ_i will result in a catastrophic change in the solution.
2. The system is chaotic from the start. Then there may be many similar sets of λ that are equally stable (including combinations allowing for life)—this is the argument of the *many-worlds theory.*[31]

There is still more evidence for this "universal fine tuning." For

[31] See Everett, 1957, DeWitt and Graham, 1973. For a more recent review, see Barrow and Tipler, 1988, p. 472ff.

example, the mass-size diagram shows that on a logarithmic scale the ratio of the mass of objects in space to their radius is a linear relation. Thus, all objects from the proton on up to stars, galaxies and the universe itself lie on a straight line (humans taking a position in the center, by the way).[32] Similarly, the well-known Hertzsprung-Russell diagram[33] exhibits localizable structures such as the "main sequence" of stars, within which the complete evolution of stellar matter takes place.

It is possible to interpret this evidence in two ways, according to either the *stability picture* (the world is such that only certain types of structure are permitted to exist for long periods of time), or the *selection picture* (some types of structure are not detectable by man). Obviously, the first picture is a purely physical argument, stating factually how the world has evolved. The second appeals to the weak anthropic principle, and involves human properties of perceiving that lie outside the physical domain. In accordance with traditions of rationality and Occam's Razor, it appears reasonable to choose the first rather than the second interpretation, especially since we have no conclusive evidence as to the abilities of humans in regard to their individual or collective means of perception (we'll come back to this argument in Section 6).

The fine tuning itself, however, is very strict. In the case of the Hertzsprung-Russell diagram, for example, it can be shown that planetary formation is closely associated with the development of convective stars and their relatively low angular momentum, implying that stellar angular momentum is lost during the process of planet formation and now resides in the orbital motion of the planets. This result may be obtained from the requirement of the stability of stars and planets, such as the gravitational argument put forward by Carter[34]: The numerical coincidence, namely, that

$$\alpha_G \approx \alpha^{12}(m_e/m_p)^4$$

up to a very small factor (of order of magnitude 10^{-39}) characterizes the dividing line between convective and radiative stars of mass M_*, which is the minimum mass necessary for hydrogen ignition:

$$M_* \geq \eta^{3/4}(\alpha/\alpha_G)^{3/2}(m_p/m_e)^{3/4} m_p \approx 10^{33} \text{ grams},$$

[32] Barrow and Tipler, 1988, p. 290.

[33] Ibid, p. 337. See also, Scheffler and Elsässer, 1974, p. 73 and p. 83. The diagram displays the luminosity of stars divided by their temperature on a logarithmic scale.

[34] cf. Barrow and Tipler, 1988, p. 335ff.

where η is a material parameter. In fact, $\eta(H)$ is approximately 0.025 for hydrogen combustion. According to an anthropic conclusion, the fact that α_G is just a bit larger than $\alpha^{12}(m_e/m_p)^4$ ensures that the more massive main sequence stars are radiative, while the smaller members of the main sequence are almost all convective. Note again that this formulation is due to the weak anthropic principle, and states actual facts only. The numerical coincidence ensures the validity of a condition that is particularly relevant for the formation of planetary structures. Therefore, it provides a reason for the processes involved, but **not** a meaning. Nor does it give any hint of a way that this fine tuning might be initiated in accordance with a plan or goal.

Similarly, from M_* alone it's possible to deduce consistency relations for planets, characterizing certain parameters such as a typical "medium" temperature of 470° Kelvin for thermal equilibrium, a typical distance between planet and star of around 1.5×10^{13} cm, and so forth, from which similarly significant values may be deduced—like the typical year and the typical day—by arguments involving purely graviational considerations.[35]

4. ETI Arguments

There is no doubt that the evidence of "universal fine tuning" is a significant result in its own right. But as such it does not really point to any mysterious coherences in the world. On the contrary, it reminds one more of well-known results of nonequilibrium thermodynamics on which the formation of structure is based.[36] Therefore, arguing in favor of a teleological structure of the universe is quite arbitrary in view of the fact that fine structure balancing is quite common for many evolutionary processes. For examining more closely the features of arguments of this type, it's of value to discuss one of the anthropic aspects in more detail.

A prominent example is the problem of extraterrestrial intelligence (ETI), a topic that has been raised within the framework of projects such as OZMA, CYCLOPS, and other SETI radiotelescope programs,[37] which are mainly associated with the names of Drake and Sagan.[38] In contrast to the opinion about ETI leading to the research activity of the early 1960s, in recent years the tendency has been to come back

---

[35] Barrow and Tipler, 1988, p. 335ff.

[36] Nicolis and Prigogine, 1977.

[37] Barrow and Tipler, 1988, p. 577.

[38] Drake, 1960, Shklovskii and Sagan, 1966.

to a strictly anthropocentric view.[39] Barrow and Tipler, for example, argue in favor of an absolute density of ETI in the universe of one per galaxy (which rules out any civilization in the Milky Way other than our own).

To begin with, let's note the following: All those who support the argument of Barrow and Tipler would actually like to show that there are no other civilizations in the *universe.* But what they actually do is estimate the number of ETI *for our galaxy* (by means of the Drake equation, which will be discussed in a moment). They then invoke what is basically a space-travel argument to show that this low estimate of one ETI per galaxy is confirmed by empirical evidence (saying that if there were ETI somewhere, they would have arrived here on Earth by now).[40]

But this line of argument is inconsistent: An estimate of one ETI per galaxy (leaving aside for a moment whether it is plausible or not) would be equally true for **any** galaxy in the universe, therefore giving approximately 10^{12} civilizations in the universe! Although the distances (that is, the "mean free path" between any two planets having civilizations) are considerably increased, this conclusion would contradict the space travel argument, since then one would equally well expect (after having suitably expanded the timescale for transport and/or communication and reduced the relative frequency of direct contact) that "they" would have already arrived. Conversely, if those who follow the Barrow and Tipler argument preferred to state that the distances are **too large** for civilizations to establish any contact in a reasonable amount of time, then this would *again* contradict their original statement ("Where are they?")[41], this time by answering their question directly, since by again adjusting the timescale and frequency, one would have to conclude that already within our own galaxy distances are quite considerable, and the probability for "a hit" is very low.

Thus, we see immediately that anthropic arguments (especially when involving any of the strong versions of the principle) run into inconsistencies. This is principally because the introduction of additional aspects into the discussion means leaving the purely physical evidence behind and distorting the interpretation by creating a new, nonphysical

---

[39] In fact, in West Germany well before 1980 Breuer of the Munich school had already publically (though not correctly) argued in favor of an open universe and the strong anthropic principle. This led him to the view that humans are the only intelligence in the universe.

[40] Cf. Casti, 1989, p. 411ff.

[41] Fermi, as quoted in Casti, 1989, p. 340.

context for the original evidence. On the other hand, the *mediocrity* view is far more plausible: ETI being almost equidistributed over the universe, but meeting infrequently because the mean free path separating it from others is too large. This is actually the original line of argument taken in the 1960s.[42]

The immanent inconsistency of anthropic arguments can also be seen when looking at the Drake equation itself, upon which most of the estimates of the number of ETI are based. This equation[43] gives the relative frequency for ETI within a galaxy.[44] A common version of the Drake equation has the form

$$p = f_p n_\ell f_\ell f_i f_c,$$

where p stands for the relative frequency of intelligent life that attempts interstellar communication. The factors f_p, f_ℓ, f_i, and f_c refer to the relative (estimated) frequencies that a given star will have planets, that, if so, life evolves on a habitable planet, that intelligence then develops, and that this intelligent life form attempts interstellar communication within 5×10^9 years after planet formation. The quantity n_ℓ represents the number of habitable planets, provided a stellar system possesses any planets at all.[45]

Casti[46] gives a slightly modified form of the equation which is basically equivalent, only that a factor R^* is added representing the anuual rate of star formation in a galaxy.[47]

The main difficulty with this equation is that f_c depends mainly on sociological assumptions. Usually the criteria for estimating f_c are based on historical experience gathered from the exploration and/or colonizing of Earth! Clearly, it's very unlikely that the motivational

[42]See Ref. 38.

[43]Sagan, 1973.

[44]In fact, data are actually adapted to the Milky Way, so that it is our own galaxy with which we are concerned here.

[45]Barrow and Tipler, 1988, p. 586ff.

[46]Casti, 1989, p. 344.

[47] In my opinion, the inclusion of the rate of star formation is not quite plausible, because we can expect that it will result in a small value for p, further tending to reduce the final number of ETI. But what we want is an estimate of the number of ETI, not a prognosis. Hence, for the present number of ETI, the rate of star formation should be negligible. On the other hand, the number of G2 stars is already taken into account by the factors f_p and n_ℓ. This is because of the fine tuning needed to obtain a habitable planet at all.

structure and the means by which ETI explores interstellar space is at all comparable to what has been used by humans when exploring the Earth. Under this assumption, all possible variations in the level of historical processes, mentalities, group behavior, and individual socialization are neglected. The same is true for the time scale assumed to be necessary for exploring the galaxy, since this is also very much dependent on the ideas a civilization has arrived at about exploring in the first place. Since f_c strongly influences p,[48] it appears to be more plausible to dispense with this term altogether. This would decisively increase the value of the product $f_\ell f_i f_c$, estimated by Barrow and Tipler as 10^{-10} (p then referring to intelligent life without saying anything about interstellar communication). More precisely, since f_c refers to the probability that ETI attempts communication after a certain period of development, dispensing with f_c means that it is left completely unclear whether ETI attempts communication at all and, if it does, whether it stops trying after a while. Stopping interstellar communication and/or exploration after say a couple of thousand years (which is a lot in historical terms, cf. the development of space technology) would considerably reduce the number of contacts, even if f_c were almost 1! Because we do not know which of these courses to take with regard to f_c, Occam's Razor suggests we restrict our attention to physical and biological arguments only.

What about biological arguments? Usually f_ℓ and f_i are also taken to be small, the idea being that it is rather unlikely that life like ours will develop very often.[49] But given that there is a habitable planet somewhere, we can rely on the line of argument introduced by Prigogine:[50] The "order through fluctuation" concept says that methods of equilibrium thermodynamics are inadequate to remove the apparent contradiction between the biological order of complex structures and the (entropic) laws of physics. As Prigogine has shown, the time necessary to produce a comparatively small protein chain of around 100 amino acids by *spontaneous* formation of structure is much longer than the age of the Earth. Hence, spontaneous formation of structure is ruled out. And so is *equilibrium* statistics. This means that according to the modern theory of self-organizing systems, classical arguments concerning the "coincidental" realization of a complex living system cannot be employed. On the contrary, powerful autocatalytic and self-organizing cy-

[48]Barrow and Tipler, 1988, p. 589.

[49]Cf. Casti, 1989, p. 407 for the case of random DNA formation.

[50]Nicolis and Prigogine, 1977, p. 23.

cles of production come into play once a certain initial eco-scenario has been established during a stage of evolution. And this is in fact another point mentioned by Casti:[51] It's quite unlikely that the relative frequencies in the Drake equation are independent of one another. Hence, a product form for p doesn't appear plausible. On the contrary, the product should be replaced by a more complex expression $F(f_p, f_\ell, f_i)$, one that is presumably nonlinear (and, unfortunately, completely unknown at present). Because any substantial value for f_p (it is probably about 0.1 according to the frequency of G2-type stars in our galaxy) means a lot of fine tuning and matching conditions in the universe, and because the transition from nonliving matter to living and from life to intelligent life is subjected to autocatalytic processes of the aforementioned type, it is to be expected that the existence of habitable planets alone already implies many details of the evolution that follows from these conditions. Or, put another way, f_p influences $F(\cdot)$ nonlinearly. Presumably, once f_p exceeds a minimal critical level, F will do the same (connecting a local and a global argument by means of the fine tuning).

Table 2 shows a number of estimates for p, both by those who favor the strong anthropic view and the idea of man's special role, and by those who favor man's mediocrity. The first row gives the estimate of Barrow and Tipler, one that is strongly on the anthropic side. Note that we have left out the estimate for f_c except in the case of Barrow and Tipler. But since it is not possible to determine f_c alone from their text, instead we insert the overall value 10^{-10} into the table. The other estimates basically favor the mediocrity view, and range from the very optimistic picture of Shklovskii and Sagan to an amalgamated view taken from a variety of sources published in the 1960s.[52]

Factor	f_p	n_ℓ	f_ℓ	f_i	$f_\ell f_i f_c$	p	$E(N)$
Barrow & Tipler (1988)	0.1–1	1	–	–	10^{-10}	10^{-10}	1
Shklovskii & Sagan (1966)	1	1	1	0.1	–	0.1	10^9
Hart (1980)	0.2	0.1	0.1	0.5	–	0.001	10^7
"Amalgamated" (1960s)	0.1	1	0.1	0.1	–	0.001	10^7

Table 2. Values for p from the Drake equation. Here p is defined to be the relative frequency of planets having intelligent life, while $E(N)$ is the estimate of the absolute number of ETI civilizations, where $E(N) = pN^*$, N^* being about 10^{10} stars per galaxy.

[51]Casti, 1989, p. 344.

[52]See Ref. 38.

Referring to standard astrophysical values,[53] we can fix the number of galaxies at around 10^{12}. About 55% (5.5×10^{11}) of them are of type S_b/S_c, which is basically the same as that of our own galaxy. Thus, there are about 10^{10} stars, 10% of which are of type G2, per galaxy.

There is still another point we should bear in mind: So far we have limited our discussion to estimating the possible number of carbon-based intelligent lifeforms in the galaxy. Because of the trickiness involved in getting a handle on the factor f_c, we have dispensed with it. We have also eliminated the question of communication and/or direct visitation. What we now find is that the universe appears to be heavily populated. But this does not tell us anything about the probability of meeting ETI or communicating with them. We have already mentioned the problem of the mean free path between galaxies or between any two planets in a galaxy, and the problem of ETI (or ourselves) stopping exploration one day for one or another reason. Under these conditions, we can expect to be able to communicate with ETI only when its evolution takes place in a period roughly comparable (and therefore accessible) to our own historical period (or periods). Hence, the number of possible contacts is dramatically decreased by the fact that only within a period of about $\pm 2.5 \times 10^5$ years bracketing our own historical time frame can we expect that communication would be possible at all. It is thus not so much the question of whether communication as such is possible[54] (we can settle this with the principle of mediocrity mentioned above), but whether ETI's evolutionary cycles roughly coincide with our own!

What do we actually learn from this example of the ETI problem? We see clearly that two levels of argument are mixed when invoking strong anthropic concepts for application to detailed questions. Those who argue in favor of strong anthropic concepts and in favor of the special position of humans show that *in principle* no intelligent life could have developed. But they take the fact that at least one form of intelligent life (namely, ourselves) has indeed developed, as a signal

---

[53] Scheffler and Elsässer, 1974.

[54] Casti, 1989, p. 412ff. See also the "chauvinisms" listed more generally on p. 363ff. Note that besides the problem of having the mentioned "gross features" of culture adjusted among each other, we are left with the problem of interpreting the information offered and attaching a certain meaning to it. But this problem (as Casti himself emphasizes) is one which is not really new for us, because we encounter it already everywhere on Earth itself. This general problem of achieving an adequate hermeneutics which ensures a minimum level of communication is clearly a universal one.

for some kind of vocation which is to be associated with this seemingly impossible fact: Man has the task of giving a meaning to the universe, and by this activity he ensures the universe's stability (for eternity, in fact). From this conclusion they derive the anthropic principle, using it as a post hoc foundation for their surprise. And while always arguing strictly within the physical domain, they often mix arguments taken from the metaphysical domain thus established. By this problematic procedure they give a hazy philosophical foundation to physical facts, rather than taking into account from the outset the possibility that man *as observer and participant* cannot be separated from nature in the first place.

5. Self-Reference and Knowledge

The true problem is indeed based on this misjudgment of the world's totality. This problem is the traditional separation of the components of the world that cannot be separated without losing their synthetic power of integration (which is mandatory for their understanding): The fact is that man is not only *produced* by nature, but at the same time produces nature as well. This observation lies at the heart of what self-reference is all about. So coming back to our original question (on the use and purpose of anthropic principles), we should try to discover what the anthropic concept can offer to this view of the world as a self-referential totality, of which nature and man are two categories of reciprocal domains capable of being visualized as *forms of extracts* (*Auszugsgestalten*) in the sense of Bloch.[55]

The traditional separation of man and nature in modern science is due mainly to the fact that science appears to be more easily understood because the objects of nature can apparently be observed "from the outside," while social objects are always obscured by the deficiency of distance.[56] But this is an illusion; in reality, there is no such distance between man and nature. Although it's possible to locally instrumentalize physical processes on a comparatively objectified and thus generally acceptable level, we have already seen that explicit results are typically arrived at *by consensus;* that is, by a *subjective* process of constituting opinion.[57] Of course, this is a process based on

[55] Bloch, 1985e, p. 68. Cf. Raulet, 1982, p. 183.

[56] Bloch, 1985e, p. 18.

[57] Kuhn has discussed explicit examples for this in his 1970. Cf. Thom, 1979, p. 8: "I think that the physical laws exist, essentially because of the fact that physicists can and have to communicate between themselves."

a *collective,* not on some individual. But we are nevertheless confronted with the difficulty that everything seems to start with man, although we know—and we can prove this by empirical evidence—that man appeared on Earth a long time after its formation (and that the universe came into existence a long time before the very first civilization could possibly have arisen). This is one of the chief aporetic difficulties that we cannot solve within the framework of this paper, since it touches on the very foundations of several branches of philosophy.

What we *can* do, however, is concentrate on the further clarification of this interactive dialectic of man and nature, starting from sober physical facts (as established by the weak form of the anthropic principle), taking care not to eventually mix them with remote, quasi-philosophical impulses. Hence, according to this perspective, we find the following: Evidence shows nothing more than that there is in fact a very detailed balancing of structure constants in the universe that determines its explicit form as a necessary and sufficient condition for the evolution of intelligent life. It is *necessary* because it is plausible that a small variation in the structure constants causes a catastrophic change of structure in the universe. As these special conditions do appear to be less likely—in view of the fine tuning and the many-worlds arguments discussed above—than those that avoid the formation of life, the implication is that any change in the structure of the world would mean *no life* rather than the realization of some other, equally unlikely, life form. It is *sufficient* because we note that indeed life has developed, and that the free play of possible variations leaving the present structure of the universe invariant is far too small to suggest a small shift to conditions that are structurally stable, yet empty of life. Note that the above does not hold just for **our** solar system; there is no privileged region in the universe in the sense that life would preferentially develop in them (according to the mediocrity view adopted here). The cosmological principle assures us that there is an equidistribution of possible locations for evolving life, since the universe at its present stage is homogeneous and isotropic in the large. So what we should do is rephrase the WAP in a modified form, taking this view into account. We call this the "Realistic Anthropic Principle (RAP)":

REALISTIC ANTHROPIC PRINCIPLE (RAP). *(1) There may be a set of possible universes, each of which might have been realized at the initial singularity. This set must contain at least one element.*

(2) There is a subset of universes in this set of possible universes (with at least one element) that is determined by a finite number of

structure constants. This subset admits the evolution of intelligent life.

(3) Once a universe is realized belonging to the aforementioned subset, the research activity of emergent intelligent life, once undertaken, is cognitively adapted to the frame of reference of the scenario provided by that universe's structure.

The final point calling for clarification is one relating the RAP to self-reference. This is reflected in the fact that *observation restricts knowledge,* in the sense that any observation includes a biased self-observation (the observer is always actor and participator in the whole system), and it is not possible to comprehend both of these constitutive components completely. This is in fact what Gödel's Incompleteness Theorem and Turing's Halting Theorem tell us. This is also true for the RAP. It is only in this sense can we say that the universe is observing itself.[58]

Hence, evolution shows up as a kind of motion that produces the observer and the observed at the same time (though apparently in a well-defined succession). It is therefore a self-reflective motion, which stochastically tests itself in view of possible innovative states (or structures). This can be modeled in terms of fractal motions, as has been discussed by Eisenhardt, Kurth, and Stiehl.[59] At the same time, the phenomenological structure of the world offers itself for interpretation by its own participators. In the same way that observer and observed are coevolving and thus *irreducible,* comprehension of the world coevolves with the system itself. Hence, the research **on** and the description **of** the world depend on the context offered **by** the world:

Adam: "Let's call that one a hippopotamus."
Eve: "Why?"
Adam: "Because it looks more like a hippopotamus than anything else we've seen today."[60]

This circle cannot be fully encompassed by any hermeneutical technique (nor by any formal technique); there is always a "bad space" sep-

[58] This is basically what Wheeler intended to claim, cf. Patton and Wheeler, 1975. But it is not quite clear whether this claim has been correctly derived, because Patton and Wheeler originally started with an argument from quantum theory. Although according to the general model of Wheeler, fluctuating microgeometries merge to form spacetime as an average macrogeometry, there is still no consistent way to superpose distributed geometries to give their expected value.

[59] Eisenhardt, Kurth, and Stiehl, 1988 and Callan, 1960.

[60] Solomon, 1983, p. 415

arating the irreducible components constituting observer and observed, respectively, very much in the sense of Ernst Bloch. With respect to the painting of a landscape, he notes that the foreground (which is to be painted in accordance with its own nature) is a sort of aberration, something approximate. "Between painter and landscape, subject and object, there is a harmful (detrimental) space."[61] This causes what Bloch calls "non-simultaneousness" (*Ungleichzeitigkeit*)[62], due to the two tendencies of the components of the world—one according to which nature as background or harmful foreground is a mere frame of reference; the other according to which nature means itself (*bedeutet sich selbst.*) While nature in the first case is subjected to the meaning to which it contributes, in the second it forms its own meaningful context that should be accepted by human praxis, constituting what Bloch later calls "technique of alliance," the method by which this dialectic between man and nature is taken into account, in this case being Bloch's *objective-real hermeneutics.*[63]

The main objective of this theory is to deny from the very beginning any separation of man's history and the evolution of nature. It's necessary to investigate the "human project" and the "project of nature" in their *indissociability.*[64] Hence, objective-real hermeneutics means to accept nature in its own subjectivity, which for us shows itself as productive potential and physical process.[65] This argument by Bloch follows another traditional line that ranges from the "Aristotelian left" to Giordano Bruno and the early Schelling, and which is mixed together in the more recent philosophical development going from Feuerbach to Marx and on to Bloch himself. The objective-real hermeneutics appears here as a double secularization: first, of the interpretation of religious texts by a kind of subversive hermeneutics, and secondly of the real in view of man as a producer of signs who is confronted with the opacity of natural order, although this opacity is no longer of divine origin.[66] Hermeneutics in this generalized sense means the recognition of the

---

[61] Bloch, 1985e, p. 16ff. Cf. Bloch, 1982, p. 929ff and "Verfremdungen II" in Bloch, 1984.

[62] Bloch, 1985a, p. 104ff.

[63] Raulet, 1987, p. 29, 59ff. For parallels to the philosophy of symbolical forms in the tradition of Cassirer, see Bourdieu, 1974 and Zimmermann, 1988c, 1988d, and 1989.

[64] Raulet, 1982, p. 124.

[65] Ibid, p. 138.

[66] Ibid.

generic symbolism of nature. Insofar as nature signifies itself and those who interpret and use these signs, self-reference shows up as a process of autosignification. The *deus sive natura* of Spinoza is replaced by its inverse: *natura sive deus*.[67] We have seen in the preceding section that one difficulty in the interpretation of anthropic principles arises from the absence of an integrating concept for which is substituted a vague sort of metaphysics—possibly of psychological origin (being caused, for example, by the threat to the world of ecological ruin or nuclear war)—which is uncritically transferred to an otherwise purely physical argument. The concept of objective-real hermeneutics offers the opportunity to develop a consistent basis for a dialectics of nature which would turn out to be a *metaphysics of matter* at the same time:[68]

> Matter becomes a moment of Spirit, but Spirit becomes also a moment of (this dialectic) matter. In one word: subject and object are constantly—on always new levels of mediation—bound to each other in a substantial dialectic, in the motion and evolution of substance.[69]

Thus, what the RAP shows is simply that self-reference is a concept that reflects the process of evolution and the process of comprehending that evolution in their mutual reciprocity. In this sense, *nature appeals to man* in that it offers its phenomena for interpretation. A science that does not take this dialectic into account explicitly remains exterior (and harmful) to nature and thus to man.

6. *Poetical Praxis and Speculative Philosophy*

The philosophy of Schelling was the first modern system to take into account explicitly the aspect of totality in a straightforward and consequent development of thought. In this respect, Schelling has extended Hegelian dialectics into a generalized historical concept that surpasses its material conditioning towards self-projected goals without ever being able to extinguish this basis.[70] This "non-abolishability" of being—as Frank calls it—is the immediate condition for rendering possible an infinite and principally open motion, which points to the problem of the structure and function of thermodynamically open systems in terms of

[67] Bloch, 1985d, p. 302. Cf. Raulet, 1982, p. 138ff, 156.

[68] Raulet, 1982, p. 184.

[69] Quoted according to Raulet, 1982, p. 187ff from Bloch's unpublished "matter manuscript," Leipzig, 1957.

[70] Frank, 1975, p. 17ff.

the recently developed theories of self-organization and the formation of structure. Schelling, as well as Hegel, already thinks of subject and object as being *open systems* in this very sense.[71] Schelling, especially, defines nature *in toto* as the contradiction moving itself and in itself, the structural formula of which coincides with its law of motion.[72] History is nature in a higher power, as its subjects are free (within the law of motion) to do the necessary.[73] Nature is therefore reflected in its own subjectivity:

> *Nature* as mere *product* (natura naturata) we call nature as object (to which alone all empirics is pointing (*auf das allein alle Empirie geht.*)) *Nature* as *productivity* (natura naturans) we call *nature as subject* (to which alone all theory is pointing (*auf das allein alle Theorie geht.*))[74]

By this construction (still on an idealistic basis, but already preparing the necessary steps towards materialism) Schelling is able to achieve a unification of metaphysical speculation and scientific research, "speculative physics" taking the part of "theoretical physics" in our sense, but being integrated into a *generic* philosophical context.[75] The evolutionary steps recovered by modern theories of self-organization exhibit a large number of similarities with Schelling's "theory of powers," as they have also been included in a somewhat modified form in the ontology of Nicolai Hartmann. Especially with a view to the problems we face at present with evolutionary theories (in particular when trying to deal with the two significant jumps from inorganic to organic matter and from biological to cultural systems), the "including of human thinking and knowing in the process of being as such" offers a basis for most of the recent developments of science.[76] The changing of nature by human work and the changing of man by working *with* nature corresponds within this picture to the historical process leading from hominization to humanization.[77]

The relevance of Schelling's philosophy for those modern scientific concepts which emerged in the late 1960s, and which already deal with the notions of self-reference, self-organization, autocatalysis, au-

---

[71]Warnke, 1977, p. 13.

[72]Ibid, p. 114.

[73]Sandkühler, 1984, p. 35.

[74]Schelling, 1856–1861, III, p. 284. Cf. Sandkühler, 1984, p. 27.

[75]Hartkopf, 1984, p. 85, 100.

[76]Ibid, p. 113ff, 122ff.

[77]Sandkühler, 1973, p. 131.

topoiesis, and so forth, has been shown recently in the works of Marie-Luise Heuser-Keßler.[78] Her thesis is basically that reviving the speculative philosophy of nature and generalizing it in the light of present scientific results, we have not lost the systematic meaning that appears to offer promising possibilities for a new approach to the conception of a general theory of self-organization.[79] By explicitly discussing Schelling's texts, she demonstrates that within his conception nature is a historical process, the production of which shows up as the potential to create innovative and *unpredictable* forms of organization.[80] Self-reference is included in this picture when Schelling states that, "Nature has its reality in itself—it is its own product—a totality which is organized in itself and which is organizing itself."[81] The crucial aspect is a methodological one that is also discussed by Schelling: "Doing philosophy of nature means to create nature."[82] As Heuser-Keßler emphasizes: "The objects of nature have to be reinvented in order to recognize them in their underlying productivity."[83] This relationship between ontology, the theory of cognition, and the creation of forms creates a deep and fundamental basis of thought, which in turn relates philosophy and science to the domain of aesthetical production. Bloch has identified the important passage in the works of Schelling for that end:

> After all, arts is therefore the highest for the philosopher, because it opens for him the Holy of Holies so to speak, in which there burns in One Flame ... in external and original unification (all) what is segregately specified in nature and history, and what must escape itself permanently in life and acting as well as in thinking.[84]

From this Bloch asserts[85] out of the will to unity of production follows the steady assimilation of nature to its reflection in the production and imagination of poetry on the one hand, and the fantasy of the romantic philosophy of nature on the other, which can often be understood only in terms of poetry. The "organizing nature" (called by Schelling the "soul of the world" (*Weltseele*) sometimes) is here at the heart of *nature in the sense of poiesis* (πὁίησιζ) altogether, mediated and

---

[78]Heuser-Keßler, 1986.
[79]Heuser-Keßler, 1989, p. 1.
[80]Ibid, p. 2. I quote according to an earlier version of the paper.
[81]Schelling, 1856–1861, III, p. 17.
[82]Ibid, III, p. 13.
[83]Heuser-Keßler, 1989, p. 5.
[84]Schelling, 1856–1861, III, p. 628.
[85]Bloch, 1985b that p. 220.

fluctuating between organic and anorganic levels of nature. The arts, being connected in Bloch's system with the notion of *Vor-Schein,*[86] couples imagination and reality as two dialectically mediated components of the world, in that *Vor-Schein* assures that mere appearance is transcended towards the realization of what Bloch calls "aesthetics without illusion" (*real Vor-Schein).*[87]

> Not only arts now (Bloch emphasizes),[88] but in particular philosophy has now the consciously active function of *Vor-Schein,* and insofar of *Vor-Schein* as of *objective-real Vor-Schein* as process-world, the real world of hope itself.

I have shown elsewhere[89] that there is an interesting parallel here relating arts to philosophy in the philosophy of Jean-Paul Sartre: Although dealing with social philosophy exclusively, Sartre's dialectical *progressive-regressive* method can be taken as a paradigm for the methodological consequence of the approach Bloch advocates when discussing nature:[90] In the case of Sartre, when discussing social groups it can be shown that the contextual language with which each individual tries to achieve a state of coexistence with the group (as its Other) in an effort of conceptualization will differ considerably from all other languages, because the context of any of the individuals in a social group is different from that of the others. It is necessary therefore to form a consensus in order to guarantee a sufficiently stable communicational basis. Nevertheless, this consensus can only represent a sort of "cognitive average" of the group, and it will soon diffuse at the group's boundaries. In fact, the permanent effort to secure such a basis and its stability is nothing but a general sort of aesthetical production. Indeed, a truly poetical group context is achieved, being defined subsequently as the group's reality. Hence, as the approach to consensus (about reality) is always an approach to meaning, the building up of a consensus proper is a permanent creation of meaning (in that case of social systems achieved by the progressive mediation of biography and social

---

[86] *"Vor-Schein"* means "bringing to light," "producing," and "appearing beforehand" (that what is announcing itself, but not yet arriving).

[87] Bloch, 1985e, p. 59, p. 197.

[88] Bloch, 1982, p. 1627.

[89] Zimmermann, 1988c.

[90] I have discussed the relationship of Sartre's method to aesthetical production in my paper Zimmermann, 1988a. Another paper—discussing the relationship of the systems of Sartre and Bloch—is in preparation (Zimmermann, 1990).

process). Meaning is drawn (*geschöpft*) from the redundancy of the communicative system that originates in the various nonoverlapping parts of the individual contexts. This redundancy opens a space of free play out of which meaning can be drawn (*geschöpft*) by consensus and in which meaning can be created (*geschöpft*) by production. This reflects the productivity of a social group as a common effort to recover lines of orientation in the process of reciprocal mediation of individual and group on the one hand, and of group and society as social totality on the other. This general sort of aesthetical production is condensed (*verdichtet*) in rare cases by the artist who is able to focus the general structural aspects of this meaning that is being produced. This is exactly what we nowadays call "autopoiesis"[91] The social system's autocatalytic organization proceeds by permanent aesthetical production of the reciprocally mediated constituents of the system (the individuals whom Sartre calls "singular universals"). The characteristic (and progressive) property of Sartre's theory is that he does not leave this autopoietic process to an abstract system of mediated structures, but gives explicit hints as to the interactive formation of these structures by a dialectical process that is initiated and carried out by the individual (the subject) who plays the role of a singularity in and hence of a source for a social field generating the system as part of totality.

The ethical program implicit in this model is clear: Solidarity can be achieved by trying to establish a common basis of communication within the free play left open by the redundancy, which is produced by the permanent missing of discourses within the struggle of what we can call "identity and difference."[92] Solidarity can be approached then by the steady effort to systematically search out this free play and test possible material emerging from this space in the light of its aptitude for building a consensus. The biographical method of Sartre (the "progressive-regressive method") is in fact a method offering a theory for this paradigm, as well as a method for how to apply it. In this sense it is an epistemology of ethics.

Note that this concept appears also to be adequate to eventually

[91]: Note that we talk of "autopoietic" systems here in the sense of systems that permanently create and annihilate themselves as their mode of evolution. Cf. Jantsch and Waddington, 1976 for examples. We do not share the conservative interpretation of autopoietic systems as put forward recently by some of their "inventors," like that given in Maturana, 1982, 1987 and Maturana and Varela, 1975. For a detailed critique of this view, see Heuser-Keßler, 1989.

[92]Zimmermann, 1988b.

encompass nature itself. Despite a clear demarcation of Sartre from the dialectics of nature,[93] he himself has opened possibilities for following this route further. When speaking in terms of productivity, nature can be visualized as a subject with a history too (we have seen this above when discussing the theories of Schelling and Bloch). Sartre's method can be used for the analysis of this history; especially in view of the modern theories of self-organization referred to in this paper, it appears possible to eventually bridge the language gap by transforming this method to another component of the objective-real hermeneutics we have discussed here.

René Thom once proposed the revival of natural philosophy.[94] In the catastrophe theory that Thom developed, he saw the possibility for eventually introducing the generalized language necessary to bridge the above-mentioned gap between formal mathematics and hermeneutics, in order to connect "quantitative models with verbal conceptualization."[95] If relevant philosophy today is basically *research* from the start—as Theunissen says[96]—where the reflective method is based mainly on critique, then this appears to be a promising route to follow. Perhaps it will thus be possible to achieve that sort of dialectic hermeneutics that Sandkühler sought.[97] As we have seen, the very beginning of this project involves the basic idea of systematically combining speculative philosophy and truly *poetical* praxis.

Such a theory should satisfy the following requirements:

(1) A unified philosophy of totality should include man and nature from the very beginning.

(2) A unified ansatz of both ontology and the theory of cognition should be provided from the outset.

(3) A historical method should be developed that takes into account the progressive and regressive components of the evolutionary motion and the intrinsic irreversibility of processes in the world. Hence, what is required is a unification of (progressive) analysis and (regressive) synthetic reconstruction of processes, where the techniques of projection that determine

---

[93]Sartre, 1985. Cf. Sartre, Garaudy, Hippolyte, Vigier, Orcel, 1965.
[94]Thom, 1979.
[95]Ibid, p. 11.
[96]Theunissen, 1989.
[97]Sandkühler, 1973, p. 400.

the structural contexts of systems are of particular importance.

(4) A generalized language should be developed which might be based on topology, this being the "most qualitative" language in mathematical terms and the most verbalized formalization technique thus far achieved.

(5) Finally, an axiomatic structural design should be developed, giving explicit formalisms allowing us to transform between the domains of man and nature.[98]

Ideas in science and philosophy about the interactions of these domains of man and nature have changed in recent years, but no definitive consequences of this change can be drawn if a new epistemological basis is not developed taking these new ideas explicitly into account. In the final section of this chapter we give an example suggesting what such a new epistemology might look like, using the twistor geometry introduced by Roger Penrose.

7. Twistor Geometry

In his 1949 work *Subject-Object* Bloch noted that "the most recent physics, especially, contains a lot of unconscious dialectics," and that it is not due to the nature of different epistemological domains that this fact is not yet acknowleged, but rather because of the social limitations of the physicists involved.[99] According to this view, an abstract physics does not have much of a chance to eventually achieve a different tendency in its development. Indeed, not until about 30 years after the publication of Bloch's work did a basis for such an alternate view appear through the results of the Bourbaki school, who based mathematics on structures and qualities rather than on quantification. On this basis have been erected the very foundations of catastrophe theory, the theories of self-organization, and chaos theory. Similar results took place in theoretical physics, where techniques of differential geometry and differential topology have been applied to Einstein's theory of gravitation.[100] Some years ago, Roger Penrose introduced an alternative concept for a unified treatment of physics. We will try to demonstrate here that this new approach is very much of the type

---

[98] Cf. Zimmermann, 1990.

[99] Bloch, 1985c, p. 207ff.

[100] See for reviews Maturana and Varela, 1975, Thom, 1975, 1983, Nicolis and Prigogine, 1977, 1987, Hawking and Israel, 1979.

required for a new integrated approach to the desired philosophy of nature as we have discussed it in the preceding sections.

Penrose's starting point is the uneasiness experienced by physicists when looking at the conceptual differences between the general theory of relativity and quantum theory. This gap seems to separate modern physics into two disjoint halves, one macroscopic, the other microscopic.[101] Hence, Penrose begins with the question of whether the concept of a spacetime point is at all definable, once we start speaking of mean-free paths of the order of 10^{-12}cm and below. The question is also whether events on spacetime, characterized by points of a real Lorentz manifold of signature −2 and dimension 4, can be brought into any relation to the states of microphysics, which are characterized by points[102] of a positive-definite Hilbert manifold of a large number of dimensions.[103] This also leads to the problem of the contradiction between the continuum of macrophysics on the one hand, and the quantum discontinuum on the other, as well as to the question of what one should do with spacetime singularities where this continuum structure breaks down even in macroscopic physics.

Penrose begins his discussion with consideration of angular momentum in physics, since this is a quantity that's relevant in both the continuum and the discontinuum domain. In macrophysics, as well as in microphysics, angular momentum has a *discrete* spectrum. The idea is to use the properties of angular momentum as a sort of primordial structure from which space and time may be built as secondary quantities (instead of starting from them as elementary primitives from the outset). Thus, Penrose's idea is a radically new concept, involving the attempt to construct space and time from a primordial, *purely combinatorial* structure.

For reasons of simplicity, Penrose initially considers the unrealistic case of particles having only the property of spin, which is characterized by spin number $2i$. Here $i = n\hbar/2$, where n is a natural number. These particles do not exhibit any interaction other than the exchange of spin among themselves. Hence, one can only define a topological relationship given by the continual change of spin numbers. Thus we have a network of fictitious particles for which we assume that the well-

[101] We follow here the main line of argumentation in Zimmermann, 1988e.

[102] Note that we talk here of points when referring to any manifold. In Hilbert space notation, particles move on and are represented by lines rather than by points.

[103] Penrose, 1971, 1975.

known combinatorial rules of quantum mechanics are valid, including the conservation of angular momentum.[104]

When choosing particularly large systems as parts of this network, one is then able to define the concept of direction by using the direction of the spin axis of the respective system. If the possibility exists to define an angle between any two axes of this kind, we should look for a transformation that carries angles from the network onto three-dimensional Euclidean space. If two such systems are given, with total spin numbers N and M, say, then we can take one unit of spin number 1 (that is, an arbitrary particle of spin $\frac{1}{2}$) from either of the systems (N say) and add it to the other. Hence, we are left with a system of spin number $N-1$ and a system with spin number either $M-1$ or $M+1$. Which of these is the case depends on the rules of angular momentum combinatorics. For both possibilities we have probabilities that can give us information about the angle between the systems N and M. According to whether the systems are parallel or antiparallel, we will have probabilities 0 (for $M-1$) and 1 (for $M+1$), and conversely. Generally, for an arbitrary angle θ, we have probabilities $\frac{1}{2}-\frac{1}{2}\cos\theta$ (for $M-1$), or $\frac{1}{2}+\frac{1}{2}\cos\theta$, respectively. Hence, one obtains a general concept for angles that is very similar to that from Euclidean space. The probabilities are rational numbers and, thus, more elementary than conventional probabilities that are expressed as real numbers between 0 and 1. A rational probability can be thought of in the following way: Given $p = m/n$, the universe can select (anthropomorphically speaking) from m possibilities of one kind and n possibilities of another kind that are equiprobable. Only in the limit do we attain the entire continuum of values. This is exactly the point; the objective is to get "pure" probabilities determined by physics alone, not by conditions involving the observer. This fits in well with the intentions of Prigogine, who tries to explain spontaneous formation of structure (and selective formation especially) on the basis of initial equidistributions.[105] On the other hand, this "argument of choice" introduces explicitly the "history of a system."[106]

Hence, geometry is *defined* by purely combinatorial systems. Making no use of complex numbers, one can create a three-dimensional structure, not just a two-dimensional partial structure. Usually, for representing all possible directions of spin states for particles with half-number spin, it's necessary to introduce complex linear combinations.

[104] Penrose, 1971.

[105] Nicolis and Prigogine, 1977.

[106] Penrose, 1971, 1975, Sparling, 1975.

Penrose, however, uses rational numbers that cannot approximate complex numbers. In fact, the reason for this is somewhat unclear,[107] but the reproduction of Euclidean space directions is possibly due to the fact that the combinatorial rules for networks can be derived from representations of the symmetry group $SO(3)$. The generic group isomorphism $SU(2) \rightarrow SO(3)$ for Euclidean three-space is obviously of a constitutive character here.

In generalizing this concept for "real" particles, it is necessary to adapt it to relativistic particles that can in addition change their position relative to the network (and not only exchange their spin). The appropriate mixture of (orbital) angular momentum and spin is still a purely relational concept. It is not necessary to define a spatial concept for this. The symmetry group $SO(3)$ must now be replaced by the group $SU(2,2)$ (in order to take account of rotations as well as translations). For technical reasons that we shall not discuss here, Penrose works with a partial set of the new group isomorphisms. He calls $SU(2,2)$ the *twistor group,* and discusses the isomorphism $SL(2,C) \rightarrow O\uparrow(1,3)$.[108] Here $SL(2,C)$—the group of two-dimensional complex matrices—is the well-known group of spinors, of which the famous Pauli spin matrices are elements.

Now what is a twistor? It is basically an element of a four-dimensional complex vector space which can be interpreted as a quantity describing momentum and angular momentum (including the spin) of a particle of rest mass zero. Hence, twistors can also be defined as quantum operators. Their commutation relations constitute a basis from which the usual commutation relations for momentum and angular momentum can be derived.

Thus spinors (which, in fact, represent the original network) can be replaced by twistors. Because of the commutation relations between twistors, we can define a generalized sort of uncertainty relationship that mixes twistor position and twistor momentum at all times, thus leading to the macroscopic "impression" of curvature. Hence, operator properties of twistors determine the signature and dimension of spacetime. The latter can be *derived* from the former.

We have already mentioned that at a singularity the continuum structure of spacetime breaks down. In the twistor picture, the primordial network would "appear" at these places. But Penrose has

---

[107] Penrose, 1980, Hughston and Ward, 1979, Penrose and Rindler, 1984, 1986.

[108] Penrose, 1975.

further strengthened this relationship between singularities and primordial spacetime structure. For him, the key to the solution of the irreversibility problem for evolutionary processes can be found in the structure of the singularities themselves.[109] The basic idea is to ask for the differences between the initial singularity with which one deals in cosmology, and the local singularities of astrophysics (such as black holes). Penrose finds that the difference is essentially their respective entropy contents: He uses the time arrows in the universe from which the irreversibility of nature can be derived, showing that it is possible to associate a low entropy state with the Big Bang and a high entropy state with the Big Crunch.[110] To some degree, this contradicts our expectation: Structures in the universe should emerge out of an initial chaos that represents a thermodynamically energetic equidistribution, i.e., *high* entropy. In the case of gravitational entropy, however, one can show that this is not the case due to the anomalies of gravitation associated with its attracting nature, which cause systems to behave as if they had something like a negative specific heat. Only in the case of stellar collapse is maximal entropy achieved.[111]

If the universe had been of maximal chaos, then the initial entropy would also have been maximal. In this case—assuming time-symmetric physics—no time arrow and thus no dissipation would have emerged.[112] Hence, Penrose concludes that a boundary condition on the universe as a whole has to be invoked to ensure that the initial entropy is sufficiently low. Thus, chaos initially becomes a structure of high order, being equivalent to something that can be visualized as an initial symmetry breaking. Chaos thus attains the nature of Bloch's *prima materia*.[113] And this boundary condition does not act upon any primordial distribution of matter, but rather acts immediately upon geometry itself.[114] Hence, we have two basic results: On the one hand, singularities of this kind can be related to dynamical aspects of evolution and to the recent results on attracting chaos.[115] On the other hand, with a view to the anthropic principles discussed earlier, this avenue of research opens new perspectives on the interaction of geometry

---

[109] Penrose, 1979, p. 581.
[110] Ibid, p. 611, p. 629.
[111] Ibid, p. 612.
[112] Ibid, p. 629.
[113] Bloch 1982, p. 748ff.
[114] Penrose, 1979, p. 630. Cf. Kibble, 1982, pp. 391–408.
[115] Prigogine, 1986, p. 24ff.

and physics, perception (and cognition), and science.

Penrose's theory also opens up a new path to a consistent philosophical treatment of the totality problem. We can see this in the following way: As I have shown elsewhere, the dynamics of evolution can be philosophically rephrased in terms of the dialectic ansatz discussed in the preceding sections.[116] This follows because fractals can be identified with a pure form of motion that contains motion as potential and the unfolding of structure at the same time. This is the generic analogue to what Hegelian theory calls *motion in itself (Bewegung an sich).* Hence, network structure and spacetime structure are related to each other in the same way as are motion *in itself* and motion *for itself*,[117] very much in the sense of Blochian *potentia* (δύναμις) and *actus* (ἐνέργεια).[118] Motion is thus in deferment (in adjournment),[119] which again unifies the ontological level with the phenomenological one and the domain of cognition.

In terms of the anthropic concept it's interesting to note that twistor theory, as being originally geometry, is based on the reconstruction of the world. But it leaves free play for the inclusion of history, as well. In contrast to the isolating tendency of anthropic conceptions, twistor theory points to the acceptance of nature as difference, hence to an intersubjective approach to nature. Taking into account the philosophical implications arising from its physical modeling, this appears to be an adequate approach in the sense of the principles which a new philosophy of nature should contain, and which have been listed in the preceding section.

References

The references are given according to the editions actually used.

Adler, R., M. Bazin, and M. Schiffer (1965): *Introduction to General Relativity,* McGraw-Hill, New York.

Barrow, J. D. and F. J. Tipler (1988): *The Anthropic Cosmological Principle,* Oxford University Press, Oxford.

Bjorken, J. D. and S. D. Drell (1964): *Relativistic Quantum Mechanics,* McGraw-Hill, New York.

[116] Zimmermann, 1988e.

[117] Hegel, 1970, §261, Z.

[118] Bloch, 1985b, p. 242ff, p. 235 and 1985c, p. 207, p. 219ff.

[119] Zimmermann, 1988c.

Bjorken, J. D. and S. D. Drell (1965): *Relativistic Quantum Fields,* McGraw-Hill, New York.

Bloch, E. (1982): *Prinzip Hoffnung,* 3 volumes, Suhrkamp, Frankfurt.

Bloch, E. (1984): *Literarische Aufsätze,* Suhrkamp, Frankfurt.

Bloch, E. (1985a): *Erbschaft dieser Zeit, Werke, Vol. 4,* Suhrkamp, Frankfurt.

Bloch, E. (1985b): *Das Materialismusproblem, seine Geschichte und Substanz, Werke, Vol. 7,* Suhrkamp, Frankfurt.

Bloch, E. (1985c): *Subjekt-Objekt. Erläuterungen zu Hegel, Werke, Vol. 8,* Suhrkamp, Frankfurt.

Bloch, E. (1985d): *Atheismus im Christentum. Zur Religion des Exodus und des Reichs, Werke, Vol. 14,* Suhrkamp, Frankfurt.

Bloch, E. (1985e): *Experimentum Mundi. Frage, Kategorien des Herausbringens, Praxis, Werke, Vol. 15,* Suhrkamp, Frankfurt.

Bourdieu, P. (1974): *Zur Soziologie der symbolischen Formen,* Suhrkamp, Frankfurt.

Callen, H. B. (1960): *Thermodynamics,* Wiley, New York.

Carter, B. (1974): in *Confrontation of Cosmological Theories with Observation,* M. S. Longhair, ed., Reidel, Dordrecht, Netherlands.

Casti, J. L. (1989): *Paradigms Lost: Images of Man in the Mirror of Science,* Morrow, New York.

DeWitt, B. S. and N. Graham (1973): *The Many-Worlds Interpretation of Quantum Mechanics,* Princeton University Press, Princeton.

Drake, F. D. (1960): *Intelligent Life in Space,* Macmillan, New York.

Eisenhardt, P., D. Kurth, and H. Stiehl (1988): *Du steigst nie zweimal in denselben Fluß,* Rowohlt, Reinbek, West Germany.

Everett, H. (1957): *Reviews of Modern Physics,* 29, 454.

Frank, M. (1975): *Der unendliche Mangel an Sein,* Suhrkamp, Frankfurt.

Gasiorowicz, S. (1966): *Elementary Particle Physics,* Wiley, New York.

Habermas, J. (1981): *Theorie des kommunikativen Handelns,* (In two volumes), Suhrkamp, Frankfurt.

Hartkopf, W. (1984): *Denken und Naturentwicklung. Zur Aktualität der Philosophie des jungen Schelling,* in *Natur und geschichtlicher Prozeß,* H. J. Sandkühler, ed., Suhrkamp, Frankfurt.

Hawking, S. W. and G. F. R. Ellis, (1973): *The Large-Scale Structure of Spacetime,* Cambridge University Press, Cambridge.

Hawking, S. W. and W. Israel, eds. (1979): *General Relativity: An Einstein Centenary Survey,* Cambridge University Press, Cambridge.

Hegel, G. W. F. (1970): *Enzyklopädie II, Werke, Vol. 9, Theorie Werkausgabe,* Suhrkamp, Frankfurt.

Heuser-Keßler, M.-L. (1986): *Die Produktivität der Natur. Schellings Naturphilosophie und das neue Paradigma der Selbstorganisation in den Naturwissenschaften,* Duncker & Humblot, West Berlin.

Heuser-Keßler, M.-L. (1989): *Wissenschaft und Metaphysik. Überlegungen zu einer allgemeinen Selbstorganisationstheorie,* in *Selbstorganisation in den Wissenschaften–Selbstorganisation der Wissenschaft,* G. Küppers, ed., Vieweg, to appear.

Hughston, L. P. and R. S. Ward, eds. (1979): *Advances in Twistor Theory,* Pitman, San Francisco.

Jantsch, E. and C. H. Waddington, eds. (1976): *Evolution and Consciousness: Human Systems in Transition,* Addison-Wesley, Reading, MA.

Kibble, T. W. B. (1982) in *Quantum Structure of Space and Time,* M. J. Duff and C. J. Isham, eds., Cambridge University Press, Cambridge.

Kramer, D., H. Stephani, M. McCallum, and E. Herlt (1980): *Exact Solutions of Einstein's Field Equations,* Deutscher Verlag der Wissenschaften, East Berlin.

Kuhn, T. S. (1970): *The Structure of Scientific Revolutions,* University of Chicago Press, Chicago.

Mandelbrot, B. (1982): *The Fractal Geometry of Nature,* Freeman, San Francisco.

Maturana, H. R. and F. Varela (1975): *Autopoietic Systems,* Biological Computing Laboratory, Report 9.4, University of Illinois, Urbana, IL.

Maturana, H. R. (1982): *Erkennen: Die Organisation und Verkörperung von Wirklichkeit,* Vieweg, Braunschweig, West Germany.

Maturana, H. R. (1987): *Kognition,* in *Der Diskurs des Radikalen Konstruktivismus,* S. J. Schmidt, ed., Suhrkamp, Frankfurt.

Maturana, H. R. (1987): *Biologie der Sozialität,* in *Der Diskurs des Radikalen Konstruktivismus,* in S. J. Schmidt, ed., Suhrkamp, Frankfurt.

Newhouse, S., D. Ruelle, and F. Takens (1978): *Communications in Mathematical Physics,* 64, 35.

Nicolis, G. and I. Prigogine (1977): *Self-Organization in Nonequilibrium Systems: From Dissipative Structures to Order Through Fluctuations,* Wiley, New York.

Nicolis, G. and Prigogine (1987): *Die Erforschung des Komplexen. Auf dem Weg zu einem neuen Verständnis der Naturwissenschaften,* Piper, Munich.

Patton, G. M. and J. A. Wheeler (1975): "Is Physics Legislated by Cosmogony?" in *Quantum Gravity,* C. J. Isham, R. Penrose, and D. W. Sciama, eds., Oxford University Press, Oxford.

Peebles, P. J. E. (1971): *Physical Cosmology,* Princeton University Press, Princeton.

Penrose, R. (1971): "Angular Momentum: An Approach to Combinatorial Space-Time," in *Quantum Theory and Beyond,* T. Bastin, ed., Cambridge University Press, Cambridge.

Penrose, R. (1975): "Twistor Theory: Its Aims and Achievements," in *Quantum Gravity,* C. J. Isham, R. Penrose, and D. W. Sciama, eds., Oxford University Press, Oxford.

Penrose, R. (1979): "Singularities and Time-Asymmetry," in *General Relativity: An Einstein Centenary Survey,* S. W. Hawking and W. Israel, eds., Cambridge University Press, Cambridge.

Penrose, R. (1980): "A Brief Outline of Twistor Theory," in *Cosmology and Gravitation,* P. G. Bergmann and V. DeSabbata, eds., Plenum, New York.

Penrose, R. and W. Rindler (1984): *Spinors and Space-Time, Vol. 1: Two-Spinor Calculus and Relativistic Fields,* Cambridge University Press, Cambridge.

Penrose, R. and W. Rindler (1986): *Spinors and Space-Time, Vol. 2: Spinor and Twistor Methods in Space-Time Geometry,* Cambridge University Press, Cambridge.

Prigogine, I. (1979): *Vom Sein zum Werden,* Piper, Munich.

Prigogine, I. (1986): "Natur, Wissenschaft und neue Rationalität," in *Dialektik 12, Die Dialektik und die Wissenschaften,* Pahl-Rugenstein, Cologne.

Prigogine, I. and I. Stengers (1981): *Dialog mit der Natur,* Piper, Munich.

Raulet, G. (1982): *Humanisation de la nature—naturalisation de l'homme. Ernst Bloch ou le projet d'une autre rationalité,* Klincksieck, Paris.

Raulet, G. (1987): *Natur und Ornament. Zur Erzeugung von Heimat,* Luchterhand, Darmstadt, West Germany.

Ruelle, D. and F. Takens (1971): *Communications in Mathematical Physics,* 20, 167.

Ryan, M. P. and L. C. Shepley (1975): *Homogeneous Relativistic Cosmologies,* Princeton University Press, Princeton.

Sagan, C., ed. (1973): *Communication with Extraterrestrial Intelligence,* MIT Press, Cambridge, MA.

Sandkühler, H. J. (1973): *Praxis und Geschichtsbewußtsein,* Suhrkamp, Frankfurt.

Sandkühler, H. J. (1984): "Natur und geschichtlicher Prozeß. Von Schellings Philosophie der Natur und der Zweiten Natur zur Wissenschaft der Geschichte," in *Natur und geschichtlicher Prozeß,* H. Sandkühler, ed., Suhrkamp, Frankfurt.

Sartre, J.-P. (1985): *Critique de la raison dialectique précédé de Questions de méthode, 2 Vols.,* Gallimard, Paris.

Sartre, J.-P., R. Garaudy, J. Hippolyte, J.-P. Vigier, and M. Orcel (1965): *Existentialismus und Marxismus. Eine Kontroverse,* Suhrkamp, Frankfurt.

Scheffler, H. and H. Elsässer (1974): *Physik der Sterne und der Sonne,* BI, Mannheim, West Germany.

Schelling, F. W. J. (1856–1861): *Sämmtliche Werke, 14 Vols.,* Stuttgart.

Shklovskii, I. S. and C. Sagan (1966): *Intelligent Life in the Universe,* Dell, New York.

Solomon, R. C. (1983): *In the Spirit of Hegel,* Oxford University Press, Oxford.

Sparling, G. (1975): "Homology and Twistor Theory," in *Quantum Gravity,* C. J. Isham, R. Penrose, and D. W. Sciama, eds., Oxford University Press, Oxford.

Theunissen, M. (1989): "Möglichkeiten des Philosophierens heute," in *Was ist Philosophie?* Vorlesungreihe, Institut für Philosophie, Free University of West Berlin.

Thom, R. (1975): *Structural Stability and Morphogenesis,* Benjamin, Reading, MA.

Thom, R. (1979): "Towards a Revival of Natural Philosophy," in *Structural Stability in Physics,* W. Güttinger and H. Eikemeier, eds. Springer, Berlin.

Thom, R. (1983): *Mathematical Models of Morphogenesis,* Horwood, Chichester, UK.

Warnke, C. (1977): "Systemdenken und Dialektik in Schellings Naturphilosophie," in *Dialektik und Systemdenken—Historische Aspekte,* H. Bergmann, U. Hedtke, P. Ruben, and C. Warnke, eds., Akademie Verlag, East Berlin.

Zel'dovich, Ya. B. and I. D. Novikov (1983): *The Structure and Evolution of the Universe,* University of Chicago Press, Chicago.

Zimmermann, R. E. (1986a): "Poetik des sozialen Seins: Sartres Auffassung vom Unbewußten," in *Dossier Sartre,* R. E. Zimmermann, ed., Lendemains, 42, 61–69.

Zimmermann, R. E. (1986b): "The Transition from Town to City: Metropolitan Behavior in the 19th Century," in *Disequilibrium and Self-Organization, 2nd and 3rd International Whitsun Meetings, International Study Group on Self-Organization,* C. W. Kilmister, eds., Klosterneuburg, Austria, Reidel, Dordrecht, Netherlands.

Zimmermann, R. E. (1988a): "Imagination und Katharsis. Zum poetischen Kontext der Subjektivität bei Sartre," in *Sartre. Ein Kongreß,* T. König, ed., Rowohlt, Reinbek, West Germany.

Zimmermann, R. E. (1988b): "Authenticity and Historicity. On the Dialectical Ethics of Sartre," in *Inquiries into Values,* S. H. Lee, ed., Mellen Press, Lewiston, ME.

Zimmermann, R. E. (1988c): "Poetics and Auto-Poetics: Sartre's "Flaubert" as an Epistemology of Ethics," in *Sartre Round Table,* W. McBride, ed., XVIIIth Congress of Philosophy, Brighton, UK (in press).

Zimmermann, R. E. (1988d): "Ordnung und Unordnung—Zum neueren Determinismusstreit zwischen Thom und Prigogine," *Lendemains,* 50, 60–74.

Zimmermann, R. E. (1988e): "Zur Dialektik der Naturwissenschaften," in *Hegel Jahrbuch 1989,* W. Lefèvre, H. Kimmerle, and R. Meyer, eds., to appear 1990.

Zimmermann, R. E. (1989): "Neue Fragen zur Methode. Das jüngste Systemprogramm des dialekitischen Materialismus," in *Jean-Paul Sartre,* R. E. Zimmermann, ed., Junghans, Cuxhaven, West Germany.

Zimmermann, R. E. (1990): "Seinsmangel und Produktion. Zur Signifikanz der Entwurfskonzepte bei Sartre und Bloch," *Prima Philosophia,* to appear.

Zimmermann, R. E. (1990): "Selbstreferenz und Poetische Praxis. Zur Grundlegung einer axiomatischen Systemdialektik," in preparation.

CHAPTER 3

Problems of Explanation in Physics

PETER T. LANDSBERG

Abstract

The problem of explanation in physics is studied by reference to four experiments (the falling of a body, the emission of a photon, cold fusion, the spreading of a gas into a vacuum) and two theories (the free-electron theory of metals, the uncertainty principle). From these some conclusions emerge regarding the nature of explanation in physics of which the most important perhaps are the crucial role of the background knowledge that is assumed to be valid, the ability to invent new general concepts, and the need to mutate reality.

1. Introduction

The program for this meeting assigned me the subject *Problems of Explanation.* I am an obedient man and so I adopted it. But my main occupation is physical theory and so I limited the scope of the title. The subject is best studied I feel by discussing a few cases where explanations are required in physics. One may hope that some broad conclusions can be reached by this procedure.

I shall briefly consider what is involved in explanations of four types of experiment (§§2–5). Then I shall consider two theoretical results (§§6–7). For the latter, explanations enter twice: first to explain the theoretical result itself, and secondly to consider the explanatory power that the result has in its applications in physics. In §§6–7 we shall be concerned mainly with the second of these.

2. Case 1: The Falling of a Body

The great feat of Newton's theory of gravitation was the reduction of planetary orbits and the falling of a body to the same force: universal gravitation. It was a magnificent achievement for it unified a diversity of phenomena and it made possible statements about properties believed to be obeyed everywhere in the universe. How do you explain that an apple drops? It is due to gravitation!

At a deeper level this explanation is inadequate. How is it that the force acts instantaneously through vacuum without any intermediaries?

In that sense the theory is nonlocal, for it uses action at a distance, and such action has a built-in mystery.

Along came relativity. I will not rehearse here how it amended our concept of simultaneity. More important for us is that in a uniformly-accelerated box (without windows!) the motions to be observed are indistinguishable from those one would find if a *uniform* gravitational field were acting (this is the equivalence principle). The Earth, and indeed any gravitating body, does not produce a *uniform* field, since the field is always directed towards the center of the body. The idea is now that this nonuniformity warps spacetime and this then may be thought of as the cause of the gravitational force. Ergo, we have a local theory: the warp of spacetime induced by the mass causes the force. So one is satisfied—I am bound to add "for the time being." This allows for future and better explanations!

Concerning the nature of explanation, we learn: (1) that we can explain only in terms of background knowledge, and (2) to achieve our explanation we need the freedom to discover, or invent, laws.

3. Case 2: The Emission of a Photon

Radiation from stars has been known for centuries, but only quantum theory has supplied some background which goes in the direction of an explanation of photon emission: An excited atom drops back to its ground state and the energy is released in the form of radiation energy $h\nu$. The transition probability per unit time can be worked out by the usual rules using the wavefunction ψ and its expansion into a set of appropriate orthonormal states. This is the superposition principle of quantum theory. We thus recover the requirements for an "explanation" already encountered in Case 1: One needs to have background knowledge and one has to be able to invent new concepts (ψ)—or at least use other people's inventions of new concepts.

Is this level of explanation satisfactory? Although quantum theory works, some people might want to have a precise prediction of *when* a *given* atom will decay, and they would not regard a mere transition *probability* per unit time as an adequate description. The fact that quantum theory is a statistical theory is then seen to be an imperfection. One might also want to know what is going on inside an atom to produce a quantum jump. Nobody knows the answer. Will people know more about this in the future? Probably so; science is always imperfect and capable of development. Quantum theory gives one an "explanation" of photon emission at some level, but I am afraid it is not entirely satisfying.

The description of the atomic decay in terms of the states of the atom means that it is assumed that the atom is relatively stable with respect to the decay. One can then use perturbation theory so that the transition probability is relatively small. If these conditions were not satisfied the complete system of atom plus electromagnetic fields would not even approximately be described in terms of the states of the atom.

The decay of our atom can be arranged to cause a detonation, the firing of a gun, the killing of a cat ("Schrödinger's cat"). Here, and in many other cases, a macroscopic system, which we may call generically a detector, interacts with a quantum system. This means that a system that is governed by the superposition principle and by quantum mechanical rules for calculating probabilities interacts with a system for which the *ordinary* rules of probability apply. One can analyze this only by mutilating reality in the following way: quantum system and detector have to be regarded as separate systems, whereas they are in truth just part of one larger system. This is just one aspect of the unsolved measurement problem of quantum mechanics. As part of the explanation of photon emission resulting in a measurement one has to mutilate reality!

This is not accepted by everybody. It can be argued that these deep questions are meaningless or irrelevant. One should just use the quantum formalism and be satisfied that it yields correct answers, as indeed it does in all the thousands of cases that have been investigated. This is *perhaps* the Copenhagen interpretation of quantum theory (it has never been properly defined). In a more general philosophical context this attitude can be regarded as positivist. However, this also has its shortcomings, well known to philosophers. Whitehead talks of complicated rules for working out the motion of the planets before Neptune was discovered and being satisfied with that as a good positivist would be. If you are **not** satisfied and make a deeper analysis, you would discover a new planet in the form of Neptune! This would be the result of a strongly nonpositivistic attitude [1].

As regards the problem of measurement, we note the view that for a rapid sequence of measurements made at times which are integral multiples of some small time τ, a system will show no change at all. It seems as if time has stood still for it. This effect can be understood in terms of quantum mechanical perturbation theory, which gives for small times a transition probability *per unit time* proportional to τ and so vanishes in the limit $\tau \rightarrow 0$ (for larger values of time, the usual *constant* transition rate is found). This result is not controversial, but it needs to be explained away if quantum theory is to continue to make sense. The most obvious explanation is that measurements require a time period $\Delta\tau$ and the paradox becomes ineffectual if it can be shown that $\Delta\tau \geq \tau$. A number of investigations of this point exist in the literature and at present one book per year can be expected on the general problems of quantum theory [2–7] and the fact that it is a nonlocal theory.

4. Case 3: Cold Fusion

For some months (March-July 1989) the general public, and therefore the scientific establishment, were puzzled by the possibility of cheap energy by nuclear fusion using electrochemical means. Hopes and counterhopes swayed hither and thither and a generally accepted explanation of the new experiments continues to elude us. There was a considerable amount of heat evolution [8a], but was it nuclear fusion [8b]? The matter is not entirely resolved, though the consensus seems nearer the view that electrochemical cold fusion using deuterium-laden metals has not been established at a useful level (muon-catalyzed 'cold' fusion is also being investigated [9]).

What are the implications for our subject? Most important is the fact that there is a delay in producing a widely accepted reduction of new results to accepted principles—for that is what an explanation is in the present context. Case 3 is one where an "explanation" is hard to achieve and of wide interest. Also over any explanation there hangs the threat of new experiments that might cause a revised explanation to be developed.

The question of priority is also raised, and in a rather amusing way. Fritz Paneth and Kurt Peters used a vaguely similar setup already in 1926 at a time before the neutron, which plays a key role in the typical fusion reaction, was discovered:

$$\text{deuterium } (\mathrm{D} = {}^2\mathrm{H}) + \text{tritium } (\mathrm{T} = {}^3\mathrm{H}) \rightarrow {}^4\mathrm{He} + \text{neutrons} + 17.6\mathrm{MeV}$$

We have two protons and three neutrons on both sides of this equation, which amounts to a conversion of hydrogen to helium. This is precisely what Paneth and Peters were trying to achieve—however there objective was to produce helium for airships! (Remember, the neutron had not been discovered yet.) In any case they had to withdraw at least some of their claims owing to the fact that sources of experimental error had been ignored in their first publication [10]. So the advance of science, which sometimes seems to proceed by a big leap, can by careful historical study often be seen to actually be the result of a slow step-by-step advance. Even Shockley's field-effect transistor could not be patented by the Bell Laboratories in 1948 due to preexisting patents by the somewhat neglected Julius Lilienfeld [11].

5. *Case 4: The Spreading of Gas into a Vacuum*

This most natural occurrence becomes problematical if one recalls that molecular collisions (insofar as they are elastic) are reversible in time. Thus the entropy is constant and states of the system will recur again and again (Poincaré recurrence). For example, a three-particle gas started off in the left half of a box will come back to essentially this same state again and again. The principles of mechanics should thus make it rather difficult for a gas to expand into a vacuum, since it should contract back again from time to time. To overcome this difficulty, one talks of an increase of entropy with time resulting from using a coarse-grained description of the system. One then talks about macrostates, each of which is an umbrella state for many distinct microstates; their number is referred to as their "weight." Thus one may stipulate that, when a macrostate corresponding to equilibrium (or other given constraints) has an overwhelmingly larger weight than any other macrostate (compatible with the given constraints), the system is found in this state with a corresponding overwhelmingly high probability. This principle also gives a direction to time, since it tells you the state which a system assumes if a constraint is removed. For example, it tells you that diffusion into a vacuous space occurs on withdrawal of a partition. Of course a super intelligence with maximum knowledge about all the molecules does not need coarse-graining, and the entropy is then constant for it (Liouville's Theorem). But it can converse with humans only if it can use our rough concepts such as pressure, wall, pointer, apparatus, etc. It can do this only if it deliberately sacrifices information at its disposal and performs some time averages over molecules, thus deriving our coarse sense data [12]. This will lead to an increase in the entropy as it corresponds to the passage from a fine-grained to a coarse-grained description.

Case 4 teaches us again that the invention of new concepts such as coarse-graining and the introduction of new variables such as entropy is an important mechanism for arriving at an explanation. There are still discussions going on about other ways of making reversible mechanical processes and irreversible thermodynamic processes compatible, so that even in this comparatively mature branch of physics explanations are provisional.

6. Case 5: The Free Electron Theory of Metals

This theory goes back to the late 1920s. It very roughly pools the closed shell valence electrons of atoms into a band of energy levels that form the valence band. The single valence electrons of the atoms of sodium metal, say, form another band of energy levels that form the conduction band. If there are unoccupied electron states in a band, then electrons can gain energy from an externally-applied electric field by being promoted to an energetically close level and, hence, it is possible to have reasonably good electrical conductivity. This explanation of the good electrical conductivity of metals and of appropriately doped semiconductors works quite well. It also accounts for the poor electrical conductivity of insulators in which the bands are full so that electrons cannot easily gain energy from an applied electrical field.

This explanation of the properties of metals and insulators is better than it ought to be, since the Coulomb interactions between the electrons have been largely neglected. One would have thought that they really have to be considered. Wigner has remarked on this point in his "Unreasonable Effectiveness of Mathematics in the Natural Sciences" [13].

In the present context it is a reminder that scientific theories are almost always approximate and, in that sense, wrong. Approximations can improve, but "truth" recedes and eludes [14–16]. Where does this leave "explanations"? I am afraid they are normally only roughly correct. Fortunately, they can be good enough to put a man on the Moon and bring him back again, so that what is strictly correct and what is wrong, but roughly correct, becomes a rather subtle matter of making distinctions.

7. Case 6: The Uncertainty Principle

In the general formulation given to it by Robertson [17]

$$\sigma_A \sigma_B \geq \tfrac{1}{2} < [A, B]_\psi > \tag{1}$$

where $[A, B]$ is the commutator of the operators A and B, and $< \cdot >_\psi$ denotes the expectation value for a state ψ. The quantity σ_X is the standard deviation of the values of an observable X as worked out according to the probability rules of quantum mechanics. The original uncertainty relations were however derived by a detailed discussion of certain experiments [18, 19] and they led to

$$\Delta x \Delta p \geq h$$

where $\Delta x, \Delta p$ are genuine least uncertainties in the values of x and p which are attainable in the experiment. However, these two concepts, Δx and σ_x, cannot be identified, as can readily be seen by taking a simple example [20]. In addition, the right-hand side of (1) vanishes if ψ is an eigenfunction of A or B, leaving the result (1) as a purely mathematical inequality to which there are no contributions from physics. This has led quite recently to alternative formulations of (1), for example in [21], following some other attempted reformulations.

This long delay reminds us that the human mind is blinkered and will repeat transmitted wisdom sometimes uncritically. We are in fact liable to be brainwashed to accept as explanations notions which are not entirely satisfactory. It is part of the point we have made repeatedly, namely, that explanations are in terms of background knowledge—it is just too bad if this itself is defective, as it is always liable to be.

8. Conclusions

To summarize our observations on this difficult subject:

1. An explanation requires the reduction of the topic considered to background knowledge, which for this purpose has to be *assumed valid.* It could be the theory of gravitation, quantum theory, fusion reactions, classical and statistical mechanics, etc.
2. Depending on the time in the history of science at which the explanation is required, one must be able to invent new concepts. These might be a gravitational theory, the wavefunction, coarse-graining, the entropy concept, etc.
3. Easily and often neglected, but noted by the conscientious researcher, are not only the well known but also the little known earlier studies. The big steps in the advance of science are often big only because the little intermediate steps have been forgotten or ignored.

In the course of the work one may encounter certain difficulties which may make it hard, though not impossible, to proceed:

4. Positivism (see §3).
5. The realization that science is imperfect (§§2, 3, 6).
6. The need to mutilate reality (§§3, 6).
7. The nonuniqueness of the explanation (§§4, 5).
8. The threat of explanation by new experiments. This is, of course, always present (§4). The need for control experiments.

9. The fact that the human mind is brainwashed by accepted ideas, and so is blinkered.

I have not gone to the philosophers for an explanation of "explanation," for I am not expert enough to do so. Fred Wilson [22] quotes J. S. Mill's *System of Logic* (1858):

> An individual fact is said to be explained, by pointing out its cause, that is, by stating the law or laws of causation of which its production is an instance . . . a law or uniformity in nature is said to be explained, when another law or laws are pointed out, of which the law itself is but a case, and from which it could be deduced.

This seems in agreement with what has been said here. However, we then come to Wilson's own introduction to the subject, which runs:

> An explanation of certain facts is a set of sentences, including a sentence about the fact or facts (individual or general) to be explained. The sentences are *about* facts, including the fact or facts to be explained. The sentences which *are* the explanation constitute an explanation of the facts because sentences in the explanation are about those facts.

This seems too cumbersome for our purposes; so I did not use this approach.

A worthwhile volume to look at is [23] which is also written from a philosophical point of view, but has good points of contact with physics and probability theory. Although I did not use it in the preceding discussion, one can observe, for example, that if B is statistically relevant to A, then

$$s(A,\ B) \equiv p(A|B) - p(A) > 0 \tag{2}$$

This states that the probability of A is enhanced if B holds. The function $s(A,\ B)$ is a kind of support function for A from B, and its properties have been discussed recently by reference to a certain controversy regarding induction (see [24] and references cited therein). In any case, it follows from (2) that

$$p(A \cap B) \equiv p(A|B)p(B) > p(A)p(B) \tag{3}$$

The fact that the probability of $A \cap B$ occurring is greater than the probability that they occur together, if they were independent, needs "explanation." How about a common cause C being responsible? Then

$$p(B|C) > p(B) \tag{4}$$

One finds indeed from the definition of $p(A \cap B)$ and (2) and (4) that

$$p(A \cap B|C) = p(A|B)p(B|C) > p(A)p(B)$$

which is essentially (3). Arguments of this kind are also relevant, but I have tried to keep close to physics and so I have not employed them here.

References

1. Whitehead, A. N. *Adventures of Ideas,* Cambridge U. Press, Cambridge, 1933, p. 161.

2. Herbert, N. *Quantum Reality,* Rider & Co., London, 1985.

3. Rae, A. *Quantum Physics: Illusion or Reality?,* Cambridge U. Press, Cambridge, 1986.

4. Rohrlich, F. *From Paradox to Reality,* Cambridge U. Press, Cambridge, 1987.

5. Krisp, H. *The Metaphysics of Quantum Theory,* Clarendon Press, Oxford, 1987.

6. Forrest, P. *Quantum Metaphysics,* Basil Blackwell, Oxford, 1988.

7. Landsberg, P. T. in *Physics in the Making,* A. Sarlemijn and M. Sparnaay, eds., Elsevier, Amsterdam, 1989.

8a. Fleischmann, M., S. Pons, and M. Hawkins, *J. Electroanalytical Chemistry,* 261 (1989), 301.

8b. Lewis, N., et al. *Nature,* 340 (1989), 525.

9. deW van Siden, C. and S. Jones, *J. Physics C,* 12 (1986), 213.

10. Dickman, S. *Nature,* 338 (1989), 692.

11. Sweet, W. *Physics Today,* May 1988, p. 87.

12. Landsberg, P. T. *Thermodynamics and Statistical Mechanics,* Oxford U. Press, Oxford, 1978, p. 145.

13. Wigner, E. P. *Comm. Pure & Appl. Math.,* 13 (February) 1960.

14. Landsberg, P. T., "The Search for Completeness," *Nature and System,* 3 (1981), 236.

15. Rescher, N. *The Limits of Science,* U. of California Press, Berkeley, 1984.

16. Schlegel, R. *Completeness in Science,* Appleton-Century-Croft, New York, 1967.

17. Robertson, H. P. *Physical Review,* 34 (1929), 163.

18. Heisenberg, W. *Z. f. Phys.,* 43 (1927), 172.

19. Heisenberg, W. *Die Physikalischen Prinzipien der Quantentheorie,* Hirzel, Leipzig, 1930.

20. Landsberg, P. T. *Foundations of Physics,* 18 (1988), 969 and in *The Measurement Problem of Quantum Theory,* M. Cini and J. M. Lévy-Leblond, eds., Adam Hilger, Bristol, UK, 1990.

21. Maassen, H. and J. B. M. Uffink, *Physical Review Letters,* 60 (1988), 1103.

22. Wilson, F. *Explanation, Causation, and Deduction,* Reidel, Dordrecht, Holland, 1985, p. 2.

23. *Explanation,* S. Körner, ed., Basil Blackwell, Oxford, 1975 (note the article by W. C. Salmon).

24. Landsberg, P. T. and J. Wise, *Philosophy of Science,* 55 (1988), 402.

CHAPTER 4

Boscovich Covariance

OTTO E. RÖSSLER

Abstract

A general principle of mathematical equivalence was proposed by the eighteenth-century physicist Roger Joseph Boscovich: Whether the observer is in a state of motion relative to the world, or vice versa, are equivalent propositions. Similarly, whether the observer is subject to an internal motion (of periodic type, say) or whether the world is, are also equivalent. This unexpected generalization of relativity follows from Boscovich's own example of a breathing world. He invented the principle to explain the fact that the empirical world is rigid. He was the first to discover, and explain, a nonclassical feature of the world. The potential scope of his principle remains to be determined—whether or not, for example, all quantum features can be derived from it. To facilitate the task, a new translation of Boscovich's seminal 1755 paper is given as an appendix to this chapter.

1. Introduction

Kittens who are prevented from exploring the world actively but rather are carried around passively in accordance with the free motions of another kitten (to which they are linked by means of a system of levers), never learn the motion-independent invariant features of the world [1]. Covariance, which is the lawfulness present in all the different "co-varying" (with displacement) aspects of the world [2], can only be discovered actively by an intelligent animal. This lends credence to a conjecture by Poincaré [3] to the effect that regardless of the group structure the world may possess, there is a mechanism already present in the brain that will allow this structure to become second nature to us after a sufficient amount of exploration. Humans, however, can go even one step further and discover through armchair reflection a second covariance based not on position, but on state of motion. In analogy to the position-specific perspectives in an invariant world (geometry) of the previous case, we then have frame-specific versions of the world in an invariant hyperworld (hypergeometry)—as Einstein [4] discovered. Surprisingly, the series can be continued. In this chapter we will describe a third type of covariance, one which was discovered by Boscovich [5].

2. Boscovich's Principle of Equivalence

Boscovich is already known as one of the forerunners of Einstein (after Leibniz) in the discovery of relativistic covariance [6]. However, the fact that Boscovich was unaware of the motion-independence of the speed of light necessarily diminished the impact of his insight that the state of motion of an observer relative to his world constitutes a primary reality. No similar qualification appears to be necessary when it comes to evaluating his second discovery in the present context, a discovery which moreover appears thus far to have gone unnoticed by the scientific community.

Boscovich, who is the inventor of Laplace's demon [7], was the first physicist to apply Newton's theory to the microscopic realm. He founded theoretical chemistry by giving the first potential function for an atom (Figure 1, p. 42, of his book [5] and also [28]). This was apparently the source of his insight that, contrary to appearances, matter (and the world) cannot be rigid but must be more malleable than meets the eye. At any rate, he proposed a new general principle stating that the world must be described relative to an observer.

Specifically, Boscovich claimed that the observer can never observe the world as it is—he can only describe the interface (or difference) between himself and the world. One consequence of this principle is that a state of external motion of the observer relative to the world is *equivalent* to a state of motion of the whole world relative to a stationary observer. This insight forms the basis of both Einstein's special and general theories of relativity [6]. Boscovich's seminal paper of 1755, which he added as a supplement to all editions of his textbook, is reproduced in a new translation in its entirety in the appendix to this chapter. It's of special interest to note the similarity of Boscovich's own example of the ship and the shore to Einstein's example of the two trains standing in a train station.

Being a discoverer of kinetic theory, however, Boscovich apparently saw a second implication. The same equivalence principle applies also with respect to the state of *internal* motion of the observer relative to the world. For example, if both the observer's internal microscopic motions and those of his environment are time-inverted—a possibility of which Boscovich was aware—nothing will change for the observer because the interface will not be affected. While this may sound artificial, there is an interesting corollary: If *only* the observer's internal motions are time inverted, this is *equivalent* to no change having taken place in the observer, but rather to the external world having been time-inverted instead. This means that any change inside the observer that can in principle be exactly compensated for by some external change in the environment (so that the net effect on the interface would be zero), is equivalent to that compensatory change having occurred objectively in the environment with no change having taken place in the observer.

This is a very unexpected result. It means that the observer does not see the world as it is. Since the existence of such compensatable-for changes in the observer can never be excluded a priori, the observer *never* sees the world objectively. Only a "cut" (or "transform") depending on his own state of internal motions will be accessible to him. Moreover, since the interface is the only reality he knows, to the observer the world objectively possesses all those features that in actuality it acquires only relative to his own state of internal motion.

Thus it appears as if Boscovich discovered a generalization of Einsteinian relativity in a direction not foreseen by Einstein himself. The new situation is very reminiscent of that surrounding Lorentz covariance. There, too, it is a directly inaccessible hyperreality (Minkowski's absolute world) that determines the actual reality. At the same time, however, there also exists a major difference between Boscovich covariance and Lorentz covariance.

3. Opacity

In the Boscovich case, even an indirect route to the hyperreality (the invariant description) appears to be blocked. The difficulty is that the observer cannot manipulate his own internal motions in the same way that he can, for example, manipulate his position or his own state of external motion. This puts him into an unresolvable situation similar to that of a kitten that has been deprived of active exploration of its world.

One might object at this point that nothing in fact prevents the observer from actively manipulating his own state of microscopic internal motions. For example, he might put his head into a 4-Tesla medical imaging machine—with the consequence that a measurable distortion of the outside world would occur. However, the point is that the observer cannot manipulate his interface in a specific, planned-in-detail fashion. To do so, he would in effect have to be able to temporarily leave the interface and enter another one in order to compare the two. In the previous two cases of covariance, such alternative aspects could indeed be entertained because a "combined" interface here is still an admissible option. Two interfaces can be combined through communication, either across space (ordinary communication) or across time (memory) or in a combined manner. However, an internal-motions-dependent interface is time-dependent on a time scale so small that the interface cannot be deliberately interfered with and, moreover, cannot be excited. Whatever part of the world is to be looked at, close by or distant, external or internal to the observer's body, in temporal proximity or far into the past, it will be subject to the difference principle. That is, it will be affected by the momentary state of the observer's internal motions. Since the resulting "distortions" affect the whole world, not even memory provides an outside vantage point. Any change of the whole world, however, is imperceptible by definition: Two whole worlds can never be compared.

4. *Falsifiability*

An important counterargument poses itself at this point: Is Boscovich covariance indeed a falsifiable (and hence scientific) concept? At first sight, a covariance of which never more than a single element is on hand does not appear to lend itself to detection, since the larger invariance cannot be constructed out of a single building-block. Unexpectedly, the answer to the question of falsifiability is nevertheless affirmative. While the particular variation remains hidden, its very existence is accessible. Otherwise, the method of analogy used by Boscovich could never have been discovered in the first place.

Boscovich's method involves imagining an entire lower-level world, one completely transparent to a privileged onlooker. Relative to such a world, the existence of the interface principle and its opacity toward being breached from the inside is understandable. Nevertheless, this still doesn't prove anything because of the privileged external position assumed. However, the inhabitants of that world *themselves* are able, at least in principle, to build a still lower-level world and look at it in detail since the latter is totally decoupled from their own world. Therefore, they may by happenstance stumble across a world that is isomorphic to their own as far as the formation of the interface is concerned. Even this may still not be enough. However, if "unreasonable coincidences" showed up between lower-level distortions and upper-level facts, they might start getting suspicious. If so, they could then embark on a systematic program of studying model worlds that are different in type and sophistication (like number of forces incorporated), thereby accumulating a catalog of particular non-objective features of opaque type that could, in principle, arise. Although nothing guarantees completeness or even overlap, they possess a nonzero chance of arriving at the correct identification of some of the laws governing their own world that are not accessible directly. Thereafter, only the individual facts within the identified classes of phenomena would remain forever beyond correction. Such a program has recently been initiated for our own world in [8].

5. Von Neumann's Argument

The fact that Boscovich covariance will be detectable with a nonzero probability if it plays a role in our own world implies nothing about its actual existence. The structure of our own world may well be such that there is simply no room for Boscovich's principle. In particular, the fact that this principle was discovered in a purely classical context by Boscovich gives rise to the question of whether *quantum mechanics* may not invalidate the principle. This important question was first seen and (so it seemed) answered in the affirmative by von Neumann in his classic work [9].

Von Neumann, who was unaware of Boscovich's work, rediscovered the Boscovich principle in the following form: "The state of information of the observer regarding his own state could have absolute limitations by the laws of nature." He proposed the specific hypothesis that this principle, if correct, might explain the quantum-mechanical fact that "the result of the measurement is indeterminate." Thereafter, it only remained for him to show that the laws of quantum mechanis are such that the state of information of the observer regarding his own state is a matter of complete irrelevance. He was able to do this in one line. The conclusion was that if quantum mechanics is the most fundamental description of nature containing all others as special cases, "the attempted explanation breaks down." The envisaged principle of Boscovich would then play no role in our own world.

Von Neumann was convinced that quantum mechanics *is* the most fundamental description of nature. Therefore, the question was settled for him. However, his own hypothesis had explicitly stated that quantum mechanics might be explicable by a more fundamental principle. This more fundamental principle, if it exists, will presumably not need to be couched in terms of the most fundamental description of nature available—quantum mechanics. Thus it turns out that the question is actually *not* settled. There exists a remaining "gap" in von Neumann's proof. To close it will not be easy, since quantum mechanics, the only theory available that's relevant to his proof, will need to be transcended in an unknown direction in order for the proof to be completed.

Therefore, it's defensible to consider the hypothesis that Boscovich covariance does play a role in our own world and, specifically, that the quantum phenomena owe their existence to it. However, even if one opts for this position, it remains true that von Neumann's result represents a major step forward. It shows that should Boscovich's principle reign in our own world, it cannot do so on a quantum-mechanical basis: The "hyperreality" must be classical. Thus, Boscovich's own way of thinking can be adopted without quantum-mechanical modifications.

6. The Problem of Nonlocality

At this point, a second objection necessarily arises. There exists a well-known theorem due to Bell [10], showing that one feature of quantum mechanics, its nonlocal nature, can never be explained by any classical (local) theory. Boscovich's classical Newtonian view is, however, local. Unexpectedly, it turns out that Bell's result, too, leaves Boscovich covariance untouched.

Bell [10] presupposed explicitly that the quantum correlations to be explained are objective. One possible way to interpret this assumption is to say that they might be a characteristic of the Everett [11] world of one particular observer (each observer)[12]. In an explanation based on Boscovich covariance, the Bell correlations would *also* be objective for each observer in question. However, there would exist a second level of objectivity, valid from the outside, where the observed correlations would cease to be objective. Since Bell was unaware of this possibility, he didn't mention it as an exception to his theorem. Specifically, two objectively equal classical particle motions will be affected in the same way by an explicit observer's interface. Nothing more than the assumption of such a "symmetry" is required in order to be able to formally arrive at the Bell correlations [13]. The latter, therefore, are indeed compatible with a deterministic local hidden-variables theory—of the Boscovich type.

7. How It Could Be

Boscovich's view is reminiscent of Bohr's saying that the observer "participates" in the creation of his world (cf. [14]). More specifically, it is very close to the Everett [11] interpretation of quantum mechanics in the version formulated by Bell [12]. To the internal observer, there is always but one Everett universe existing at any moment in time. However, according to Bell [12], the different Everett worlds existing for every possible alternative outcome of any measurement, each containing its own version of the same observer, no longer exist simultaneously, but rather exist successively. In this view, the (single) observer is "hopping" amongst all the different internally consistent universes in a very rapid fashion. Then the shielding principle, which takes the place of state reduction in Everett's formulation [11], no longer cuts across the different universes "vertically" (across a non-physical new dimension), but rather "horizontally"—across the time axis of the observer. But since the shielding remains equally effective, there is no difference for the observer: At every moment he finds himself in a consistent universe with a consistent history. The fact that he is taken out of each universe very quickly and placed in another is "censored out." This view of Bell's [12] is, admittedly, counterintuitive. However, it certainly is less so than the equivalent canonical version of Everett's [11] in which the foliation causes an indefinite multiplication of worlds along many new lateral dimensions.

The classical Boscovichian view proposed above implies a very similar result. The internal motions of the observer "transform" the real modes of existence (the hidden level of objectivity) in a manner that necessarily changes as rapidly as the changes in the internal motions themselves. However, again these changes must be imperceptible. For if they were not, the true reality, from which they all deviate by either a small or a large amount, would have to be perceptible, too—despite the fact that this is impossible by virtue of the equivalence principle (Section 2).

This result is robust. It is independent of the nature of the particular changes occurring in the interface. The latter are contingent on the particular properties of the classical universe assumed. These details will determine whether or not more specific properties of quantum mechanics (like Nelson's diffusion coefficient [15]) are implied as well. In particular, we can imagine a case in which Nelson stochasticity—and hence the Schrödinger equation—is implied, but with a wrong and nonconstant value of the parameter h (Planck's constant) contained in the diffusion coefficient [8]. Unexpectedly, such a "defective" reproduction of quantum mechanics will be perfectly adequate. In each momentary universe, the distortions of objective reality will be such that the resulting "quantum world" is consistent. This applies not only to the individual measurement results, but also to all records—including their statistical properties. Therefore, the value of "h" will also be perfectly well defined within the universe in question. The fact that there is no objective reality behind it will be just as inaccessible as is the fact that the underlying individual quantum events also do not exist objectively.

Thus we find that a Boscovichian world will vary imperceptibly along *more* parameters than is the case in the Everett-Bell picture. In particular, the value of h (and therefore the length scale) will not be conserved across universes, in general. This additional free parameter will be negligible relative to the number of foliation dimensions already contained in the Everett picture. At the same time, however, a prediction made by Boscovich (see the appendix) is unexpectedly confirmed: Not even length scales can be trusted from an objective point of view.

Thus the above view, when compared with Boscovich's, amounts only to a refinement. The two time-*in*dependent quantum effects that Boscovich already singled out (invariance of scale and existence of solidity) no longer stand alone. Time-dependent quantum effects (or rather their analogues) make their entry, too. They comprise, in the Nelsonian case [8], measured eigenstates (like clicks of a geiger counter and plotter excursions) and everything else that can be made dependent on these (like cats and explosives), as well as derived features like nonlocal correlations and, to the observer, universal constants. Boscovich's observer-centered picture can therefore be confirmed qualitatively from a modern perspective.

8. Main Counterargument

The main counterargument can be lifted directly from Boscovich's own writing as well. Toward the end of his essay (see the appendix), he is forced to admit that common sense (the opinion of the many) is *not* ready to accept his views, and that even amongst his colleagues he is completely isolated. There is little doubt that—unless something has changed fundamentally in the scientific attitude—the view presented above is as unacceptable today as his own views were almost a quarter millenium ago.

On one hand, Everett's observer-centered quantum mechanics is fully accepted today by only a small minority of physicists. On the other, the "improvement" achieved by Bell is virtually unknown. Most importantly, however, the explanation offered above involves an unfamiliar element—the introduction of a "second level of reality."

Other "hidden-variables theories" have been introduced in the past, e.g., that of Bohm [16]. Nevertheless, they did not go so far as to specify an entire hidden reality. Maybe if the Boscovich proposal had been more alien in type, containing nonlocal elements like Bohm's theory, it could have been assimilated somewhat more easily. The assertion made above that the world adheres to an outdated, nineteenth-century type of structure, without this fact being accessible from the inside, is particularly hard to accept.

On the other hand, something *has* changed since Boscovich's time. It is not so much a change in psychology, or an enhanced emphasis on experiment, or a new degree of mathematical proficiency, or even an increase in the size of the scientific community; rather, it is the advent of the computer. About a century ago, it still took a major flight of the imagination to arrive at a picture of an outside operator to a kinetic world. Only a "demon" could interfere with the Second Law from the outside. Today, since the invention of the molecular-dynamics simulation by Alder and Wainwright (who put billiard balls into a computer [17]), all connotations of sorcery have vanished completely.

9. Second-Level Experiments

Still, it is not the computer alone which makes the above new type of explanation acceptable today. It is not even the fact that new predictions can be arrived at using this instrument, predictions that concern our own world. Rather, what is decisive is that *we* are ready to test these predictions.

The notion of an experiment was "invented" in the Renaissance. Even though the Greeks had used it earlier with great mastery, its decisive function as an escape hole from the darkness of speculation remained undiscovered in their time, due to the subconsciously held belief that experimentation is childish. Until very recently, it appears that a similar haughty prejudice kept theory subservient to direct experiment. The motivation for experiment had to be "serious." Second-level experiments would have been frowned upon, not only by the ancient Greeks, but also by the contemporaries of Boscovich and Maxwell—and even in the age of mainframes, too.

A second-level experiment is an experiment performed in our own world motivated entirely by the levels-jumping paradigm. The underlying rationale is the discovery (Section 4) that the inhabitants of a toy world have access in principle to *actions* whose outcome would teach them facts about their own world that are ordinarily hidden to them. Of course, to embark on these very actions, the inhabitants would first need to be "cued." And the only thing that can cue them is a still lower-level world.

From the point of view of traditional theory, it would be irresponsible to take this fact at face value in our own world and start experimenting right away. Probability stands in the way. More specifically, the nonuniqueness principle (cf. Section 4) makes the approach appear hopeless from the beginning. The endeavor therefore has only the character of "mathematical playwork."

On the other hand, nothing guarantees that experimentation is impossible. Only an "adult" would not even try because he is too well aware of the pitfalls. The point is that it costs nothing to try. This mental lightness is, perhaps, in itself a consequence of the computer age (which lets having your cake and eating it too no longer appear impossible [18]). Explanation, prediction *and venture,* taken together, appear to form a new post-Renaissance triad: Second-level experiments are there to be tried.

10. An Example

Let us not be too daring at this point. Since it is too early to get a concrete hint from a particular lower-level universe, all that can be done with confidence is to exploit the fact that the outlook on our own world has changed. New questions concerning "standard experiments" come into view that would have been impossible to consider outside the toy-worlds paradigm.

There exists an experiment which combines Einsteinian relativity and Boscovich relativity in a nontrivial fashion. Unfortunately, there is no way to predict its outcome using Boscovich's theory, since a relativistic molecular-dynamics model world will not be available for the foreseeable future. On the other hand, there is virtually no disagreement about its outcome. In fact, the agreement is so unanimous that the experiment will probably never be carried out, despite the fact that it is feasible with current technology [19].

The problem was apparently first seen by Susan Feingold in unpublished 1978 notes, quoted by Peres [20]. It is a standard correlated-photons experiment, in the sense of Bell, with one unusual feature. The two measuring stations (analyzer, detector, observer) do not belong to the same inertial frame but rather recede from each other. The idea is to have each side reduce the singlet-like superposition state of the photon pair "first," in its own frame. A very simple space-time diagram [19, 20] shows that this is possible. Suppose that current theory [20] is correct in predicting a negative outcome (no change in correlation relative to an ordinary Bell-type experiment). What does this outcome *mean*?

It means that each observer can stick to the assumption that the other's measurement was done *after* the singlet state had already been reduced by herself. The photon impinging on the other's measuring device already possessed a fixed spin. The subsequent registration of the latter was a matter of course (since, according to relativistic quantum mechanics, a receding measuring device will measure an unchanged spin). Therefore the fact that there is no difference relative to an ordinary Bell experiment will be in complete agreement with what each observer expects.

Moreover, these two interpretations, held symmetrically by the two observers, are at first glance "compatible." Bell's correlations are the same no matter which side reduces the singlet state [10]. In an ordinary Bell experiment, therefore, each side can *pretend* to bear the responsibility for the other's correlated outcome. Nevertheless, since the two observers are mutually at rest, it is always possible in principle to find out which side accomplished the reduction (passing to a different reference frame might change this result, but not without additional hypotheses). Thus, ordinarily no more than one side can be proven wrong. In the present case, however, either side can always be proven wrong by the other. Still, since each of these two allegations when taken alone is compatible with the observed facts, it appears straightforward to conclude that both taken together are compatible, too.

This opinion appears tenable. Each observer's world adheres to the rules of nonrelativistic quantum mechanics (that is, obeys Bell's theory). Therefore, everything is all right. Nevertheless something is subtly wrong. One has the feeling that there is *too much* symmetry present.

A very similar situation held sway once in the past. In that earlier situation, it was also the case that no one took offense until a whole new branch of physics had to be developed as a remedy. After the constancy of the speed of light had been discovered experimentally, and after Lorentz had formulated his contraction law which explains the phenomenon, ten more years would pass before the decisive question was put forward by Einstein: Shouldn't each observer be concerned that it's always the other side that contracts so conveniently?

Einstein realized that too much symmetry must be taken as a cue in its own right. Whenever two observers have a consistent view of the world, yet each in such a way that the two views contradict each other in the sense that the same facts are explained differently by the two sides even though both make use of the same theory, we have an example of "empirical covariance."

Once one has been alerted to such an empirical state of affairs in a concrete case, it is natural to search for a deeper—invariant—picture from which the two covariant facts have sprung. In the present case there are two ways open as to how to arrive at an invariant solution. The first possibility is to postulate that quantum events are "cooperative" events: The future—the other observer—participates in the creation of the presently observed outcome on an equal-rights basis.

This is the message taught by field theory and relativistic quantum mechanics. A nonuniqueness (overdetermination [19]) in space-time allows each frame to pick out the facts consistent with nonrelativistic quantum mechanics. The commutation relations can be violated, but only in retrospect [19].

The alternate possibility takes seriously Bohr's view that the decision of the observer as to what to measure determines critically how the world is going to look. All other observers, linked to him by linguistic communication, are bound to comply. Everett's interpretation says exactly the same thing. Nevertheless it contains an added element of "free play" that is ordinarily unnecessary, since isomorphism to Bohr's world suffices under all thus far known conditions. In the present case, however, this element unexpectedly becomes crucial. While every Everett world is consistent, one would be mistaken in assuming that the Everett worlds of two observers are equal, in general. Not even two Everett worlds of the same observer are equal.

Thus, unlike Bohr's quantum mechanics, Everett's *can* be brought into a covariant form. In analogy to Einstein's solution in the earlier case ("either side contracts in the frame of the other"), it becomes possible to say: "either side complies in the world of the other." Just as the existence of more than one "frame" had to be acknowledged in the previous case, so more than one "world" is needed in the present one.

While in the first possiblity above the invariant world could be indicated immediately (although it turned out to be overdetermined), in the present case the necessary second step (pinpointing the invariant absolute world that gives rise to the two covariant worlds) is open. None of the currently discussed hidden-variable theories (including stochastic ones, cf. [21]), appears to be up to the "enlarged" task of having to accomodate more than one quantum world. The only candidate theory is the above proposal. Thus, an *empirical* reason why a theory as unfamiliar in structure as Boscovich's may be needed appears to have been found.

11. Discussion

A new "rationalist" theory of the physical world has been presented. It revives a proposal made almost a quarter-millenium ago by Boscovich. The idea belongs in a long tradition in science which includes the names of Heraclitus and Maimonides, to mention only two. True reality is different from the way it appears. Nevertheless it is possible to find out about this fact, such that some or even all of the laws involved can be discovered, with only the contingent details remaining hidden forever.

The approach in a sense generalizes Gödel's [22] "valid from the inside" limitation principle. Unlike previous attempts to capitalize on Gödel's mathematical insight for physics (cf.[8] for references), it looks at explicit observers in a continuous, reversible context. It still suffers from the drawback that fields, even though eventually needed to explain the discrete spectrum of whatever particles are assumed, are not included. Therefore a complementary program like Wheeler and Feynman's [23], which allows one to get rid of the infinitely many degrees of freedom of a classical field, will be needed in the long run.

Boscovich in his own time was able to identify only two properties of our empirical quantum-mechanical world—scale and solidity—as being incompatible with the consistent classical picture he himself had developed. For these properties he was forced to find a new type of explanation. To date, many more quantum features, all derived from time-dependent state reduction, would have to be included. Unexpectedly, Boscovich's approach does not crumble under the enlarged task. The basic idea required—making his approach time-dependent—is already contained in his own work (see the appendix). The interface principle is couched there entirely in terms describing the relation of the outside world to the state of the "organs" of the observer. The latter's state by definition is time-dependent, in general.

In this way, a qualitative explanation of the three major new phenomena of quantum mechanics—stochasticity, nonlocality and state reduction—can apparently be arrived at. What is open is whether quantitative details (like the numerical value of Planck's constant or its degree of isotropy in space) can also be captured eventually. A precondition for tackling such questions is complete success in the task of calculating and visualizing the simplest case (namely, how a third particle, acting as an "environmental forcing," impinges on a two-degree-of-freedom chaotic subsystem if all necessary symmetries [8] are taken into account). Afterwards, τ, the mean interval between effective time reversals in a multiple-particle explicit observers (cf. [24]), will become calculable, and from it h_{obs}.

Even more important, however, than qualitative explanation and quantitative prediction of things already known, may be the fact that the present approach alerts us to new possibilities. New covariances should exist empirically in which the observer plays a privileged role.

In this context, let's consider the most extreme observer-centered phenomenon possible—the phenomenon of nowness. Bell was the first to realize [12] that the sealing-off principle employed by Everett [11] has the same structure as the sealing-off that occurs phenomenologically in the transition from one momentarily valid world to another. Subjective nowness was thereby elevated from the status of a codimension one phenomenon to that of a codimension two phenomenon (with no longer only the time axis but also Everett's parameter Ξ participating in the unfolding). However, no rationale for this elegant hypothesis could be offered.

The present approach to quantum mechanics not only unexpectedly confirms and explains Bell's version of Everett's theory as shown above, but at the same time also "explains" nowness. Since the previous states of the interface are not represented within the interface itself, they are extinguished, in effect, along with everything they entail, from one time slice to the next. Hence there exists a "cross inhibition" between neighboring elements on the time axis that has no corollary along the other axes. The present approach thus explains more than it was meant to. The formulation of nowness as a covariance principle unexpectedly arises.

Nowness, on the other hand, belongs to subjective phenomenology. So far, only macroscopic brain states have been known to possess subjective correlates. For example, the microscopic states by definition are irrelevant for the macroscopic workings of a computer. Could they nevertheless be relevant to the "world" of such a machine, should the latter possess any? Here, by the interface principle, the answer is yes.

Does this mean that subjective experience may perhaps become predictable scientifically? The answer is no. We can never hope to find anything more than objective "correlates." On the other hand, a theory admitting an observer-private interface into the picture may allow for much *finer* correlates to be found than has been possible before.

The present approach therefore in a sense revives Cartesian rationalism. As Descartes knew, rationalism would be dead (falsified) if a single instance of sorcery were reliably reported [25]. Equivalently, it would be sufficient if a single example of an in principle inexplicable arbitrariness were found to reside within the world. This has indeed been the situation since the advent of quantum mechanics. To give a recent example, a Dehmelt atom [26] emits a Morse-code-like time series of light pulses that is inexplicable in its structure according to modern science. Only the individual transition events' probabilities can be indicated. Therefore it is no use, even in principle, to look at this "telegraphic noise" from the point of view of empirical investigation. Any particular sequence by definition has the same interest as a paranormal phenomenon. It makes no difference whether or not a sorcerer of high standing like Don José (Don Juan) is able to prove that he can influence the code. The inexplicable, by definition, not only invites sorcery, it *is* sorcery.

Boscovich's approach combines Einstein's proposal to look for a mechanistic explanation with Bohr's suggestion to look for a holistic one. The "bad dream" that both insisted would pertain in the absence of a rational explanation would thereby be avoided. However, would the price—a return to old-fashioned rationalism—be too steep? The answer appears to be: On the contrary.

Cartesian rationalism consists of a single falsifiable hypothesis. The world, which to the individual has the character of a Big Dream, may be *relationally consistent.* The universe therefore must be a giant dynamical system ("machine"). The first major implication is determinism, while the second implication is exteriority.

Determinism is sometimes considered undesirable because it seems to imply the lack of free will. This is incorrect since, according to Descartes, one particular partial machine would be exempt. Its dynamical states would reflect the content of the whole Dream, while the latter could not be said to be determined by this internal character. Nothing more than consistency with the whole is required for the observer's brain states. Nevertheless it remains true that the brains of all other individuals would indeed be "just machines" within the first's universe. To endow them with free will, too, becomes a problem. Great cunning is required here if one is to arrive at a satisfactory solution as Descartes knew. An Everett-like new dimension of objective type is one of several possibilities (see the discussion of such a dimension by Boscovich [5], p. 234). However, the real problem that poses itself in this context is not metaphysics, it is ethics.

The *exteriority principle* has been rediscovered and named only in our own time by Lévinas [27]. Lower-level worlds by definition pose an ethical problem because of one's being completely exterior to them. However, this "vertical" exteriority is but a minor special case to the "horizontal" one described above. If the other individuals are just machines within the first's universe, he is equally exterior—and hence privileged—relative to them as he is relative to a lower-level world. Absence of sorcery, paradoxically, conveys the ultimate power. This was a desirable state of affairs to Descartes. Thereby he felt that the arbitrariness of the whole dream becomes acceptable. If the victim of the dream cannot exclude, within the dream, that he is endowed with as much power once more as is being brought to bear against him, he is even. For he can then *refrain* from misusing this infinite power. Lévinas says the same thing without the theological connotations. The unlimited power of exteriority can be used to create one thing even if it had never existed before—fairness. Rationalism thus unexpectedly is a most desirable option. It *had* to be created anew after science had abandoned it.

The present attempt at restoration is, therefore, not really crucial. Many details still need to be filled in before it can be called a qualitative success. Alternatively, the "gap" in von Neumann's counterargument may turn out to be bridgeable. The more vigorous the attempts at falsifying the Boscovichian position, the greater are the chances that a more reliable position will be found.

To conclude, a scientific idea proposed 234 years ago and rediscovered 57 years ago has been re-examined. No fault could be found with it. Boscovich's proposal belongs within a class of physical principles whose independent standing becomes fully visible only now, that of covariance principles. It is the most general example discovered so far. In fact, it is so general that it even ceases to be demonstrable directly. Maybe it does not exist. Alternatively, it may represent a new way of decoding the world.

12. Summary

A new generalized covariance principle has been offered. Even motions *inside* the observer transform the world. This new principle may allow for an alternate explanation of some known facts and for the prediction of some new facts. Quantum mechanics, exteriority and nowness provide candidates. A new sensitivity to observer-centered covariant phenomena is called for. A new type of experiment may become possible. The proposed invariant "higher-level" description of the world is close to the traditional classical picture. Some connections to the history of science on the one hand and to philosophy on the other were pointed out.

13. Acknowledgments

I thank John Casti and Bob Rosen for stimulation. Discussions with Peter Erdi, Karl-Erik Eriksson, George Kampis, Martin Hoffmann, Peter Weibel and Jürgen Parisi were also very helpful. Werner Kutzelnigg first mentioned Ruder Boscovich's name to me in 1988. Jens Meier dug up the 1758 textbook and contributed to the discussion of the Feingold experiment. The Abisko conference under the midnight sun acted like a focus.

14. Appendix

On Space and Time as They Are Recognized by Us

R. J. Boscovich [5]

§ 1. (It will be shown that we cannot have absolute recognition of the local modes of existence, nor of distances and magnitudes.)

After our having in the previous paper (supplement) dealt with space and time as they are in themselves, it remains—if I am to attain the goal to say something pertinent about them—to talk about how they are being recognized. We by no means immediately recognize through our senses the real modes of existence, nor can we discern them from each other. We do perceive, from the discriminate impressions which through the senses are excited in our mind, a determinate relation of distance and position which originates from any pair of local modes of existence. However, the same impression can arise from innumerable pairs of modes, such as real points of location which induce relations of equal distance or of similar position — whereby the positions are so between them as they are relative to our organs and to the rest of the circumjacent bodies. This is because a pair of material points which at some place have by one pair of modes of existence been given a distance and an induced position, can at some other place under the influence of another pair of modes of existence again possess a relation of equal distance and similar position, whereby the distances of course exist in parallel. If now these points, and we ourselves, and all circumjacent bodies, changed their real locations (but) such that all distances remained equal and parallel to the former; then we would (will) have exactly the same impressions. And we would still have the same impressions if, while the magnitudes of the distances remained the same, the directions were all turned through an equal angle such that their mutual inclinations remained the same as before. Furthermore, even if all those distances were diminished while their angles and mutual ratios remained unchanged, but such that the forces would not be affected by that change in distances since their scale—namely, that curve whose ordinates express the forces—would be changed in proportion; then we would still have no change in our impressions.

§ 2. (It will be shown that a common motion of ourselves and the World cannot be recognized by us, not even if the latter as a whole were increased or decreased in size by an arbitrary ratio.)

From the above follows now that if this whole visible World was moved forward in a parallel motion in some arbitrary planar direction and was simultaneously turned through some arbitrary angle, we would not be able to feel this very motion and rotation. Similarly, if the whole alignment of the room in which we are with the plains and the mountains outside was in the same sense rotated through some arbitrary common motion of the Earth, we cannot feel such a kind of motion because impressions that are equal to the senses are excited in the mind. In fact it might happen that this whole visible World contracted or expanded in a matter of days while the scale of the forces involved contracted or expanded in unison. Even if this happened, the impressions in our mind would not be altered and we would have no feeling of this kind of change.

§ 3. (It will be shown that if either our own position changes or that of everything that we see, our impressions will not be different, and that therefore we cannot legitimately ascribe motion either to ourselves or to the rest.)

In case it is either the external objects or our own organs that change their modes of existence — in such a way that neither the above equality nor the above similarity are preserved — , then indeed the impressions are altered and we do have a sense of the change. However, *the impressions are completely the same, no matter whether it is the external objects that undergo the change, or our organs,* or both in a nonmatching fashion. Always our impressions report the difference between the new state and the previous one, not the absolute change which does not fall under the senses. Thus, whether it is the stars that are moved about the Earth or whether it is the Earth with ourselves on it that turns about itself in the opposite direction, the same are the impressions, the same are the feelings. The absolute changes we can never feel, the difference (discrimination) from the previous form we feel. Whenever there is nothing present to admonish us of the change of our organs, we indeed judge ourselves unmoved in accordance with the saying (common prejudice) that we take for nothing what is not in our mind since it is not recognized. Moreover, we attribute the whole change to objects that are situated outside ourselves. Such errs who enclosed in a ship judges himself motionless while the shore and the mountains and the wave are taken to be moving.

§ 4. (It will be shown that from the way we judge the equality of two from the equality with a third, it follows that we can never have a congruence in length or in time that is not indirectly inferred.)

A most important implication of the above principle of unchangeability of those things whose change we do not recognize through our senses is the method to which we adhere when comparing the magnitudes of intervals between themselves. For we take that which we use for the measure to be unchangeable. Specifically, we make use of the principle that "what are equal to a third are equal among themselves," from which follows the corollary that "what are multiples or submultiples of a third are equal among themselves" and "what are congruent are equal." Let us assume the example of a ten-foot rod made of wood or iron. If we find the latter congruent to a first interval after applying it once (or a hundred times), and then congruent to another interval after one (or one hundred) applications, we call these intervals equal. Furthermore we take that wooden or iron ten-foot rod to be identically the same standard of comparison after the translation in space. If the rod consisted of perfectly continuous and rigid (solid) material, it could justly be taken to be identically the same standard of comparison. However, according to my above theory on the mutual distance of points, all points of the rod continually change their distance in reality during the transfer. This is because distance is constituted by real modes of existence, and the latter change all the time. Even if they change in such a way that the modes before and after the transfer establish real relations of equal distances, the standard of comparison will not be identically the same in length. Nevertheless it will be equal, and the equality of the measured intervals will be correctly inferred. We are certainly no better able to directly compare the length of the rod, as it is constituted in the previous situation by the previous real modes, with its length in the later situation as it is constituted by the later real modes, than we are able to compare those intervals through measurement using the rod. But since we feel no change during the translation which demonstrates to us the relation of length, we consider the length itself to be identically the same. However, in reality the latter never remains totally unchanged during the translation. It could even happen that it underwent some giant change, the rod as well as our senses (which we would not feel), but that it when returned to the previous place would resume a state equal or similar to the previous one. Some slight change at any rate takes place in general. This is because the forces which connect the material points between themselves cannot remain completely unchanged when the position relative to all points of the remaining parts of the World has been changed. The same thing also holds true in the usual theory. This is because no physical body is free of little spaces interspersed within it and is therefore never totally incapable of internal compression and dilation. This dilation and compression is generally believed to occur at least on a small scale in every translation. We, however, take the measure to remain identically the same because we feel, as I have stressed above, no change.

§ 5. (It will be shown in conclusion that common sense and scientific judgement are at variance.)

From all of the above follows that we are totally unable to recognize absolute distances directly, and that we also cannot compare them with one another by means of a common standard. Rather, we have to estimate the magnitudes by using the impressions through which we recognize them. For common standards we have rulers (measures) in which according to common sense (popular opinion) no change has taken place. The scientists (philosophers), however, certainly must acknowledge the change. Moreover, since there is no reason envisionable why the homogeneity (equality) should be changed notably, they consider the change to occur homogeneously (equally).

§ 6. (It will be shown that, despite the fact that the real modes which constitute the relation of interval change with the translation of the ten-foot rod, nevertheless equal intervals can be taken to be identically the same with good reason.)

Let us add that even though it is true that, when material points change their place (as in the translation of the ten-foot rod), in reality the distance changes as a consequence of the change in those real modes which constitute the rod, we will nevertheless, in case the change is such that the later distance is exactly equal to the previous one, adopt the language that we call it the same and unchanged. Similarly, equal distances on the same standards will be called the same distance. Moreover, a magnitude is to be called the same if it is defined by those equal distances. Also, a pair of parallel directions will go under the name of same direction. And we will say in the following that neither distance nor direction have changed when neither the magnitude of the distance nor the parallelism has changed.

§ 7. (It will be shown that the same results must be applied to time, in which case even common sense knows that one and the same temporal interval cannot be transposed for the purpose of comparing two, even though it errs in this respect about space.)

What we have said about the measurement of space is not difficult to apply to time as well, for which, too, we have no certain and constant measure. Hard as we might try to derive this notion from motion, the fact remains that we have no perfectly uniform motion. Much of what is pertinent to this context, and much of what concerns the nature of our impressions and their succession in time, has been said in our notes. I add here only one point, concerning the measuring of time. Here not even common sense believes that identically the same measure of time can be transposed from one time toward another time. It is evident that the interval (measure) is a different one, although it is supposed to be equal because the translational motion is supposed to be equal. Both in spatial measurement, according to my theory, and in temporal measurement, it is equally impossible to transpose a certain length (or duration, respectively) away from its own seat toward a new one, so that two could be compared using a third. In either case a different length (or duration, respectively) is substituted which is judged to be equal to the previous one. But, of course, new real places of points constitute the new distance on the same ten-foot rod, and a new circle is generated by the same pair of compasses, and a new temporal distance applies between any pair of initial and final points. In my book "Theory," the same perfect analogy between space and time is consistently applied. According to common sense it is only in spatial measurement that the same standard of comparison can be used. Almost all scientists other than me still think that identically the *same* standard could be used for a perfectly rigid (solid) and continuous measure in space while in time only an *equal* one remains. I in either case only acknowledge the possibility of equality, never identity (sameness).

(Remarks: The numbering of the paragraphs in the original reference [5] runs from 18 to 24 rather than from 1 to 7. The word "impression" consistently stands for "idea" in the original. Italics added.)

References

1. R. Held and A. Hein (1963). "Movement-produced stimulation in the development of visually guided behavior," *J. Comp. Physiol. Psychol.*, 56, 872–876.

2. E.F. Taylor and J.A. Wheeler (1963). *Spacetime Physics*, Freeman, San Francisco, p. 42.

3. H. Poincaré (1904). *La Valeur de la Science*, Flammarion, Paris, p. 98.

4. A. Einstein (1905). "Zur Elektrodynamik bewegter Körper," *Ann. d. Physik*, 17, 891–921.

5. R.J. Boscovich (1755). De spatio et tempore, ut a nobis cognoscuntur (On space and time, as they are being recognized by us). Reprinted in his *Theoria Philosophiae Naturalis* (Theory of Natural Philosophy) of 1758, Appendix B. See the Latin-English edition of the 1763 Venetian Edition, *A Theory of Natural Philosophy,* J.M. Child, ed., Open Court, Chicago 1922, pp. 404–409. Reprinted by MIT Press, Cambridge, Mass. 1966.

6. H.G. Alexander (1956). *The Leibniz-Clarke Correspondence,* Manchester University Press, Manchester, p. xliv–xlv.

7. S.G. Brush (1976). *The Kind of Motion We Call Heat, A History of the Kinetic Theory of Gases in the 19th Century,* Book 2, pp. 594–595. North Holland, Amsterdam; see [5], pp. 141–142.

8. O.E. Rössler (1987). "Endophysics," in *Real Brains, Artificial Minds,* J. L. Casti and A. Karlqvist, eds., pp. 25–46, Elsevier, New York.

9. J. Von Neumann (1955). *Mathematical Foundations of Quantum Mechanics,* Princeton University Press, Princeton, p. 438.

10. J.S. Bell (1964). "On the Einstein-Podolsky-Rosen paradox," *Physics,* 1, 195–200.

11. H. Everett, III (1957), "Relative-state formulation of quantum mechanics," *Rev. Mod. Phys.,* 29, 454–463.

12. J.S. Bell (1981). "Quantum mechanics for cosmologists," in *Quantum Gravity 2,* C. J. Isham, R. Penrose and D. Sciama, eds., pp. 611–637. Clarendon, Oxford.

13. M. Hoffmann (1988). "Fernkorrelationen in der Quantentheorie: Eine neue Interpretation (Correlations-at-a-distance in quantum theory: a new interpretation)," Doctoral Dissertation, University of Tübingen.

14. J.A. Wheeler (1980). "Beyond the black hole," in *Some Strangeness in the Proportion: A Centennial Symposium to Celebrate the Achievements of Albert Einstein,* H. Woolf, ed., pp. 341–375. Addison-Wesley, Reading, Mass.

15. E. Nelson (1966). "Derivation of the Schrödinger equation from Newtonian mechanics," *Phys. Rev.,* 150, 1079–1085.

16. D. Bohm (1952). "A suggested interpretation of the quantum theory in terms of 'hidden variables', " *Phys. Rev.,* 85, 166–179, 180–193.

17. B.J. Alder and T.E. Wainwright (1957). "Studies in molecular dynamics," *J. Chem. Phys.*, 27, 1208–1209; *Scientif. Amer.*, Oct. 1959.

18. T.A. Bass (1985). *The Eudaemonic Pie,* Houghton-Mifflin, Boston.

19. O.E. Rössler (1989). "Einstein completion of quantum mechanics made falsifiable" in *Complexity, Entropy and the Physics of Information,* Santa Fe, NM, May 29–June 2, 1989, W. Zurek, ed., Addison-Wesley, Reading, Mass.

20. A Peres (1984). "What is a state vector?," *Am. J. Phys.*, 52, 644–649.

21. F. Guerra and M.I. Loffredo (1980). "Stochastic equations for the Maxwell field," *Lett. Nuovo Cim.*, 27, 41–45.

22. K. Gödel (1931). "On Formally Undecidable Propositions." Basic Books, New York 1962.

23. J.A. Wheeler and R.P. Feynman (1945). "Interaction with the absorber as the mechanism of radiation," *Rev. Mod. Phys.*, 17, 157–162.

24. O.E. Rössler (1989). "Explicit observers," in *Optimal Structures in Heterogeneous Reaction Systems,* P. J. Plath, ed., pp.123–138. Springer-Verlag, New York.

25. R. Descartes (1641). *Meditationes de Prima Philosophia,* (Meditations on the First Philosophy). Soly, Paris.

26. W. Nagourney, J. Sandberg and H. Dehmelt (1986). "Shelved-optical-electron amplifier: Observation of quantum jumps," *Phys. Rev. Lett.*, 56, 2797–2799.

27. E. Lévinas (1946). *Le Temps et l'Autre,* (Time and the Other). fata morgana, Montpellier 1979. See also L. Wenzler, "Zeit als Nähe des Abwesenden" (Time as presence of the absent), postface to the German-Language Edition *Die Zeit und der Andere,* Felix Meiner Verlag, Hamburg 1984, pp. 67–96.

28. E. Heilbronner (1989). "Why do some molecules have a symmetry different from that expected?" *J. Chem. Educ.*, 66, pp. 471–478.

CHAPTER 5

Entropy and Correlations in Discrete Dynamical Systems

KRISTIAN LINDGREN

Abstract

Information theory for lattices can be used to characterize the spatial structure as well as the spacetime pattern in discrete dynamical systems. In this paper we shall review some of the formalism useful for analyzing correlations and randomness in lattice systems of any dimension.

The metric entropy, the average entropy per lattice site, is a quantity expressing the degree of randomness of a system. For a spatial pattern that is a microstate in a statistical mechanics system, the metric entropy equals the statistical mechanics entropy. But for a pattern that is the state of a macroscopic system, the metric entropy measures disorder of a higher level. We shall also discuss some complexity measures that quantify the degree to which correlation information is distributed throughout the system.

The formalism is applied to deterministic and probabilistic cellular automata. The temporal behavior of the spatial metric entropy is analyzed and its relevance to statistical mechanics is discussed. For a general class of lattice gas models, the metric entropy is nondecreasing in time. The correspondence between the metric entropy and the statistical mechanics entropy shows the close relation of this result to the second law of thermodynamics.

The equivalence between n-dimensional cellular automata and $(n+1)$-dimensional statistical mechanics models is discussed. This correspondence is illustrated by a class of simple probabilistic filter automata that generate spacetime patterns similar to the configurations seen in the two-dimensional Ising model.

1. Introduction

Cellular automata are abstract simulation models that are discrete in space and time. The number of states per lattice site is finite and usually small. The interaction distance is also finite, i.e., the system evolves according to what is termed a "local rule." Nevertheless, very simple cellular automata are able to display very complex behavior [Wolfram 1983]. This behavior can be investigated using methods from, for example, formal language theory [Wolfram 1984a, Grassberger 1986, Nordahl 1988, Lindgren and Nordahl 1988] and information theory [Wolfram 1983, 1984b, Grassberger 1986, Lindgren 1987, Lindgren and Nordahl 1988]. In this paper we shall review some information-theoretic concepts stemming from the work of Shannon [1948], which are applicable to (time-dependent) spatial patterns as well as to the spacetime pattern created by a cellular automaton rule.

Cellular automata have been used to simulate a variety of physical systems, such as crystal growth, self-organizing chemical systems, and phase-transitions [Wolfram 1986]. At present, the most active applications are lattice gas simulations [Frisch, et al 1987], research stimulated by the development of computers equipped with fast, parallel processors. Also, the original idea of using cellular automata for simulating living systems [von Neumann 1966] is still of interest (see, for example, the articles on artificial life in [Farmer, et al 1986]). One example is the recent microbiological application to the internal dynamics of microtubules reported in [Hameroff, et al 1988].

For one-dimensional infinite sequences of binary symbols, information theory provides concepts for decomposing the total information of one bit per symbol into contributions from correlations of different block lengths (including the single-symbol blocks), as well as from the internal randomness of the sequence quantified by the metric entropy. The metric entropy expresses the asymptotic degree of compressibility that can be achieved by a coding scheme that exploits almost all of the correlations in the sequence.

An approach to the quantitative definition of information, different from that proposed by Shannon, was suggested by Solomonoff [1964], Kolmogorov [1965], and Chaitin [1966]. They define the algorithmic information (or algorithmic complexity) of the system as being the length of the shortest algorithm with which an (abstract) computer can generate the system. An example of such an abstract computing device is the Turing machine, a formal computer capable of performing any type of computation [Turing 1936-37]. The algorithmic approach led to a formalism, algorithmic information theory, which is very closely related to information theory in the ordinary sense. But the algorithmic information has even more general applicability, since it does not presuppose any probabilistic aspects. The concept seems to be useful in the description of living organisms [Chaitin 1979], which have very compact descriptions in the form of DNA and RNA sequences.

Most symbol sequences of a given length have no description more compact than the sequence itself, and may be said to be algorithmically random [Martin-Löf 1966]. If there is some algorithmic structure present in the sequence, not necessarily detectable as correlation information, then the algorithmic information is smaller than the length of the sequence. A quantity, the algorithmic redundancy, can be defined corresponding to the correlation information in ordinary information theory. Furthermore, it can be shown that, given an infinite stationary stochastic process, almost all sequences generated by the process have metric entropy equal to the algorithmic information [Zhvonkin and Levin 1970, Brudno 1977, Lindgren 1987]. It should be noted that the algorithmic information is, in general, an uncomputable quantity.

Bennett [1982] has shown that the algorithmic information of the microstates in a statistical mechanics system, averaged over the ensemble, equals the thermodynamic entropy of the system. The equality between algorithmic information and metric entropy can then be used to derive the following result: If a symbol sequence is a representation of a microstate in an equilibrium ensemble of a physical system, then the metric entropy equals the statistical mechanics entropy for almost all microstates in the ensemble [Lindgren 1988]. Thus it suffices to calculate the metric entropy of a single microstate, e.g., from a simulation, to get the statistical mechanics entropy as well as the other thermodynamic properties, assuming that the microstate is large enough. For one-dimensional spin models there is no correlation information from distances larger than the interaction length [Fannes and Verbeure 1984, Lindgren 1988], a fact which is not true in higher dimensions.

We have applied the information-theoretic concepts for the spatial structures discussed above to the states of infinite cellular automata, showing how the metric entropy changes in time for different classes of rules [Lindgren 1987]. Most deterministic rules decrease the accessible region of state space, so that the metric entropy decreases in time. This corresponds to a self-organizing process in which order increases. On the other hand, if the rule is sufficiently close to being reversible, we obtain a dynamics that can be used to simulate statistical mechanics, the metric entropy remaining constant. If such a reversible rule is disturbed by noise, the metric entropy increases until it reaches a stationary maximum. It can be shown that for a general class of lattice gas models for which semidetailed balance holds, the metric entropy of the spatial configuration is nondecreasing in time. In the thermodynamic limit, the metric entropy (multiplied by the Boltzmann constant) equals the statistical mechanics entropy of the system, implying that this theorem is closely connected to the second law of thermodynamics.

We shall also present a generalization of the information-theoretic formalism in dimensions two and higher [Eriksson and Lindgren 1989], including an expression for the metric entropy that converges much faster than the definition of average entropy per lattice site. The application of this formalism to the spacetime pattern of probabilistic cellular automata will be discussed, since such models are highly relevant for the statistical mechanics of spin systems [Rujan 1987, Georges and LeDoussal 1989]. We illustrate the correspondence between one-dimensional automata and two-dimensional spin models by very simple filter automata rules that generate spacetime patterns similar to the configurations seen in the two-dimensional Ising model.

2. Information Theory

As far as we know, the quantitative concept of information originated with Hartley [1928]. But a theory for information was not formulated until Shannon [1948] presented what we know today as information theory. An important branch of this field is information theory for symbol sequences [MacMillan 1953], which has been used to define quantitative measures for structure and complexity, e.g., [Watanabe 1969, Grassberger 1986, Eriksson and Lindgren 1987]. Shannon's formula for information is identical to the formula for entropy in statistical mechanics, the fundamental nature of this relation having been demonstrated by Jaynes [1957] in his information-theoretic formulation of statistical mechanics. We shall begin this section with a brief review of some basic information-theoretic concepts before turning to the analysis of one-dimensional symbol sequences and lattices of symbols.

Let $P = \{p(k)\}_k$ be a normalized probability distribution, i.e., $\sum_k p(k) = 1$ and $p(k) \geq 0$. Then the entropy of P is defined [Shannon 1948] to be

$$S[P] = \sum_k p(k) \ln \frac{1}{p(k)}. \tag{1}$$

Let $P_0 = \{p_0(k)\}_k$ be a normalized a priori probability distribution with $p_0(k) > 0$ for all k. Then the relative information (Kullback information) of P with respect to P_0 is defined to be

$$K[P_0; P] = \sum_k p(k) \ln \frac{p(k)}{p_0(k)}. \tag{2}$$

This can be interpreted as the amount of information gained when the a priori probability distribution P_0 is replaced by the distribution P. The relative information is always nonnegative, i.e., $K[P_0; P] \geq 0$, equality holding only for identical distributions, $P_0 \equiv P$.

We shall consider a one-dimensional infinite sequence of symbols and let $p(A_M)$ be the probability for a randomly chosen subsequence of length m to coincide with the block $A_M = i_1 i_2 i_3 \ldots i_M$ of M symbols. We assume that this probability distribution is well defined for all $M \geq 1$. Then the metric entropy of the sequence is defined as the average entropy per symbol,

$$s_\mu = \lim_{M \to \infty} \frac{1}{M} S_M, \tag{3}$$

where the block entropy S_M is defined as in Eq. (1):

$$S_M = \sum_{A_M} p(A_M) \ln \frac{1}{p(A_M)}. \tag{4}$$

The entropy difference $\Delta S_m = S_m - S_{m-1}$ can be written as an average entropy of a conditional probability distribution $p(i_m|A_{m-1}) = p(A_m)/p(A_{m-1})$,

$$\Delta S_m = \sum_{A_{m-1}} p(A_{m-1}) \sum_{i_m} p(i_m|A_{m-1}) \ln \frac{1}{p(i_m|A_{m-1})} \geq 0. \tag{5}$$

(We define $\Delta S_1 = S_1$.) Below we shall also see that $-\Delta^2 S_m = \Delta S_{m-1} - \Delta S_m \geq 0$. From these properties of S_m, it follows that the metric entropy can be written

$$s_\mu = \lim_{m \to \infty} \Delta S_m = \Delta S_\infty, \tag{6}$$

which converges faster than the process expressed in Eq. (3).

The average information of 1 bit, or equivalently $\ln 2$, per lattice site can be decomposed into two terms: the *redundancy* and the metric entropy. The redundancy, $k_{\text{corr}} = \ln\ 2 - \Delta S_\infty$, expresses the information k_m in correlations of all lengths m (blocks of length $m + 1$), including the single-symbol block (density information k_0) [Eriksson and Lindgren 1987]. Thus,

$$k_{\text{corr}} = \sum_{m=0}^{\infty} k_m = \ln 2 - s_\mu. \tag{7}$$

Here, $k_m = -\Delta^2 S_{m+1}$ for $m > 0$, and $k_0 = \ln\ 2 - S_1$. Note that k_0 can be written in the form of the relative information of the density distribution given by $p(i)$ and an a priori density of 1/2,

$$k_0 = \sum_i p(i) \ln \frac{p(i)}{1/2} \geq 0. \tag{8}$$

Also in the expression for the correlation information

$$k_m = \sum_{i_1,\ldots,i_{m-1}} p(i_1,\ldots,i_{m-1}) \sum_{i_m} p(i_m|i_1,\ldots,i_{m-1}) \times \ln \frac{p(i_m|i_1,\ldots,i_{m-1})}{p(i_m|i_2,\ldots,i_{m-1})} \geq 0, \quad (9)$$

the sum over i_m is the relative information of $p(i_m|i_1,\ldots,i_{m-1})$ with respect to $p(i_m|i_2,\ldots,i_{m-1})$, which explains the interpretation of k_m as the correlation information of length m.

Grassberger [1986] has introduced what he terms the "effective measure complexity" as

$$\eta = \lim_{m\to\infty} (S_m - m s_\mu), \quad (10)$$

expressing the rate of convergence of S_m/m to s_μ. This quantity can be written as a product of the average correlation length and the correlation information, i.e., $\eta = \sum_m m k_m$. Thus η serves as a complexity measure, in the sense that it is small for simple "crystal patterns" like ...01010101..., as well as for completely random patterns.

The technique used to define quantities expressing correlations and randomness in the one-dimensional case can also be used in dimensions two and higher [Eriksson and Lindgren 1989]. For simplicity, we shall assume the system to be two-dimensional, but it is easy to generalize the formalism to higher dimensions.

Consider an infinite two-dimensional lattice in which each site is occupied by either the symbol 0 or 1 (the generalization to a larger symbol set and higher dimensions is straightforward). Assume that the relative frequencies with which finite configurations of symbols occur in the lattice are well-defined. Let $A_{M\times N}$ be a specific $M \times N$-block occurring with probability $p(A_{M\times N})$. Then the metric entropy, i.e., the average information per site, is

$$s_\mu = \lim_{M\to\infty} \lim_{N\to\infty} \frac{1}{MN} S_{M\times N}, \quad (11)$$

where the block entropy $S_{M\times N}$ is defined as

$$S_{M\times N} = \sum_{A_{M\times N}} p(A_{M\times N}) \ln \frac{1}{p(A_{M\times N})}. \quad (12)$$

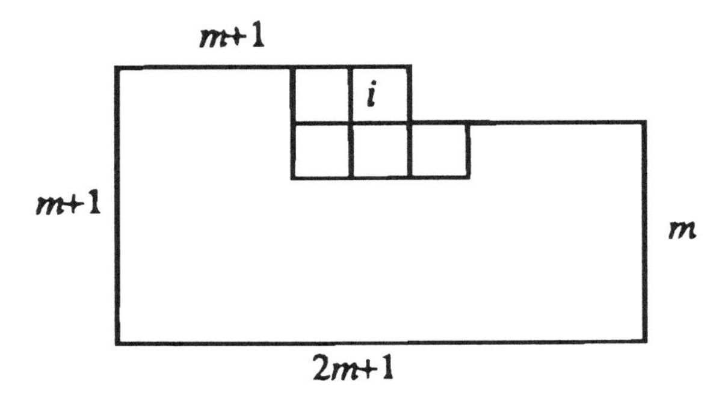

Figure 1. The block B_m used to calculate the mth approximation to the metric entropy.

Let B_m be a configuration of symbols arranged as in Figure 1, i.e., m rows of symbols, each of length $2m + 1$ are put on top of each other, and on the first $m + 1$ symbols of the top row is placed a sequence of $m + 1$ symbols. Let R be an operator that reduces a configuration B_m to a configuration B_{m-1} by removing the symbols from the leftmost and rightmost columns as well as from the bottom row. Denote the symbol in the last, i.e., the $(m+1)$th, position in the top row by i. Let X be the operator that removes the symbol in this position (it is not necessary for the block to have this shape, a point we shall comment on later). The conditional probability of the state i with respect to the "half surrounding" block XB_m is given by

$$p(i|XB_m) = \frac{p(B_m)}{p(XB_m)}. \tag{13}$$

The average entropy for this distribution is

$$K_m = \sum_{XB_m} p(XB_m) \sum_i p(i|XB_m) \ln \frac{1}{p(i|XB_m)}. \tag{14}$$

(For $m = 0$, we define $K_0 \equiv S_{1\times 1}$.) It can be proved [Eriksson and Lindgren 1989] that, in the limit $m \to \infty$, K_m equals the metric entropy, i.e.,

$$s_\mu = \lim_{m\to\infty} K_m = K_\infty. \tag{15}$$

The generalization $B_m^{[n]}$ of the block B_m in Figure 1 to n dimensions can be carried out in the following manner. Start with one symbol (the one denoted by i in Figure 1) and let its coordinate (in the block

$B_m^{[n]}$) be $(0, 0, 0, \ldots, 0)$. Denote such a single symbol block by $B_m^{[0]}$. Extend the block by adding the cells $(k, 0, 0, \ldots, 0)$, with $-m \le k < 0$, and denote this block by $B_m^{[1]}$ (this corresponds to the one-dimensional block used in Eq. (4)). Next we get the two-dimensional block $B_m^{[2]} = B_m$ of Figure 1 by extending $B_m^{[1]}$ through the addition of cells with coordinates $(j, k, 0, \ldots, 0)$, with $-m \le j \le m$ and $-m \le k < 0$. In general, we extend the block $B_m^{[n-1]}$ by adding cells with coordinates $(j_1, j_2, j_3, \ldots, j_{n-1}, k)$, with $j_1, j_2, j_3, \ldots, j_{n-1} \in [-m, m]$ and $-m \le k < 0$, to get the block $B_m^{[n]}$. The block $B_m^{[3]}$ is illustrated in Figure 2.

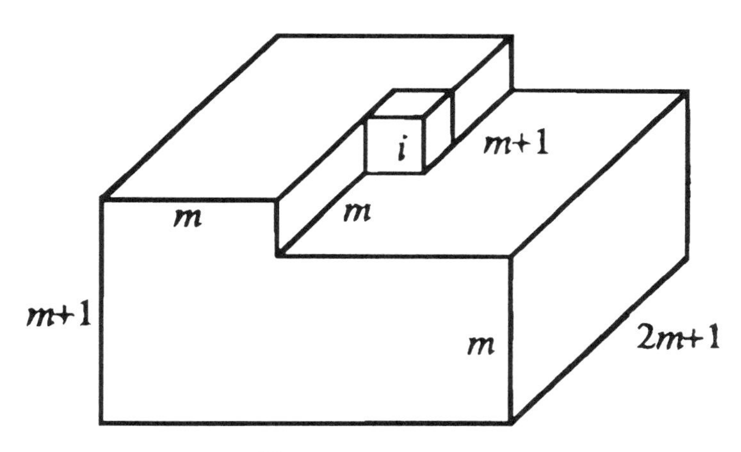

Figure 2. The block $B_m^{[3]}$ used to calculate K_m for a 3-dimensional lattice.

The blocks used to define a series of quantities converging to the metric entropy need not necessarily be based on blocks of the shape shown in Figure 2. For instance, suppose we have a series of blocks $D_k, k = 1, 2, \ldots$. It is sufficient that the size of the blocks is increasing in the series and that every block $B_m^{[n]}, m > 1$, is included in a block D_k in the series.

As in the one-dimensional case, the average information $\ln 2$ or, equivalently, 1 bit per lattice site can be decomposed into a term quantifying the information coming from correlations from different lengths (including density information) and a term measuring the internal randomness of the system,

$$\ln 2 = k_{\mathrm{corr}} + s_m. \tag{16}$$

Here

$$k_{\text{corr}} = \sum_{m=0}^{\infty} k_m = \ln 2 - s_{\mu}, \tag{17}$$

$$k_0 = \ln 2 - S_{1\times 1} = \sum_i p(i) \ln \frac{p(i)}{1/2}, \tag{18}$$

and we can write the correlation information k_m from length $m > 0$ as follows:

$$k_m = K_{m-1} - K_m = \sum_{XB_m} p(XB_m) \sum_i p(i|XB_m) \times \ln \frac{p(i|XB_m)}{p(i|XRB_m)} \geq 0. \tag{19}$$

Here, the sum over i is the relative information (2) of the probability distribution $\{p(i|XB_m)\}_i$ with respect to $\{p(i|XRB_m)\}_i$. Averaged over all possible blocks XB_m (which also include XRB_m), this expresses the average amount of information gained about the state i when a surrounding block XRB_m is extended to XB_m. This is in analogy with the definition of correlation information given in Eq. (9) for the one-dimensional case.

3. Cellular Automaton Spatial Patterns

Consider an n-dimensional cellular automaton, i.e., a mapping that simultaneously changes the states at all sites (cells) in an n-dimensional lattice according to a local rule. We restrict attention to deterministic rules of range r, which are mappings from the set of $(2r+1)^n$-blocks to $\{0, 1, \ldots, n-1\}$, when there are n possible states per lattice site and the rule determines the change of the state in the center cell.

In our simulation examples, the lattice is one-dimensional and there are only two states per cell, 0 and 1. The infinite sequence of states evolves according to a rule that depends only on the cell to be changed and its nearest neighbors, i.e., $r = 1$. Then a rule is a mapping from triplets of states $\{0, 1\} \times \{0, 1\} \times \{0, 1\} \rightarrow \{0, 1\}$, where we number the 256 possible rules using Wolfram's notation [Wolfram 1983].

As an example, the spacetime pattern for rule 110 is shown in Figure 3. At each time step, the states in a row of 400 cells with periodic boundary conditions are simultaneously updated and drawn

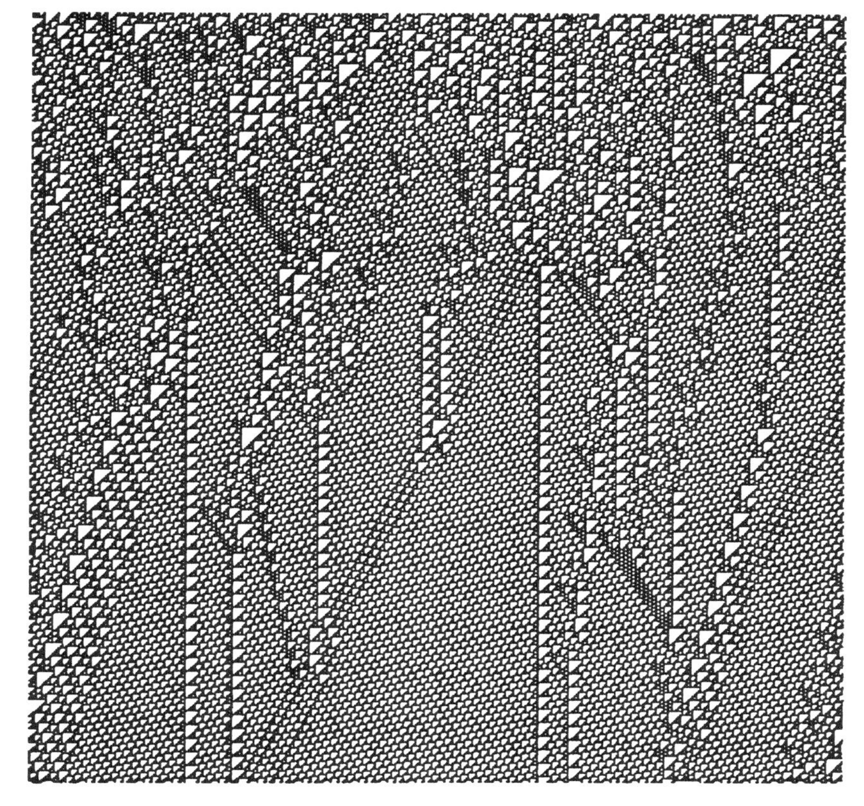

Figure 3. The spacetime pattern formed by rule 110.

as a line of white (0) and black (1) dots. The initial state is an internally uncorrelated sequence having an equal number of 0's and 1's. In the time evolution, spatial structure develops in the form of correlations in the sequence, which is seen clearly from the background pattern that forms (see also Figure 4). If the information-theoretic concepts are applied to the time evolution of cellular automata, a number of evolutionary laws can be proved.

Most of the cellular automaton rules have the common property that if applied to a "random" sequence, i.e., one that is an internally-uncorrelated infinite sequence, they develop some spatial structure or change the density of 0's and 1's. It can be shown that the metric entropy, i.e., the internal "randomness," is nonincreasing in time [Lindgren 1987],

$$\Delta_t s_\mu(t) = s_\mu(t+1) - s_\mu(t) \leq 0. \tag{20}$$

Then it follows from Eq. (16) that the correlation information is nondecreasing in time, $\Delta_t k_{\text{corr}}(t) \geq 0$. For the irreversible rule 110, this fact

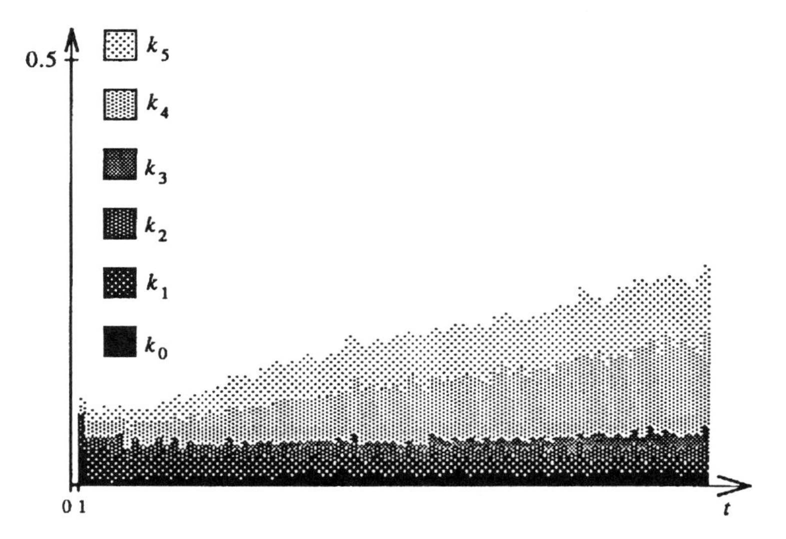

Figure 4. The evolution of correlation information (in bits) for rule 110. At each time the correlation information k_m from lengths $m = 0, 1, \ldots, 5$, i.e., blocks of length up to 6, are calculated and drawn on top of each other. Initially, there are no correlations, but during the time evolution the correlation information increases, especially for lengths 4 and 5. Actually, the background pattern developed in Figure 3 is of spatial period 14, but has no correlation information from blocks longer than 6 (lengths $m > 5$).

is illustrated in Figure 4. (We note that the inequality (20) holds also for Rényi entropies of arbitrary order [Lindgren and Nordahl 1987]).

Rules that correspond to a one-to-one mapping for infinite sequences are called reversible [4]. We call rules that result in finite-to-one mappings *almost reversible,* and for such rules the inequality (20) holds with equality [Lindgren 1987]:

$$\Delta_t s_\mu(t) = 0. \tag{21}$$

The correlation information is then also conserved in this case, i.e., $\Delta_t k_{\text{corr}}(t) = 0$. These rules lead to a time evolution in which previous (infinite) sequences can be obtained from the current (infinite) sequence by using only a finite amount of extra information.

Although the metric entropy is conserved for almost reversible rules, complex dynamical behavior for the information in correlations of different lengths k_m can arise if $s_\mu(t = 0) < \ln 2$. We illustrate this by a simulation of rule 195 (see Figure 5). The initial state is an uncorrelated sequence of 0's and 1's, with the density of 0's being 0.1. This

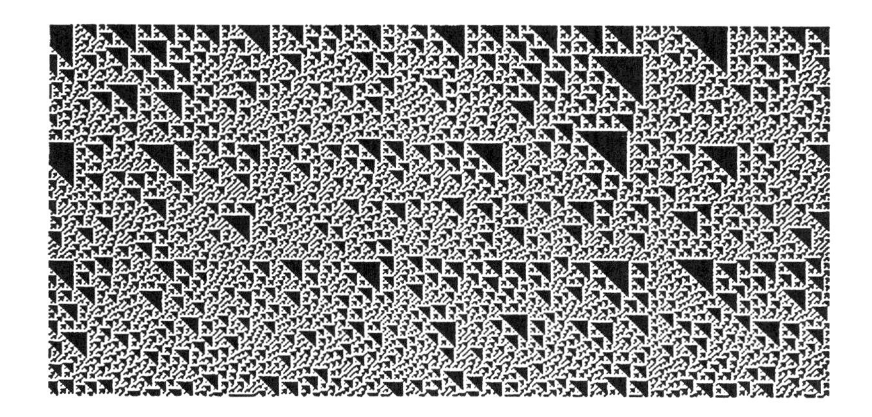

Figure 5. The spacetime pattern for the almost reversible rule 195.

gives $k_{\text{corr}} = k_0 = 0.368$ and $s_\mu = 0.325$. The metric entropy as well as the total correlation information will not change as the process unfolds, but the distribution of the correlation information over different lengths will vary, as is shown in Figure 6. The correlation information k_m is shown as a function of time (for $m = 0, \ldots, 5$). Since the total correlation information k_{corr} is conserved, the white space below the horizontal line and above the drawn contributions $k_0 + \cdots + k_5$ corresponds to information in correlations of lengths greater than 5. For time steps $t^* = 2^n \, (n = 1, 2, \ldots)$, the density of 1's is high, and the density information k_0 gives the major contribution to the total correlation information. These time steps are preceded by a series of steps in which correlation information flows from longer to shorter lengths (see Figure 6). Note that there is self-similarity in the figure, suggesting that for almost all time steps there will be no correlation information from finite distances. If the metric entropy $s_\mu < \ln 2$, it can be shown that the effective measure complexity η increases linearly in time for this rule [Lindgren and Nordahl 1987]

$$\eta(t) = s_\mu t, \tag{22}$$

with the metric entropy as the proportionality constant. (For $s_\mu = \ln 2$, the system will be completely random and there is no correlation information, giving $\eta = 0$.)

The increase in η implies that the mean correlation length increases without bound over time. Thus, the system will appear to have an increasing metric entropy. This type of process could possibly account for the apparent increase of microscopic and, hence, macroscopic entropy in a deterministic system.

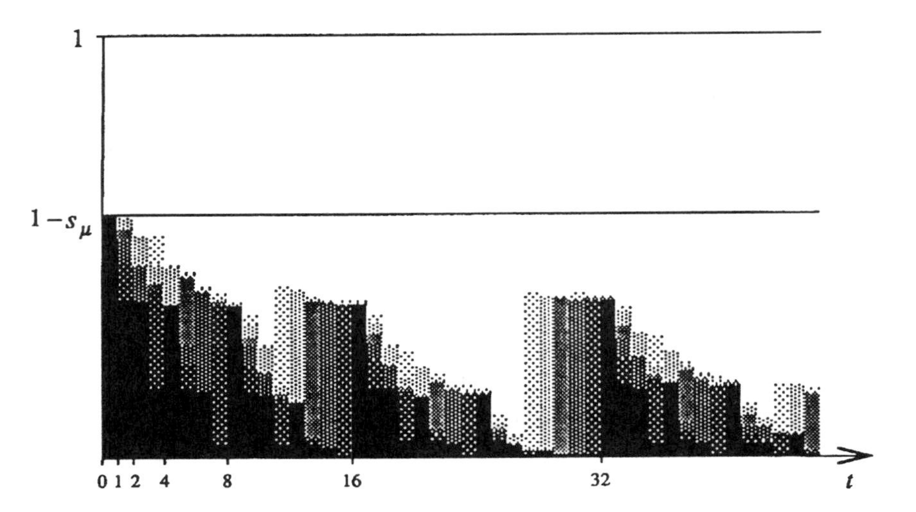

Figure 6. The contributions (in bits) to the correlation information from lengths $m \leq 5$ as functions of time for rule 195, cf. Figure 4.

If for each cell and time a rule is chosen at random according to some probability distribution, the temporal evolution of the process is nondeterministic. By introducing a constant probability q for a deterministic rule to make an error, we obtain a subclass of all nondeterministic rules. Such a rule can be specified by the corresponding deterministic rule number and the probability q. General evolutionary laws like Eqs. (20) and (21) do not exist for this class, but if one is restricted to almost reversible rules modified by noise, it has been shown [Lindgren 1987] that the metric entropy increases (for arbitrarily small $q > 0$),

$$\Delta_t s_\mu(t) > 0, \tag{23}$$

until it reaches its maximum $s_\mu(t) = \ln 2$, whereupon the system is completely randomized. In Figure 7 a spacetime pattern of rule 195 is shown, in which the evolution is disturbed by noise. Each time the rule is applied, the probability for making an error is $q = 0.01$. In Figure 8 the contributions to the total correlation information from k_m $(m = 0, \ldots, 5)$ are drawn as in Figure 6. A comparison between the figures reveals how sensitive the spatial correlations are to the noise.

If the state of the cellular automaton is interpreted as a microstate of a physical system in which the time evolution is disturbed by an arbitrarily small amount of noise, Eq. (23) is a microscopic version of the second law of thermodynamics. Here we have restricted ourselves

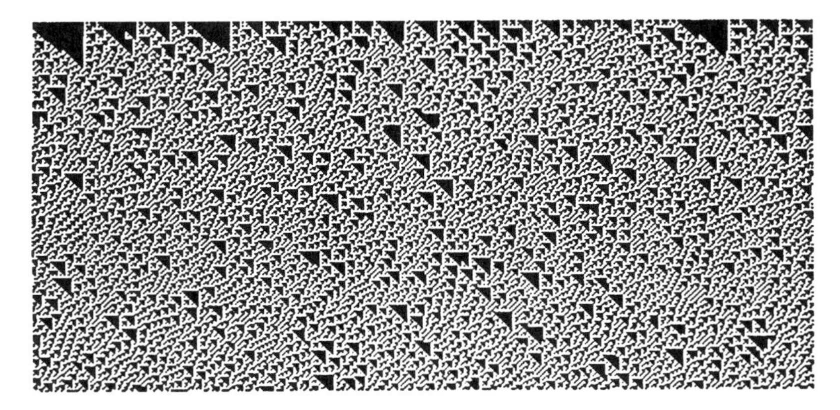

Figure 7. The spacetime pattern for the almost reversible rule 195 modified by noise.

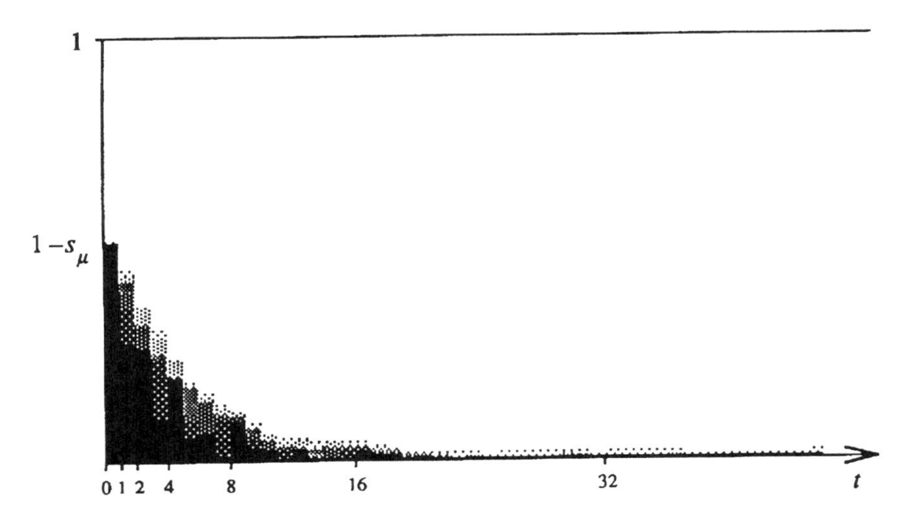

Figure 8. The contributions (in bits) to the correlation information from lengths $m \leq 5$ as functions of time for rule 195, cf. Figure 6.

to a limited class of nondeterministic rules. In the next section we give the proof of a similar relation for a general class of lattice gas models.

4. The Evolution of the Entropy in a Lattice Gas

In a lattice gas particles are moving and colliding with each other on an n-dimensional lattice. At each lattice point there may be several particles, and for each such configuration there is a corresponding state (usually not more than one particle in each direction is allowed). We shall assume the lattice to be infinite in all directions. All the states in the lattice are simultaneously updated according to a cellular automaton rule, which may be probabilistic. Each time step can be divided into two parts: in one part the particles move to neighboring lattice sites according to their velocities, while in the other part the particles change their velocities due to collisions at each node. The translational step is reversible and, hence, does not change the metric entropy of the system. Below we shall analyze the change in metric entropy due to the collisions.

Let A_M be a specific M^n-block occurring with probability $p(A_M)$. Assume that the collisions transforms A_M to a block $A_{M'}$ with probability $P(A_M \to A_{M'})$. Further, assume that this probability distribution is normalized,

$$\sum_{A_{M'}} P(A_M \to A_{M'}) = 1, \tag{24}$$

and that it obeys semidetailed balance (or semireversibility),

$$\sum_{A_M} P(A_M \to A_{M'}) = 1. \tag{25}$$

By definition, the metric entropy before the collisions is

$$s_\mu = \lim_{M\to\infty} \frac{1}{M^n} S[P_M], \tag{26}$$

where $S[P_M]$ is the entropy of the distribution $P_M = \{p(A_M)\}$. If the system is not in equilibrium, this probability distribution will change due to the collisions

$$p'(A_{M'}) = \sum_{A_M} p(A_M) P(A_M \to A_{M'}). \tag{27}$$

Using Eqs. (24–27), we can write the change in metric entropy due to the collisions as

$$\Delta_{\text{coll}}\, s_\mu = \lim_{M\to\infty} \frac{1}{M^n} \sum_{A_{M'}} p'(A_{M'}) K\left[P(\cdot \to A_{M'}); \frac{p(\cdot)P(\cdot \to A_{M'})}{p'(A_{M'})}\right] \geq 0, \tag{28}$$

where the relative information is (cf. Eq. (2))

$$K\left[P(\cdot \to A_{M'}); \frac{p(\cdot)P(\cdot \to A_{M'})}{p'(A_{M'})}\right] = \sum_{A_M} \frac{p(A_M)P(A_M \to A_{M'})}{p'(A_{M'})} \times \ln \frac{\frac{p(A_M)P(A_M \to A_{M'})}{p'(A_{M'})}}{P(A_M \to A_{M'})} \geq 0, \quad (29)$$

which assures that the change of metric entropy due to collisions is nonnegative. Since the change of metric entropy due to translations is zero, we conclude that the metric entropy is nondecreasing in time,

$$\Delta_t s_\mu(t) \geq 0. \quad (30)$$

In lattice gas models one usually has collision rules that are probabilistic. So if the system is not in equilibrium, random information will enter the system through collisions and the metric entropy increases. If the collision rule is deterministic, the metric entropy will be unaffected by the collisions (cf. Eq. (21)). In that case we expect that the information in short range correlations decreases, as was discussed for almost reversible cellular automata in the last section, making the system appear more random after the collisions.

The inequality (30) can be viewed as a "second law of lattice gas dynamics." Like the Boltzmann H-theorem, it is defined for the dynamics of a microstate. But, since the metric entropy is a quantity which takes into account all correlations in the system, this "second law" holds even if the velocities are reversed in the dynamics.

5. *Cellular Automaton Spacetime Patterns*

Consider an n-dimensional cellular automaton with range r, i.e., a mapping which simultaneously changes the states at all sites (cells) in an n-dimensional lattice according to a local rule.

A probabilistic rule of range r uses the state in the $(2r+1)^n$-block to determine with which probability $p(s)$ the state of the center cell will change to state $s \in \{0, 1, \ldots, n-1\}$. For a probabilistic rule, the metric entropy of the spacetime pattern can be given by the $(n+1)$-dimensional generalization of K_r in Eqs. (13) and (14). Assume that the block $B_r^{[n]}$ is so directed that the state i in cell $(0, 0, \ldots, 0)$ is chosen according to a probability distribution fully determined by the states in the cells $(j_1, j_2, j_3, \ldots, j_n, k)$, with $j_1, j_2, j_3, \ldots, j_n \in [-r, r]$ and $-r \leq k < 0$. In other words, the block $B_r^{[n]}$ is directed so that if the cell with state i corresponds to the center cell at time $t+1$, then it also covers the $(2r+1)^n$-block at time t. This constraint then determines the probability distribution for state i, which in turn implies that for lengths $m > r$, the correlation information k_m vanishes (see Eq. (19)). Then the metric entropy of the spacetime pattern is

$$s_\mu = K_r. \tag{31}$$

(Note that the whole block $B_r^{[n]}$ is not needed for calculating the entropy. It suffices to use a block consisting of the cell with state i and the $(2r+1)^n$-block determining the probability distribution for state i.)

If the rule is deterministic, i.e., the state i is fully determined by the block $B_r^{[n]}$, then it is clear from Eq. (14) that K_r is zero and so is the metric entropy (see also [Wolfram 1984b]). Note that the spacetime pattern formed by a one-dimensional deterministic rule has a block entropy $S_{M\times M}$ that generally grows linearly with M, and the average entropy slowly converges to zero as M increases (cf. Eqs. (11)–(12).

In the expression for K_r in Eq. (14), the entropy of the conditional probability $p(i|XB_r^{[n]})$ is given by the probabilistic rule governing the time evolution. This entropy is then averaged over all possible blocks $XB_r^{[n]}$, and the probability distribution for these blocks must usually be estimated by simulations.

Since there is an equivalence between n-dimensional cellular automata and $(n+1)$-dimensional equilibrium spin models [Rujan 1987, Georges and LeDoussal 1989], Eq. (31) provides a fast method to calculate the entropy, as well as other thermodynamical properties of spin models having a CA equivalent.

We also expect the formalism to be applicable to spin models for which a CA equivalent cannot be found. In that case expressions like Eq. (14) would be the starting point for an analysis. We note that the conditional probabilities of Eq. (13) are a generalization of the transfer matrices used by Kramers and Wannier in their analysis of the two-dimensional Ising model [Kramers and Wannier 1941]. In that situation the conditional probability of i is given by blocks of the shape shown in Figure 9a, i.e., $p(i|j_{-1}, j_{-2}, j_{-3}, \ldots; j_1, j_2, j_3, \ldots)$. In general, if the interaction length is m, one would need a block having width m in one dimension and extending to infinity in the other dimensions, as depicted in Figure 9b for $m = 2$.

Meirovitch and Alexandrovicz [1977] have used expressions based on the kinds of blocks in Figure 9a to get approximations to the free

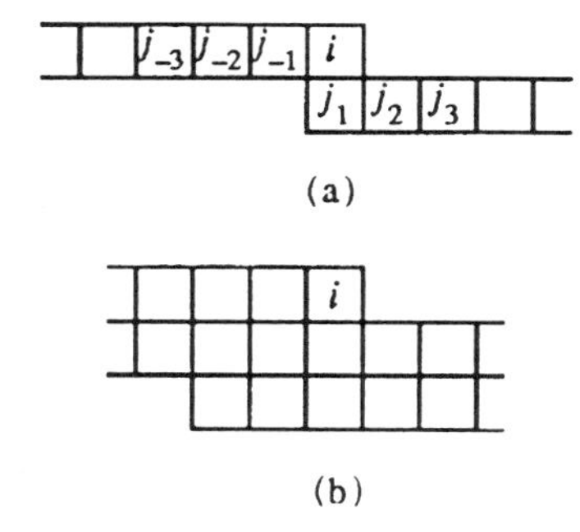

Figure 9. (a) The conditional probability that determines the entropy of the two-dimensional Ising model is given by an essentially one-dimensional (but infinite) block. (b) In general, the shape of the minimal blocks which are needed for calculating the conditional probabilities $p(i|XB_m)$ in $K_m(m \to \infty)$, Eqs. (9) and (10), has a finite width if the interaction distance is finite. The example in the figure corresponds to a case with next-nearest neighbor interactions (including diagonal interactions).

energy of the two-dimensional Ising model. Actually, for each temperature in the two-dimensional Ising model, one can construct a simple filter automaton (FA) that generates an ensemble of spacetime patterns corresponding to configurations in the Ising model (see Figure 10) [Lindgren, unpublished]. The free energy of the FA spacetime pattern does not deviate more than 1% from the exact result, although the critical temperatue is 5% off the exact value. The FA rule has the form

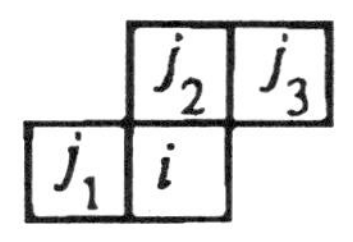

where the states j_1, j_2, and j_3 have already been generated and determine the probability $p(i|j_1, j_2, j_3)$ with which state i is randomly chosen (cf. Eq. (13)). The spatial patterns for some of these rules are shown in Figure 10. The simulation starts with a randomly chosen row of 0's and 1's, using skew boundary conditions (the last cell of row i is connected to the first cell of row $i + 1$). There is a transient period before the effect of the random initial state disappears. This corresponds to a surface effect, and is not shown in Figure 10. For each temperature the rule is derived by a minimization procedure for the free energy of the spacetime pattern that is generated.

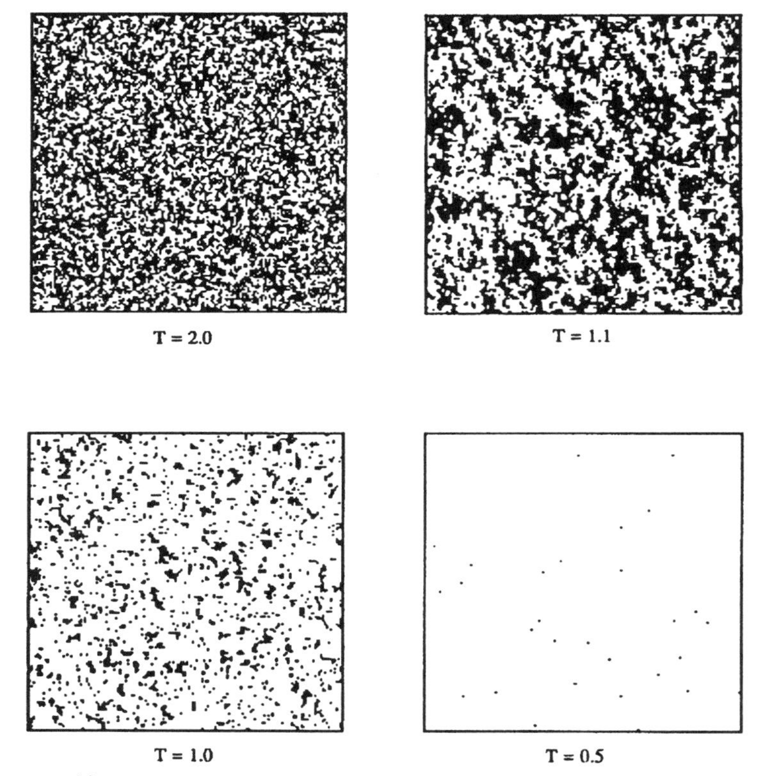

Figure 10. The spacetime pattern of simple FA-rules correpond to configurations in the two-dimensional Ising model. (Temperatures T are relative the exact T_c.)

6. *Conclusions*

Information theory provides concepts for analyzing correlations in various physical systems. Some of the results discussed here can be viewed as "second laws" for different kinds of microscopic dynamics in discrete systems. It appears that information theory, and certainly also algorithmic information theory [Zurek 1989], has more to contribute to the classical question on the origin of randomness and entropy in physical systems.

References

Bennett, C. H., 1982, "The thermodynamics of computation—a review," *International Journal of Theoretical Physics*, 21, 905–940.

Brudno, A. A., 1977, "On the complexity of trajectories in dynamical systems," *Uspekhi Matematcheskii Nauk*, 33, 207.

Chaitin, G. J., 1966, "On the length of programs for computing finite binary sequences," *Journal of the Association for Computing Machinery,* 13, 547–569.

Chaitin, G. J., 1979, "Toward a mathematical definition of life," in *The Maximum Entropy Principle,* R. D. Levine and M. Tribus, eds., MIT Press, Cambridge, MA.

Eriksson, K.-E. and Lindgren, K., 1987, "Structural information in self-organizing systems," *Physica Scripta,* 35, 388–397.

Eriksson, K.-E. and Lindgren, K., 1989, "Entropy and correlations in lattice systems," Preprint 89–1, Physical Resource Theory Group, Chalmers University of Technolgy, Göteborg, Sweden.

Fannes, M. and Verbeure, A., 1984, "On solvable models in classical lattice systems," *Communications in Mathematical Physics,* 96, 115.

Farmer, D., Lapedes, A., Packard, N., and Wendroff, B., eds., 1986, *Evolution, Games, and Learning,* North-Holland, Amsterdam, (first published in *Physica 22D,* 1–402 (1986)).

Frisch, U., d'Humières, D., Hasslacher, B., Lallemand, P., Pomeau, Y., and Rivet, J.- P., 1987, "Lattice gas hydrodynamics in two and three dimensions," *Complex Systems,* 1, 649–707.

Georges, A. and Le Doussal, P., 1989, "From equilibrium spin models to probabilistic cellular automata," *Journal of Statistical Physics,* 54, 1011–1064.

Grassberger, P., 1986, "Towards a quantitative theory of self-generated complexity," *International Journal of Theoretical Physics,* 25, 907–938.

Hameroff, S. R., Rasmussen, S., and Månsson, B. Å., 1989, "Molecular automata in microtubules: basic computational logic of the living state," pp. 520–553 in *Artificial Life,* C. Langton, ed., Addison-Wesley, Redwood City, CA.

Hartley, R. V. L., 1928, "Transmission of information," *Bell System Technical Journal,* (July), 535–563.

Jaynes, E. T., 1957, "Information theory and statistical mechanics," *Physical Review,* 106, 620.

Kolmogorov, A. N., 1965, "Three approaches to the quantitative definition of information," *Problemy Peredachi Informatsii,* 1, 3–11. (English translation in *Problems of Information Transmission,* 1, 1–7.)

Kramers, H. A. and Wannier, G. H., 1941, "Statistics of the two-dimensional ferromagnet–Part I," *Physical Review,* 60, 252–262.

Lindgren, K., 1987, "Correlations and random information in cellular automata," *Complex Systems,* 1, 529–543.

Lindgren, K., 1988, "Microscopic and macroscopic entropy," *Physical Review A,* 38, 4794–4798.

Lindgren, K. and Nordahl, M. G., 1988, "Complexity measures and cellular automata," *Complex Systems,* 2, 409–440.

MacMillan, B., 1953, "The basic theorems of information theory," *Annals of Mathematical Statistics,* 24, 196–219.

Martin-Löf, P., 1966, "The definition of random sequences," *Information and Control,* 9, 602–619.

Meirovitch, H. and Alexandrovicz, Z., 1977, "The stochastic models method applied to the critical behavior of Ising lattices," *Journal of Statistical Physics,* 16, 121–138.

Nordahl, M. G., 1988, "Discrete Dynamical Systems," Doctoral thesis, Institute of Theoretical Physics, Göteborg, Sweden.

Rujan, P., 1987, "Cellular automata and statistical mechanics models," *Journal of Statistical Physics,* 49, 139–222.

Shannon, C. E., 1948, "A mathematical theory of communications," *Bell System Technical Journal,* 27, 379–423.

Solomonoff, R. J., 1964, "A formal theory of inductive inference," *Information and Control,* 7, 1–22.

Turing, A., 1936–37, "On computable numbers, with application to the Entscheidungsproblem," *Proceedings of the London Mathematical Society,* 42, 230, and 43, 544.

Watanabe, S., 1969, *Knowing and Guessing: a Quantitative Study of Inference and Information,* John Wiley & Sons, New York.

Wolfram, S., 1983, "Statistical mechanics of cellular automata," *Reviews of Modern Physics,* 55, 601–644.

Wolfram, S., 1984a, "Computation theory of cellular automata," *Communications in Mathematical Physics,* 96, 15–57.

Wolfram, S., 1984b, "Universality and complexity in cellular automata," *Physica 10D,* 1–35.

Wolfram, S., ed., 1986, *Theory and Applications of Cellular Automata,* World Scientific, Singapore.

Zhvonkin, A. K. and Levin, L. A., 1970, "The complexity of finite objects and the development of the concepts of information and randomness by means of the theory of algorithms," *Russian Math Surveys,* 25, 83–124.

Zurek, W. H., 1989, "Algorithmic randomness and physical entropy," *Physical Review,* A40, 4731–4751.

CHAPTER 6

Dimensions of Atmospheric Variability

R. T. PIERREHUMBERT

Abstract

We consider a number of applications of the correlation dimension concept in the atmospheric sciences. Our emphasis is on the correlation dimension as a nonlinear signal-processing tool for characterizing the complexity of real and simulated atmospheric data, rather than as a means for justifying low-dimensional approximations to the underlying dynamics. Following an introductory exposition of the basic mathematics, we apply the analysis to a three-equation nonlinear system having some interesting points of affinity with the general subject of nonlinear energy transfers in two-dimensional fluids. Then we turn to the analysis of a 40-year data set of observed Northern Hemisphere flow patterns. Our approach deviates from most previous studies in that we employ time series of flow fields as our basic unit of analysis, rather than single-point time series. The main evidence for low-dimensional behavior is found in the patterns of interseasonal variability. However, even when this is removed, the streamfunction field shows clear indication of dimensionality in the range 20–200, rather than in the thousands. Due to problems connected with the nonequivalence of various norms in functions spaces, we do not claim that this represents the "true" dimensionality of the underlying system. Nevertheless, it does show the existence of a considerable amount of order in the system, a fact begging for an explanation.

From this, we turn to the matter of spatial complexity, examining the geometry of clouds of passive tracer mixed by spatially structured (atmospherically motivated) two-dimensional flow fields. Evidence is presented that the cloud ultimately mixes over a region characterized by dimension two. We also demonstrate that the correlation dimension of the tracer cloud is directly related to the algebraic power spectrum of the concentration distribution in Fourier space. From this we are led to some speculations on the role of chaotic mixing in the enstrophy cascade of two-dimensional turbulence.

Finally, we consider the spatial patterns of rate of predictability loss in the tracer problem. This information is obtained by computing finite-time estimates of the Lyapunov exponents for trajectories starting from various initial conditions. The rate of predictability loss is itself unpredictable, in the sense that it exhibits sensitive dependence on initial conditions. It is suggested that multifractal analysis could be used to characterize the spatial pattern of predictability.

1. Introduction

Of the various measures of atmospheric variability, the *correlation dimension,* introduced in dynamical systems theory by Grassberger and Procaccia (1983), is one that has recently received a lot of attention. This is a convenient and readily computable way to estimate the fractal dimension of an object traced out by a time series of points in some high-dimensional embedding space. Specifically, if we have a system whose time variation is described by state $\Psi(t)$ (where, for example, Ψ is the set of all wind and temperature values at every point in the atmosphere at time t), and a function $d(\Psi_1, \Psi_2)$ measuring the distance between states, then the correlation dimension is calculated as follows: Discretize the time series to form a sequence of states Ψ_j, and compute the distances d_{ij} between each pair of states. From this form the cumulative histogram $H(r)$ for the number of pairs of distance less than r. If $H(r)$ is well-approximated by the form $H = ar^d$, then the exponent d is said to be the correlation dimension. Grassberger and Procaccia give numerous examples showing why this quantity should be thought of as a dimension, as well as indicating how it relates to other measures of dimension.

The analysis has been applied to climatic variations on geological time scales (Grassberger 1986, Maasch 1989, *inter alia),* to daily weather fluctuations at a single station (Essex, et al. 1987), and to microscale fluctuations having time scales of seconds to minutes (Tsonis and Elsner, 1988). The analyses typically indicate low dimensionality and indeed often seem to be motivated by the (probably forlorn) hope of justifying the modelling of weather or climate in terms of a small set of ordinary differential equations. Whether or not this quest is tantamount to tilting at windmills, the correlation dimension analysis does provide useful information about atmospheric variability, information that goes beyond the bounds of traditional linear statistics. Such analyses reveal the extent to which the actual variations are concentrated on a limited subset of the space of all possible variations. It is in this spirit that we present the ideas of this chapter.

As an example of the geometric information contained in the $H(r)$ curve, consider Figure 1. The upper example shows a "thin wire grid" embedded in R^2. The object is one-dimensional viewed at small scales, but of dimension two when viewed at large scales. This would be reflected in a change in the slope of the $\log H(r)$ vs. $\log r$ curve at a distance r_0 corresponding to the average spacing of the grids. In the bottom example we consider the $H(r)$ graph for a "relaxed earthworm,"

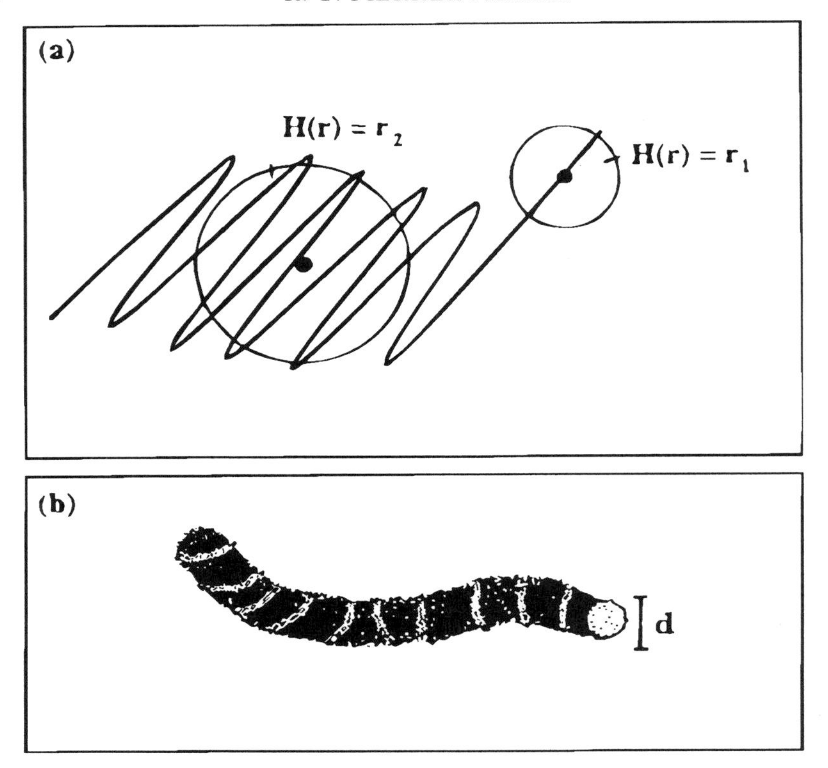

Figure 1. Geometric interpretation of the correlation dimension.

which displays a one-dimensional structure at large scales but is of dimension three at scales smaller than the thickness of the earthworm. If the earthworm should become agitated and decide to curl up into a spiral on the ground, the 1-D large-scale regime would be replaced by a 2-D regime. All objects which can be embedded in a compact subset of their phase space have $H(r)$ becoming constant for values of r sufficiently large to contain the entire object. Thus, all localized objects look zero-dimensional (pointlike) if you stand back far enough.

It is instructive to compare the correlation dimension analysis with the "empirical orthogonal function" (EOF) technique, which currently enjoys much wider application in the atmospheric sciences. This technique is a systematic way of finding a *linear subspace* of a high-dimensional embedding space, to which the trajectory of the system is approximately confined. Given a vector time series $(x_1(t), \ldots, x_n(t))$ in which each component has zero mean and unit variance, the EOF's are simply the n eigenfunctions of the $n \times n$ covariance matrix $< x_i x_j >$, where the brackets denote time averaging. The eigenvalues give the proportion of the variance of the signal explained by projection onto the corresponding eigenvector. The use of EOF's in characterizing atmospheric variability was pioneered by Lorenz, though the concept first emerged in the statistics literature through the work of Karhunen and Loéve. The minimum number of EOF's needed to account for, say, 90% of the total variance is sometimes referred to as the "Karhunen-Loéve dimension."

By way of example, consider the motion $x(t) = \cos\omega t$, $y(t) = \sin\omega t$. The eigenvectors of the correlation matrix are $(1,0)$ and $(0,1)$, and each explains 50% of the variance; the Karhunen-Loéve dimension is 2. However the correlation dimension is 1, as the motion traces out a circle in the x-y plane. The correlation dimension analysis provides a much better description of the confinement of the trajectory to a limited portion of the embedding space, because it does not need to confine the trajectory within a *linear* subspace. On the other hand, a strength of the empirical orthogonal function analysis is that it provides some information on the *patterns* of variation via the eigenvectors, which have no counterpart in the correlation dimension analysis.

An extension of this example demonstrates that the correlation dimension analysis must be interpreted with great care if one's objective is to determine the dimensionality of the underlying dynamics. Suppose that the frequency ω is time dependent (say, $\omega = \omega_0 + \alpha \sin \Omega t$, for the sake of concreteness), rather than being constant. Then a correlation dimension estimate based on the trajectory in x-y space is bound to yield a dimension of unity or less, since the trajectory is restricted to the unit circle. However, knowlege of the initial point on the circle does not determine the future course of the system, as one must also know the phase Ωt; that is, the system is actually two-dimensional. Thus, while the naively determined correlation dimension gives us a useful description of the kind of set the trajectory lives on in x-y space, it does not tell us how many equations are needed to predict the future.

The problem stems from the fact that the x-y plane in which the trajectory is being embedded is not a *phase space* for the system. The rectification of the problem is, in principal, equally simple. One forms the family of $2(N+1)$-dimensional synthetic phase spaces consisting of the x-y point and its time derivatives up to Nth order, and performs the dimension analysis for ever larger N until the answer converges. In the example above, the x-y point and all its time derivatives can be expressed in terms of the two phases $\theta_1 = \Omega t$ and $\theta_2 = \omega(\theta_1)t$, which define a two-dimensional manifold embedded in the $2(N+1)$-dimensional space. In practice, when applying the method to data one forms the synthetic phase spaces by employing time-lagged sequences $(x_n, y_n, x_{n-1}, y_{n-1}, \ldots)$, which is equivalent to computing the time derivatives by finite differencing.

In this chapter we examine the application of the correlation dimension analysis as a tool for characterizing the variability of a number of actual and simulated atmospheric time series. We begin in §2 with an analysis of a simple system of ordinary differential equations whose behavior can also be directly visualized geometrically. In §3 we apply the methods to a 40-year data set of actual atmospheric flow patterns. In contrast to most previous studies, this one deals with time series of *fields* covering most of the Northern Hemisphere, rather than single-point time series of weather or climate data. We then take up the characterization of the complexity of individual two-dimensional patterns in §4, showing how geometrically simple flow fields can nevertheless generate intricate patterns in tracers and vorticity. In this section we also discuss the relation between fractal dimensions and the spatial spectrum of concentration variance. In §5 we come up against yet another form of spatial complexity. The spatial complexity in tracer patterns arises from loss of predictability of the individual particle trajectories. It turns out that the spatial pattern characterizing the *rate* of predictability loss is itself fractal in nature, so that even the degree of unpredictability is in some sense unpredictable. Some general conclusions and speculations are given in §6.

2. A Toy Atmosphere

To illustrate the key ideas, we first treat a simple model embodying some of the energy transfers and dissipative characteristics of atmospheric flow. Consider the equations

$$\frac{dA_1}{dt} = A_2 A_3 - d_1 A_1 \tag{2.1}$$

$$\frac{dA_2}{dt} = -2A_1 A_3 + g A_2 \tag{2.2}$$

$$\frac{dA_3}{dt} = A_1 A_2 - d_3 A_3 \tag{2.3}$$

In the absence of dissipation ($d_1 = d_3 = g = 0$) these equations reduce to the conventional triad equations, and conserve the energy and enstrophy related quantities

$$E = A_1^2 + A_2^2 + A_3^2, \quad F = A_1^2 - A_3^2 \tag{2.4}$$

Hence the solutions are closed orbits defined by the intersection of the sphere E = constant and the hyperboloid F = constant. These equations are identical to the gyroscopic equations describing the evolution of the angular momentum vector of a rigid body. The inviscid triad equations can be derived from a number of *ad hoc* and formal asymptotic approximations to 2-D fluid flow, and the amplitudes A_j may be thought of as the amplitudes of three scales of motion (e.g. the amplitudes of a triplet of resonantly interacting Rossby waves, as described in Pedlosky 1979). Motion concentrated initially in the middle scale ($j = 2$) is unstable, and transfers energy to the neighboring scales. This process is at the heart of two-dimensional turbulence.

But the real atmosphere is not a closed system; it is forced by energy input at intermediate scales, and nonlinear processes transfer energy to other scales where it is dissipated. In Eq. (2.2), g represents the generation process, while d_1 and d_3 in Eqs. (2.1) and (2.3) represent dissipation. This system has the interesting property that a trajectory located initially on one of the axes remains there indefinitely. Thus the three trajectories

$$(A_1, A_2, A_3) = (a\exp(-d_1 t), 0, 0) \tag{2.4a}$$
$$= (0, a\exp(gt), 0) \tag{2.4b}$$
$$= (0, 0, a\exp(-d_3 t)) \tag{2.4c}$$

are all exact nonlinear solutions of Eqs. (2.1)–(2.3). This means that the dissipation in itself cannot be relied on to prevent the growth of energy in the system; to keep the trajectory from running off to infinity along the 2-axis, energy must be transferred out of mode 2 and into modes 1 and 3 where it can be dissipated.

This property is not an artifact of the *ad hoc* truncation; it is also encountered in partial differential equations describing fluid systems that are more like the real atmospheric flow. For example, consider the damped barotropic vorticity equation, augmented to make a single mode unstable:

$$\partial_t \nabla^2 \psi + \mathcal{J}(\psi, \nabla^2 \psi) = -d\nabla^2 \psi + g(\nabla^2 \psi)_{nm} \tag{2.5}$$

where ψ is the 2-D streamfunction, $\nabla^2 \psi$ the vorticity, $(\nabla^2 \psi)_{nm}$ the projection of the vorticity on the Fourier mode (n, m), and $\mathcal{J}$ is the Jacobian matrix defined to be $\mathcal{J}(A, B) = \{\partial_x A \partial_y B - \partial_x B \partial_y A\}$. One can think of the artificial instability as a surrogate for various instability processes (such as baroclinic instability) that cannot be captured in a single-layer model. If the system is initialized with a pure Fourier mode (n, m), the Jacobian vanishes identically, and the amplitude of the mode grows without bound. However, the mode is subject to a sideband instability rather similar to that occuring in the triad model, and so nonlinear transfer out of the mode can cause enough dissipation to limit growth. We are not aware of any integrations of Eq. (2.5) addressing the boundedness of the flow, but a similar question occurs naturally in connection with baroclinic instability in an infinite x-y plane, and there the runaway mode appears to have bounded amplitude. (I. Held, personal communication).

Now let's get a bit more mathematical. The origin, (0, 0, 0) is a fixed point of Eqs. (2.1)–(2.3). The directions (1, 0, 0) and (0, 0, 1) are stable, while the direction (0, 1, 0) is unstable. In fact, given Eq. (2.4), the *unstable manifold* of the system (defined crudely as the set of trajectories coming out of the fixed point at directions tangent to the local unstable direction) is simply the 2-axis. The *stable manifold,* which is the set of points which flow into the fixed point with time, is two-dimensional because there are two stable directions; it *contains* the 1-axis and 3-axis, but the structure in between can be (and probably is) highly contorted. A trajectory located on the unstable manifold initially will of course run directly off to infinity, but what happens if it is slightly displaced? To answer this question, we linearize Eqs. (2.1)–(2.3) about the unstable trajectory (2.4b), finding

$$\frac{d}{dt}(A_1, A_3) = (A_1, A_3)\exp(gt) \tag{2.6}$$

Hence neighboring trajectories deviate from the unstable manifold like the exponential of the exponential of time. Clearly, the unstable manifold is a very unstable creature indeed. This prevents the trajectory from shadowing the unstable manifold and running directly off to infinity. However, after deviating from the unstable manifold, the trajectory can come close to the two-dimensional stable manifold and be swept back towards the origin, where it begins the process anew. Thus trajectories are expected to funnel into the origin along the stable manifold and fountain outward from there along the one-dimensional unstable manifold.

Numerical calculations show that this is indeed what happens. The case $d_1 = d_3$ is somewhat pathological, as the quantity F then becomes monotonically decreasing and the trajectories asymptotically collapse onto the crossed planes $A_1 = \pm A_3$, and thence spiral out to infinity along a highly organized trajectory. When $d_1 \neq d_3$ the trajectory is

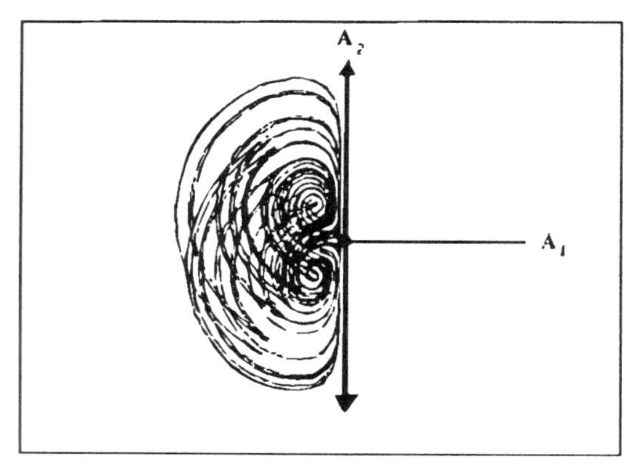

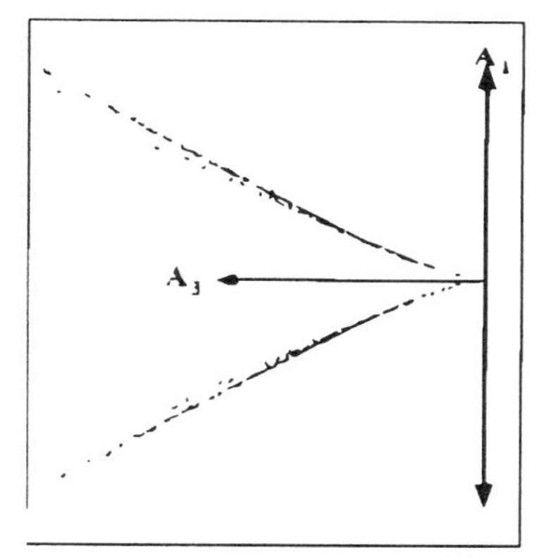

Projection on A_1 — A_2 Plane

Cross section through plane A_2=0

Figure 2(a). Phase space structure of the triad system. Left: projection of trajectories on A_1-A_3 plane. Right: cross section showing intersection of trajectories with the plane $A_2 = 0$. Parameters are $(d_1, g, d_3) = (0.5, 0.25, 1)$.

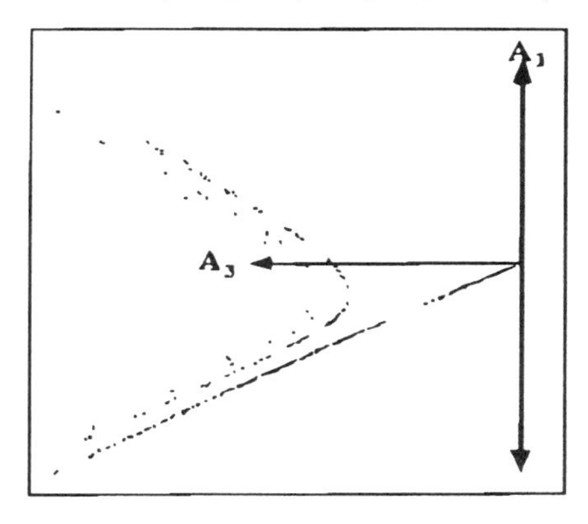

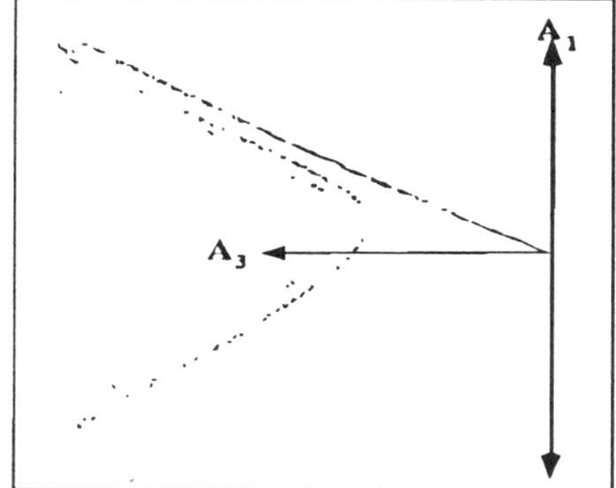

Cross section through plane A_2 = 1.5

Cross section through plane A_2 = -1.5

Figure 2(b). Cross sections as in Figure 2(a), but for $A_2 = 1.5$ and $A_2 = -1.5$.

more chaotic and unpredictable. A typical set of points tracing out the attractor for $(d_1,\ g,\ d_3) = (0.5,\ 0.25,\ 1)$ is shown in Figure 2.

Does the system remain within a bounded region of phase space as the integeration is carried out over ever longer times? This question is equivalent to asking whether the unstable fixed point (0, 0, 0) borders the attractor, since trajectories beginning arbitrarily close to the unstable manifold will attain correspondingly large amplitudes before they

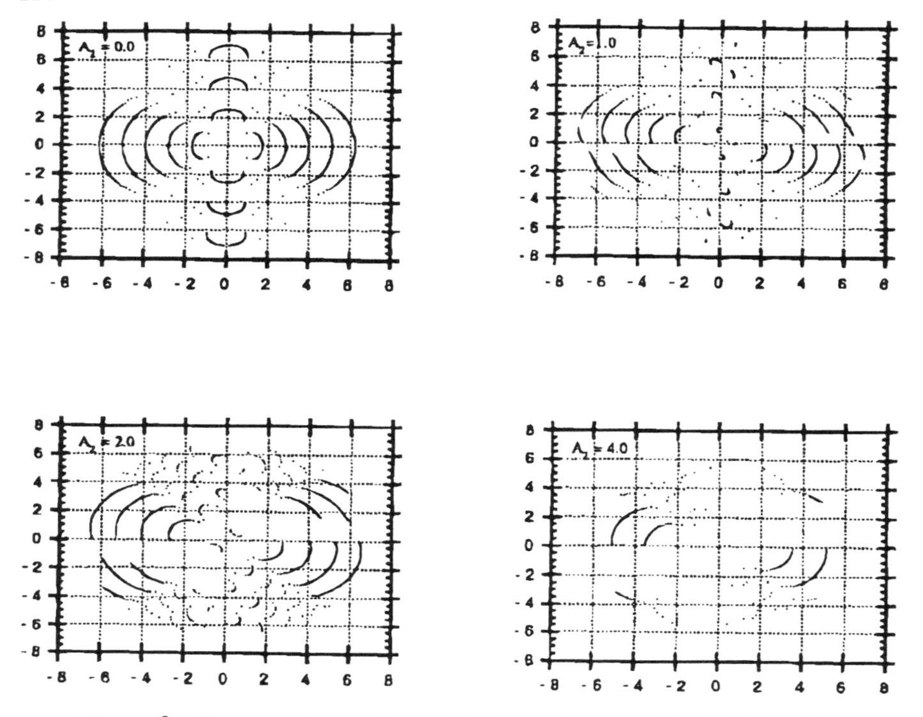

Figure 3. Cross sections through the stable manifold of the triad system for various values of A_2.

deviate. The structure of the *stable* manifold gives some clue of the likelihood of this. Some slices through this 2-D structure are shown in Figure 3; the manifold is highly contorted, rather like a crumpled ball of tissue paper. The series of shell-like structures visible in the cross-sections actually come from sheets winding helically about the A_1 and A_3 axes, giving the stable manifold the appearance of a series of four corkscrews or auger bits with points directed toward the origin along the 1 and 3 axes. Being more or less space-filling, it is rather hard to avoid, and hence trajectories should eventually come near it and be carried near (0, 0, 0). Based on the numerical results (and further analyses to be presented below), we conjecture that the amplitude is indeed unbounded. This conjecture has not yet been proved, however.

The geometric influence of the one-dimensional unstable manifold is clearly present in the trajectories. For this system, which has a three-dimensional phase space, most of what we would like to know about the geometry of the attractor can be inferred directly from looking at plots, a consequence of the fact that we are three-dimensional beings and have brains capable of interpreting three-dimensional geometric structures (or at least inferring them from Platonic shadows on the two-dimensional retina). In higher dimensions this is not possible, and so one must turn to various numerical descriptions of the geometry. The correlation dimension is the tool we shall employ here. Let us see how the geometric structure of the triad model is reflected in its correlation dimension.

We took a data set of 50,000 points from the triad model trajectories (equally spaced in time), and computed the Euclidean distance in 3-space between each pair of points. The histogram of the set of differences is shown on a log-log plot in Figure 4(a). We show results for the full data set and also for a subset consisting of the first 5,000 points. There is an extensive range of distances for which the attractor is characterized by a correlation dimension of approximately 1.5. This is consonant with what we know about its geometry: the trajectory spends much of its time clustered about the one-dimensional unstable manifold, but during its necessary excursions away from this manifold it executes a trajectory along one of the two planar "ears" sewn to the unstable manifold like pages onto the spine of a book, if we may mix metaphors. Thus, it makes sense for the dimension to be between 1 and 2.

At very small distances for the 5,000-point plot, the slope approaches unity; here we are beginning to pick up the geometry of individual trajectories, which are one-dimensional curves. For the 50,000-point plot, this behavior sets in only at still smaller distances, as the trajectory has had time to fill out the attractor more densely. At large distances the curve asymptotes to a constant, reflecting the obvious fact that for a finite set of points the distance must have an upper bound. A detailed examination of the behavior at large distances sheds some light on the question of whether the attractor lies in a bounded region of phase space, which is identical to the question of whether the histogram becomes exactly flat above some critical distance. First note that the curve for the 50,000-point data set is almost exactly parallel to that for the first 5,000 data points. To underscore this feature, in Figure 4(b) we have plotted the 5,000-point histogram rescaled by the ratio of the total number of pairs (a bit more than 100). The rescaled curve lies precisely atop the 50,000-point histogram, except at very small distances, indicating that the geometric structure of the attractor has been well defined.

A closer look at the large-distance tail (Figure 4(b)) shows that a larger maximum distance is attained in the 50,000-point case than in the 5,000-point case. In fact, the tail region of the histogram can be well fitted by a curve of the form $1 - ae^{-bd}N$, where N is the

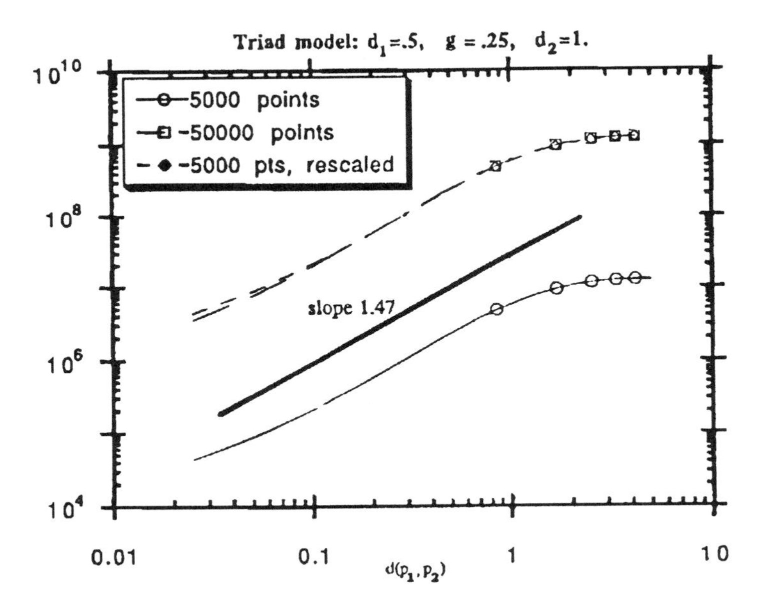

Figure 4 (a). Correlation dimension plot (log-log cumulative histogram) for the triad system.

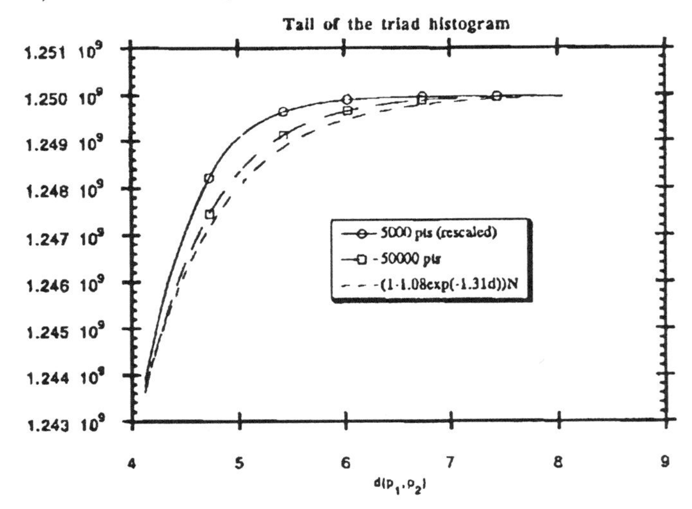

Figure 4(b). As in Figure 4(a), but showing a magnification of the tail region.

number of pairs. From this we infer a finite, though exponentially small, probability of arbitrarily large excursions. However, the distribution is *integrable* at large d, so that averages of any polynomial quantity (such as the variance) converge. This situation contrasts with $1/f$ noise.

The state of the real atmosphere is characterized by an infinite-dimensional phase space (the space of the values of wind, temperature, etc. at each point of the atmosphere), but it may be that relatively few directions in phase space are unstable. It is the signature of such a low-dimensional unstable manifold that we seek in the next section.

3. The Real Atmosphere

The data source for this study is a subset of the operational U.S. National Meteorological Center archive for 1940–1985. It is digested from the thousands of daily inhomogeneous data sources (radiosondes at several hundred stations, surface pressures, aircraft observations, and ship observations) and interpolated on a regular grid of 1,977 points covering the Northern Hemisphere poleward of 30° N latitude. We will focus on the temperature pattern on the 500 millibar (mb) pressure surface (roughly in the middle of the atmosphere, as determined by mass), and the height of the 500mb surface. The latter is of considerable dynamical interest, as it is (approximately) a streamfunction for the upper air flow. This study was made feasible by the availability of the archive on a single 600 megabyte read-only optical disk (CD-ROM) compiled by Professor Clifford Mass of the University of Washington (Seattle). The data analyses can be performed in the comfort of ones' own office with an inexpensive CD-ROM drive attached to a Macintosh II computer.

Although the archive includes data for each day, we will mostly look at the variability of the monthly means. Not only does this keep the computational requirements manageable, it also emphasizes precisely the low-frequency motions that are of primary interest in the effort to probe the feasibility of extended-range forecasting. It is generally agreed that individual storms lose predictability after a week or two. However, if the more ponderous general weather patterns embodied by the monthly means turns out to be governed by a low-dimensional attractor, that would really be news that would shake up the long-range forecasters. To be sure, the monthly mean destroys information on time scales of motion from minutes to weeks that are formally part of the underlying phase space. The premise (or pious hope) is that the *monthly mean rectified fluxes* arising from these motions are nevertheless systematically related to the monthly mean flow itself.

A single 500mb field from the archive is described by a point in a 1,977-dimensional space, and the time series moves in this space. This is not necessarily a phase space of the system, as we have neglected other data (temperature and height at other levels, moisture, etc.) needed to truly specify the future course of the system; furthermore, the monthly averaging process throws away additional information. Thus, the working space is a projection of the true phase space. Trajectories can cross in the working space, and so there is no guarantee of the existence of a flow determining future evolution from the initial conditions. As is done for the analysis of 1-D climatological time series, we seek to recover some of the true phase space structure by creating synthetic phase spaces of dimension $1,977m$, by taking m-tuples of fields $(Z_1, \ldots, Z_m)$, where each Z_i is a 1,977-point field. The correlation dimension is estimated by taking the Euclidean (ℓ_2) distance between each pair of points in this phase space, and forming the histogram.

Results for the monthly mean 500 millibar height (Z500) are shown in Figure 5 for various values of m. An immediately apparent feature is a subrange at large distances characterized by having slope 1. This indicates that the large-amplitude fluctuations are composed of trajectories that come in one-dimensional bundles. Somewhat smaller amplitude fluctuations appear to be characterized by a three-dimensional structure, and at yet smaller amplitudes the slope increases sharply, suggesting a dimension somewhere in between 20 and 40. This picture shows little sensitivity to the value of m. We have also reproduced similar results for subsets consisting of half the total data.

The 1-D structure at large amplitudes is nothing else than the seasonal cycle. For terrestrial conditions, root mean square differences in Z500 as large as, say, 150 meters occur only between two patterns belonging to different seasons. The results are telling us that the seasonal cycle is dominantly a 1-D entity; to tell what the flow pattern is to lowest order, one only needs to know the time of year. But what of the three-dimensional subrange? The origin of this feature is less apparent.

In order to remove the seasonal effects, we repeated the calculation using data only from the winter months (December, January and February). Results are shown in Figure 6. Not only has the 1-D subrange disappeared, but the 3-D subrange has disappeared as well. We are left only with the high-dimensional structure at short distances; the slight apparent decrease in its slope cannot in any way be regarded as statistically significant, a point about which we shall have more to say shortly. The results indicate that there is a component of the interseaonal variability that can be characterized by three degrees of freedom, one that is not phase-locked to the seasonal cycle. What is this motion? And what cycles of spatial patterns does it correspond to? It is one of the frustrations of the correlation dimension analysis

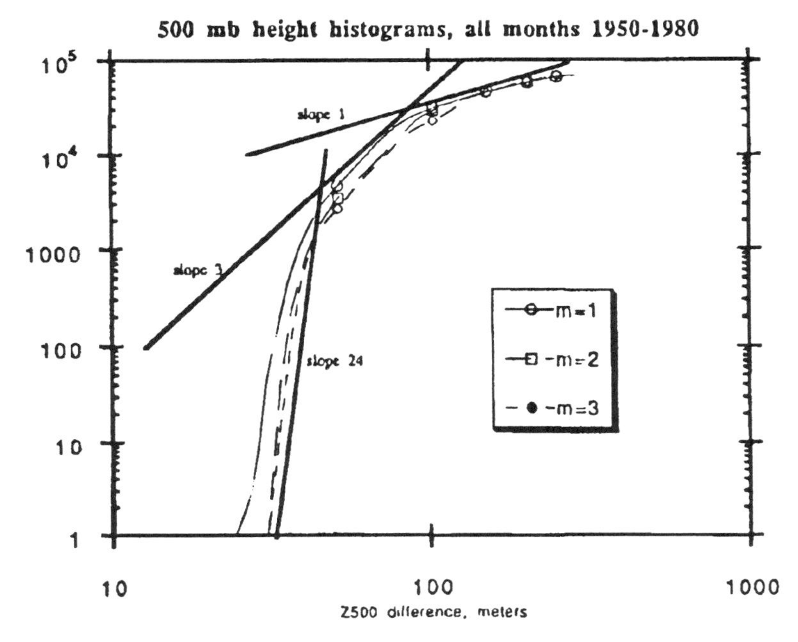

Figure 5. Correlation dimension plot for the 500mb geopotential height field of the atmosphere, using all monthly data.

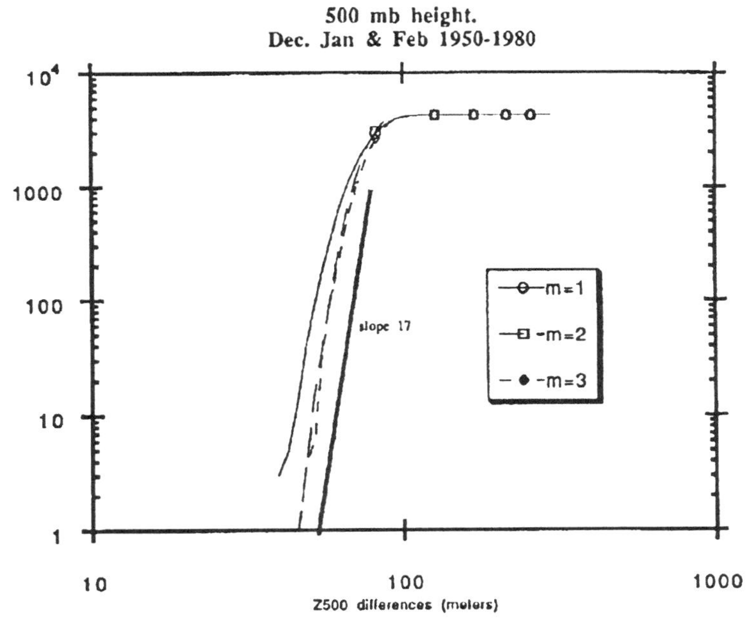

Figure 6. As in Figure 5, but using only December, January and February data.

that it tells us that the motion in infinite-dimensional function space is confined to the vicinity of a three-dimensional manifold, but it gives us no way of constructing this manifold. Specifically, there is (at least locally) a map

$$(\lambda_1, \lambda_2, \lambda_3) \Rightarrow Z(x, y|\lambda_1, \lambda_2, \lambda_3)$$

from triples of real numbers to the function space of all atmospheric flow patterns, which captures a great deal of the atmospheric variability. It would be very useful to know this map, as it would enable the characterization of much of the state of the atmosphere in terms of the specification of the triple of numbers.

An analysis of variability of winter 500mb temperatures is shown in Figure 7. The overall picture is rather similar to that for the height field; there is no indication of a very low-dimensional subrange, and slopes for small temperature differences indicate a dimensionality of around 20. Temperature is governed by very different detailed dynamics from the height field. In particular, it is (in a crude sense) advected around by the streamfunction derived from height data, and for this reason might be expected to show a markedly higher dimensionality. Evidently, the details arising from this process are averaged out in the monthly mean, leaving a temperature variability pattern of similar complexity to that for the height field.

Analysis of daily data would incorporate modes of variability that are filtered out in the monthly mean data set. Since there are 365 height fields per year, and the required number of disk accesses grows as a quadratic function of the number of samples, the rather slow access speed of the optical disk precludes an extensive multi-year study such as we have undertaken for the monthly means (though such a study will be feasible when we acquire enough fast magnetic disk space to upload an appropriate subset of the data). Still it is of interest to see what features can be discerned in the analysis of a single year's height data. Results for the year 1977 are given in Figure 8. Of course, with only one year of data we cannot pick up the 1-D seasonal cycle, as one cannot isolate a periodic signal from the sampling of a single period. Indeed, the 1-D subrange at large distances is not present in this plot. However, there is a distinct 5-dimensional subrange for moderately large distances. Probably three dimensions of this correspond to the interseasonal variability signal detected in the monthly mean analyses. We speculate that the remaining two degrees of freedom are associated with propagating features in the storm tracks. For smaller

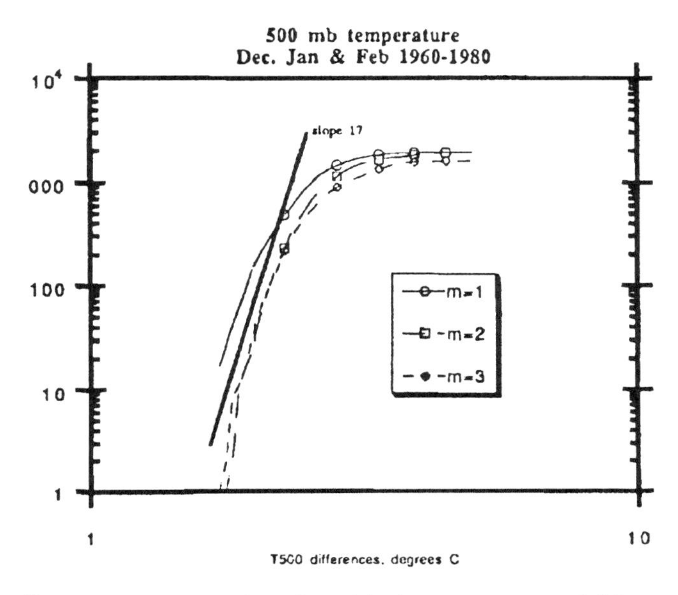

Figure 7. As in Figure 6, but for 500mb temperature fields.

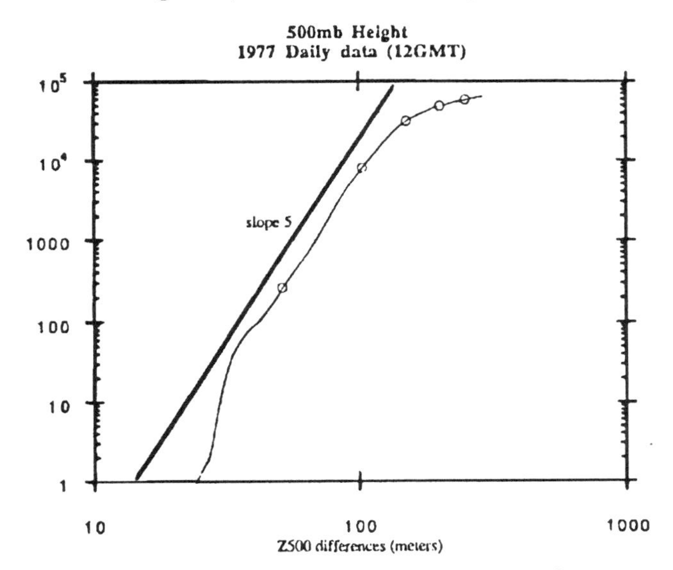

Figure 8. Correlation dimension plot for the 500mb geopotential height based on daily data at 12GMT for the year 1977.

height differences, the slope indicates a high-dimensional subrange similar to that picked up in the monthly mean analyses, suggesting that the higher frequency motion adds little to the spatial complexity of the signal; however, the number of samples in this steep subrange is rather small, and so drawing conclusions of this nature is perilous.

This leads us to the question of statistical significance of the results. Grassberger (1986) has treated the question of how much data is needed to reliably estimate dimensionality. In a nutshell, the answer is that the data requirements increase precipitously as the underlying dimension increases, since the volume of an N-dimensional sphere increases rapidly with N and one must have enough points to more or less fill up the sphere. To provide a more graphic indication of statistical reliability, which is nonetheless tied to the same idea, we have constructed sequences of random fields of N points each, filled-in using a random number generator. If the random number generator is truly random, the underlying dimensionality of the sequence will be N. We then constructed graphs of the correlation dimension from finite sequences of various lengths n, to see how well we could infer the true dimension. The results are shown in Figure 9. For a 20-dimensional underlying data set, an analysis with 120 points (similar to the size of the sample in our winter analyses) indicates a dimensionality in the 'teens, which indeed is an underestimate. Doubling the amount of data does not improve the situation much; it is not until we go to 10,000 data points and very small distances that we begin to see much improvement. Hence, we are not likely to ever know whether the true slope for the atmospheric data set is 10, 20, or 30. However, 120 point samples from data with underlying dimensionalities of 100 and 2,000 show markedly steeper slopes. From this, we conclude that the dimensionality of the atmospheric data *as present in this data set and for the ranges of difference discussed* is unlikely to be as much as 100. The issue remains as to how much of this apparent low dimensionality is in the real atmosphere, and how much is an artifact of the spatial smoothing and interpolation techniques used to derive the gridded data set from the raw observations. This question can only be answered with recourse to higher resolution data sets derived from the raw data by different methods.

In dealing with the height field (which is practically a streamfunction for mid-latitude flow), we are actually already applying a spatial filter to the underlying dynamics. To specify the future course of a two-dimensional fluid, one needs to know the *vorticity* accurately, not the streamfunction. Similar considerations obtain for potential vorticity in the atmosphere. Since the vorticity ζ and streamfunction ψ are related by $\nabla^2\psi = \zeta$, the streamfunction represents a highly smoothed picture of the vorticity. It is true that knowing the streamfunction *exactly,* one can infer the vorticity exactly. However, small amplitude

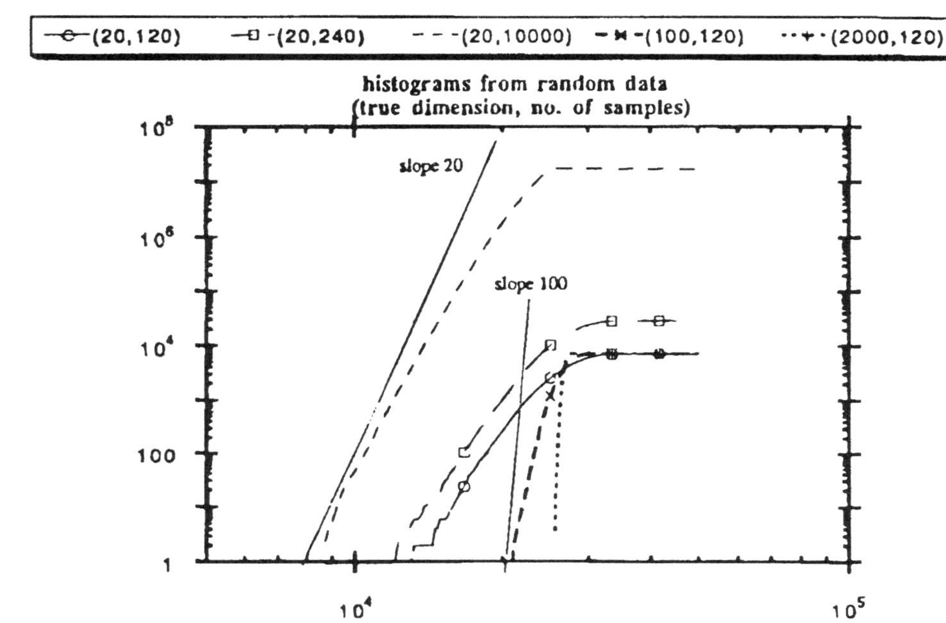

Figure 9. Correlation dimension plot for random fields of various dimensions.

streamfunction errors at small scales destroy the relation. To bring this point home, consider the functions

$$f_1(x) = \cos x, \quad f_2(x) = \cos x + \epsilon \cos\left(\frac{x}{\epsilon}\right)$$

The root-mean-square mean difference between these two functions is small if ϵ is small. However the root-mean-square mean differences of d^2f/dx^2 (which is like vorticity, if one thinks of f as the streamfunction field) diverge as ϵ approaches zero. This phenomenon is a reflection of the fact that in function spaces (unlike finite-dimensional systems) not all norms are equivalent. Distances that approach zero with respect to the streamfunction norm do not approach zero with respect to the vorticity norm. The relevance for the correlation dimension analysis, using data sampled from function spaces, is that we could get a very different dimension from vorticity data than we would from streamfunction data. We would need to have accurate data down to very small distances between streamfunctions to pick up the effects of vorticity fluctuations with small spatial scales. Yet, according to prognostic equations such as Eq. (2.5), fields with differing small-scale vorticity patterns, but nearly identical streamfunction patterns, will ultimately evolve in time along very different paths. Hence apparent low dimensionality in streamfunction data does not mean that the future course of the system can be accurately predicted by projection onto this low-dimensional manifold.

Another possible source of apparent low dimensionality is filtering of a high-dimensional process by a linear subsystem. Consider the first-order Markov process $A_n = \rho A_{n-1} + \epsilon_n$, where $\rho < 1$ and ϵ_n is uncorrelated white noise. The underlying dimensionality of the system is infinite, in the sense that no finite amount of initial data admits the prediction of the future course of the system over an indefinitely long time period. However, examination of large amplitude fluctuations will generally pick up the one-dimensionality associated with the exponential decay of A. The correlation graph will show a one-dimensional subrange for large A differences, with marked steepening as small distances are approached. However, the apparent one-dimensionality is not entirely without dynamical significance, as it does indeed reflect a certain limited measure of predictability of large amplitude fluctuations. A similar process may occur in the atmosphere, where the linear subsystem could be provided by large-scale, low-frequency Rossby waves.

The low dimensionality of the variability of the height and temperature patterns has some interesting consequences for the reccurence of weather patterns. If the dimensionality of the monthly mean fields were as high as, say, 1,000, one would have to wait $(dZ/dZ_0)^{1,000}$ months to see two fields that differed by dZ, where dZ_0 is some reference value. If this were the case, each weather pattern would for all intents and purposes be unique. With a dimensionality of 20 or so, however, recurrence of similar patterns would be rather common. In fact, long before fractal geometry had been invented, Lorenz (1969) studied recurrence in atmospheric flow. His intent was to analyze atmospheric predictability by first searching for similar initial states in the historical record, and then examining the divergence of their subsequent evolution. Lorentz's attempt was only partially successful, for reasons evident in Figures 5–8—close analogues, while not exactly nonexistent, are nonetheless rather rare. The entire 40-year historical record yields only a score or so of analogues close enough to be useful in predictability studies. Even if the dimension did not increase beyond 20 at small distances, halving the threshold dZ at which we are willing to accept two flows as analogues still would mean waiting 2^{20} times as many months (or about 40 million years) for suitable analogues to appear.

4. Genesis of Spatial Complexity: Chaotic Mixing

Regardless of the implications (or lack thereof) of the results of §3 for the presence of low-dimensional chaos in the atmosphere, it's clear that the space of all streamline patterns visited by points of the atmospheric attractor can, with considerable accuracy, be characterized by a rather small number of coordinates. On the other hand, observations of tracers such as potential vorticity (e.g., McIntyre and Palmer 1983) typically show a great deal of spatial complexity. There is no contradiction in this state of affairs. Recent results on mixing by two-dimensional, time-dependent flows show that the particle trajectories can be chaotic and exhibit predictability decay even if the velocity field moving the particles around is perfectly predictable—in fact, even if it is periodic in time (Ottino et al. 1988). In this section, we analyze an example of this kind of mixing in an idealized atmospheric flow. The correlation-dimension analysis will reappear here as a measure of the spatial complexity of the tracer field.

Consider the two-dimensional incompressible velocity field defined by the streamfunction

$$\psi = A\cos(k_1(x - c_1 t))\sin(l_1 y) + \epsilon\cos(k_2(x - c_2 t))\sin(l_2 y) \qquad (4.1)$$

This represents a primary traveling wave with amplitude A and phase speed c_1, disturbed by a perturbation with speed c_2 and (small) amplitude ϵ. Most of the results shown below are insensitive to the specific kind of waves chosen, but for the sake of concreteness let's consider solutions of the barotropic β-plane equation. This equation is a crude approximation to planetary-scale flows in a shallow atmosphere, and consists of Eq. (2.5) with $g = d = 0$, the vorticity being replaced by the potential vorticity $\nabla^2\psi + \beta y$. It is essentially a tangent-plane approximation to incompressible 2-D flow on the surface of the sphere, with x being approximately the longitude and y being approximately the latitude. The waves in this system are known as Rossby waves, and satisfy the dispersion relation

$$c = \frac{-\beta k}{k^2 + l^2} \qquad (4.2)$$

when the equations are linearized about a state of rest. In fact, with c_1 given by Eq. (4.2), Eq. (4.1) is an exact nonlinear solution of the equations for arbitrary A provided $\epsilon = 0$. The quantity c_2 is also taken to be governed by the Rossby wave dispersion relation, though there is no particular mathematical justification in doing so. The mixing results are governed primarily by the streamline geometry of the unperturbed flow and the overall magnitude of the perturbation, so this crude assumption on the form of the perturbing wave isn't particularly dangerous.

In the comoving frame defined by $X = x - c_1 t$, the velocity field has the streamfunction

$$\psi = A\cos(k_1 X)\sin(l_1 y) + c_1 y + \epsilon\cos(k_2 X - k_2(c_2 - c_1)t)\sin(l_2 y) \quad (4.3)$$

and corresponds to a steady velocity field disturbed by a perturbation that's periodic in time, the Poincaré period being $T = 2\pi/(k_2(c_2 - c_1))$. The problem of particle motion in this flow field is a perturbed planar Hamiltonian system. Figure 10 shows the unperturbed streamlines in the comoving frame for $A = 1$. The stagnation points P and P' form what is known as a "heteroclinic cycle," being connected by the upper arc PP' (the unstable manifold of P and the stable manifold of P') and the lower segment $P'P$ (the unstable manifold of P' and the stable manifold of P). The latter is preserved under perturbation, but powerful theorems of dynamical systems theory imply that the former breaks up generically into a chaotic set. It is the coexistence of closed and open streamlines that is at the heart of the chaos, leading to the inevitability of chaotic mixing. Indeed, Knobloch and Weiss (1987) have presented evidence of chaotic mixing in thermosolutal convection, which involves modulated traveling waves having an identical streamline geometry to that arising in the Rossby wave case.

In Figure 11 we show the time evolution of an initially small cloud of particles placed near the bottom of the recirculating eddy at the lower left of Figure 10. The particles were moved around in the velocity field defined by Eq. (4.3), taking $A = 1$ and $\epsilon = 0.15$. The number of time periods T that have passed are indicated on each frame. The cloud is stretched out rather gently for a time, but then "finds" the unstable manifold and expands in a burst. Thereafter, the mixing becomes roughly diffusive, displaying rapid stretching and folding and efficient cross-streamline mixing. The final state of a mixed cloud of 10,000 particles is plotted in Figure 10, in order to show the intimate association between the mixed region and the heteroclinic streamline structure.

It is common practice to characterize observed or simulated concentration distributions by their Fourier power spectra. What is the power spectrum of a distribution such as shown in Figure 10? An el-

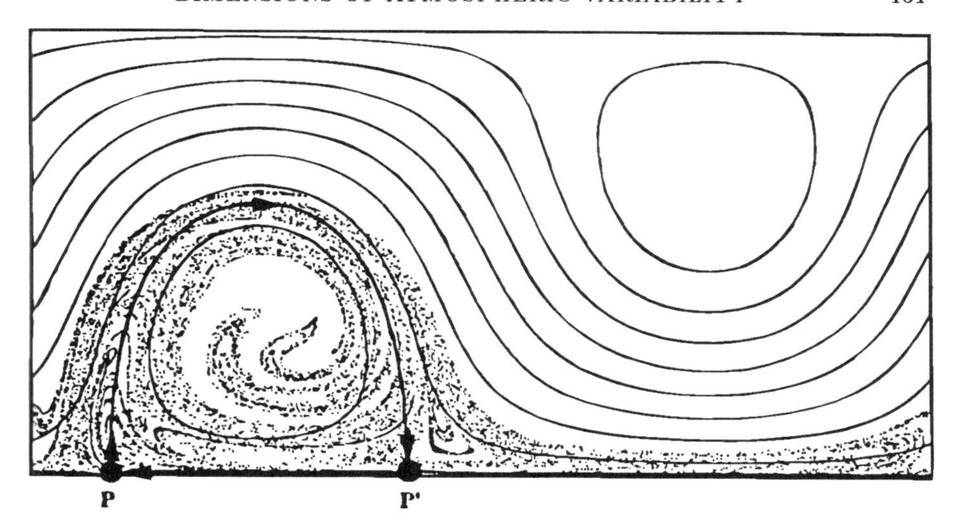

Figure 10. Streamline pattern of the primary Rossby wave in the comoving reference frame. The domain goes from 0 to $2p$ in x and from 0 to p in y. The cloud of points illustrates the dispersal of an initially small cloud of particles when they are advected by a perturbed Rossby wave with $e = 0.25$.

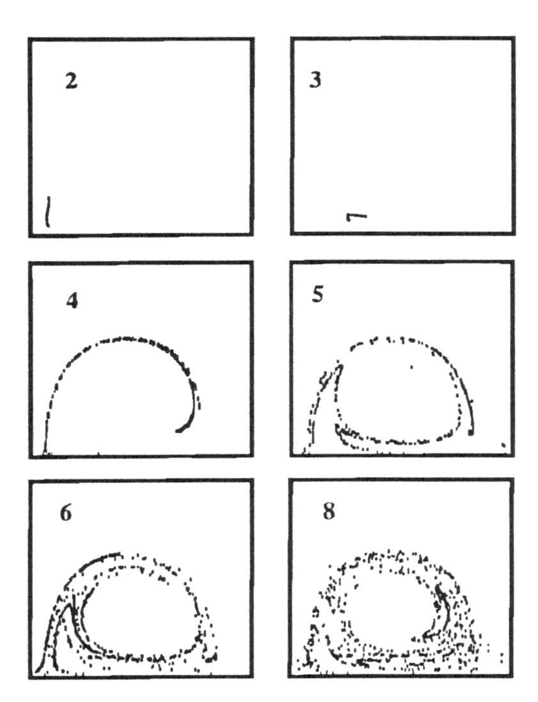

Figure 11. Time sequence of particle dispersion.

ementary argument reveals a direct relation between spectra and the correlation dimension of the particle cloud.

Consider a concentration pattern made up of an assemblage of N circular blobs located at points $\mathbf{r}_j$ in the two-dimensional plane, with $j = 1, \ldots, N$. Each blob contributes a concentration $G(|\mathbf{r} - \mathbf{r}_j|)$, where $G(0) = 1$. Thus, the concentration is

$$C(\mathbf{r}) = \sum_{j=1}^{N} G(|\mathbf{r} - \mathbf{r}_j|) \tag{4.4}$$

and its 2-D Fourier transform is

$$c(\mathbf{k}) = \frac{g(|\mathbf{k}|)}{2\pi} \sum_{j=1}^{N} e^{i\mathbf{k}\cdot\mathbf{r}_j} \tag{4.5}$$

where $g(\cdot)$ is the Fourier transform of $G(\cdot)$. Thus, the concentration variance spectrum is

$$|c(\mathbf{k})|^2 = \frac{|g(|\mathbf{k}|)|^2}{2\pi} \sum_{j,j'=1}^{N} e^{i\mathbf{k}\cdot(\mathbf{r}_j - \mathbf{r}_{j'})} \tag{4.6}$$

Now suppose that there are a large number of particles, and that the distribution of interparticle distances is governed by a probability density $P(\delta\mathbf{r}) = P(\delta x, \delta y)$, where $\delta\mathbf{r} = (\delta x, \delta y)$ is the distance between particles. Then the number of particle pairs separated by vector distance $\delta\mathbf{r}$ is approximately $N^2 P(\delta\mathbf{r})$, whence Eq. (4.6) becomes

$$|c(\mathbf{k})|^2 = N^2 \frac{|g(|\mathbf{k}|)|^2}{2\pi} \iint P(\mathbf{r}) e^{i\mathbf{k}\cdot\mathbf{r}}\, dx dy \tag{4.7}$$

in which the double integral extends over all space. For wavelengths much longer than the individual blob radius, $g = 1$. Thus, the concentration variance spectrum is proportional to the Fourier transform of the probability distribution of the interparticle distance. If we further assume the probability distribution to be isotropic, then $P(\mathbf{r}) = P(r)$, where $r = |\mathbf{r}|$, whence, upon integrating over angles θ, Eq. (4.7) becomes

$$|c(\mathbf{k})|^2 = \text{constant} \cdot \int_0^\infty r P(r) J_0(kr)\, dr \tag{4.8}$$

where J_0 is the zero-order Bessel function of the first kind. This gives us the desired connection with the correlation dimension, since

$rP(r) = dH/dr$, H being the cumulative histogram we have been dealing with in our dimension computations. If there is an extensive range of distances for which $H = r^d$, then letting (kr) be the integration variable in Eq. (4.8), we find

$$C(k) = k|c(\mathbf{k})|^2 = \text{constant} \cdot k^{1-d} \tag{4.9}$$

where $C(k)$ is the isotropic power spectrum as conventionally defined.

As the concentration pattern is embedded in the two-dimensional plane, we must have $0 < d < 2$. The case $d = 0$ corresponds to clustering of the concentration into a few isolated pointlike regions, and yields a power spectrum that increases toward short waves. The situation $d = 1$ corresponds to well-separated strands, and yields a flat spectrum. The steepest possible spectrum is k^{-1}, corresponding to mixing of the tracer over a two-dimensional region. Steeper spectra cannot be associated with self-similar fractal concentration patterns, since there is not enough variation at small scales. A more traditional way of looking at this derives from the fact that k^{-1} has a divergent integral from (any) k_0 to ∞, corresponding to an infinite total concentration variance. Steeper spectra, however, have convergent variance, whence one can define the integral scale

$$L^2 = \frac{\int_{k_0}^{\infty} k^{-2} C \, dk}{\int_{k_0}^{\infty} C \, dk} \tag{4.10}$$

This defines a preferred scale for structures in the concentration pattern. Obviously, such distributions cannot exhibit self-similarity.

It is a simple matter to apply this calculation to particle dispersion data from the Rossby wave problem. A computation of the correlation dimension for a cloud of particles like that shown in Figure 10, but recomputed with 40,000 particles, shows an extensive range of scales characterized by $d = 2$ (or perhaps a bit less). This implies a subrange with a k^{-1} spectrum. In Figure 12 we show the spectrum, calculated from a Bessel transform of the histogram. It indeed shows the predicted spectrum for the intermediate length scales. This spectrum is the same as the classical Bachelor spectrum predicted for tracers in homogeneous, isotropic, two-dimensional turbulence, on the basis of scaling arguments (e.g., see Rhines 1979).

How universal is this result? All parameter settings we have tried in the Rossby wave case ultimately lead to a mixed cloud characterized by $d = 2$. We have also found $d = 2$ for chaotic mixing by the

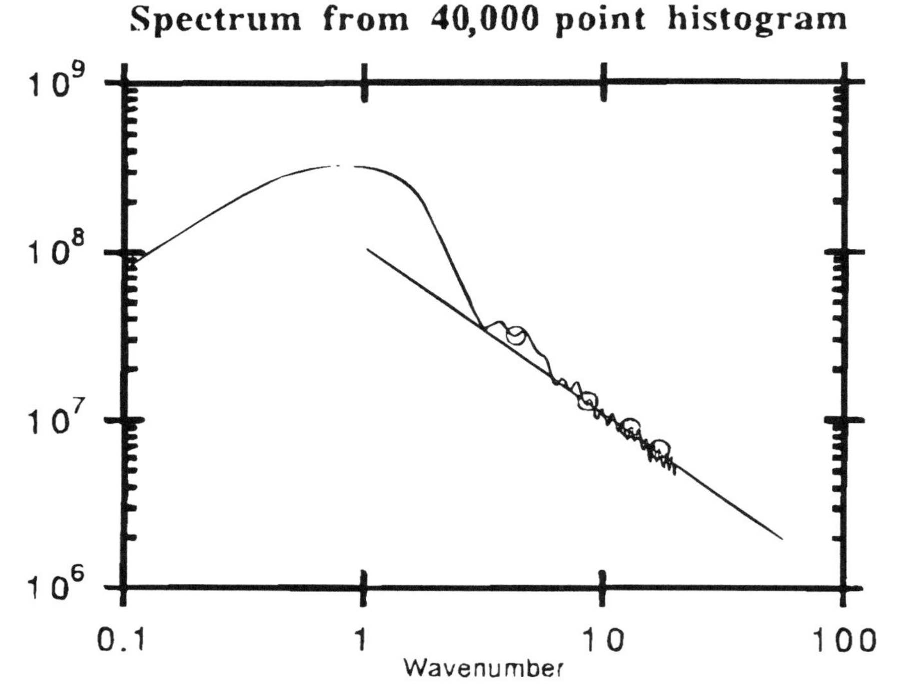

Figure 12. Power spectrum of the concentration distribution in Figure 10, computed by Bessel transform of the histogram.

Kida vortex (Polvani and Pierrehumbert 1989). Visual analysis of the numerous cases considered by Ottino and others reveals no obvious contradictions to the $d = 2$ law. While it seems plausible that chaotic mixing should always proceed to $d = 2$, we have been unable to produce a proof of this conjecture. If we take this conjecture as true, however, it implies an ubiquity for the Bachelor spectrum that transcends the restrictive assumptions of the original scaling arguments. The k^{-1} spectrum would arise naturally in any flow with fluctuating large eddies, regardless of whether the eddy field is itself homogeneous or isotropic. The spectrum can be obtained as a result of advection *by the large scale component of the flow field alone.* This is a dramatic illustration of the well-known nonlocality of two-dimensional turbulence in spectral space. The concentration spectrum at length scale L (no matter how small) is governed by the velocity field of the large eddies, rather than by advection of eddies having scales near L.

Our results on concentration spectra have some implications for the dynamics of two-dimensional turbulence as well. Vorticity (or more generally potential vorticity), is a tracer in an inviscid fluid, as it is conserved following a fluid parcel. The power spectrum of vorticity is distinguished by the name "enstrophy spectrum." Owing to the relation between vorticity and velocity, a $k^{-\alpha}$ enstrophy spectrum corresponds to a $k^{-(\alpha+2)}$ energy spectrum. The same scaling arguments that predict a k^{-1} concentration power spectrum in homogeneous, isotropic 2-D turbulence yield a k^{-1} enstrophy spectrum and corresponding k^{-3} energy spectrum. Assuming vorticity acts qualitatively like a passive tracer, the chaotic mixing properties imply that the k^{-1} enstrophy spectrum can arise in much more general circumstances. How does vorticity differ from a passive tracer?

Unlike a passive tracer, it changes the velocity field. From our standpoint, the most important manifestation of this is that as a circular vorticity blob is stretched out into an ellipse, its self-induced velocity causes it to rotate or roll up, both of which act to hold the blob together. Thus, we expect that isolated strong vortices will survive, while sufficiently weak vortices will be dispersed in the manner of a passive tracer. Chaotic mixing is generic in the sense that it is not particularly sensitive to the details of the fluctuating velocity field. This fact leads us to conjecture that the strength of "weak" vortices need not be *infinitesimal* in order to be dispersed like a passive tracer. It should be sufficient that they not be strong enough to hold together.

For many years the k^{-1} shortwave enstrophy spectrum was assumed to be a fact of life in 2-D turbulence. However, calculations by McWilliams (1984) and many others since tend to show a steeper spectrum. The steepness is generally associated with the freezing out of isolated eddies that have individually smooth vorticity distributions. The incompatibility of this situation with self similarity is in accord with the analysis presented above. When the tracer in question is vorticity, the integral scale of Eq. (4.10) represents a characteristic vortex size. Since the k^{-1} spectrum appears to be an inevitable consequence of chaotic dispersal, the steeper spectra are expected to occur only when the initial condition contains small eddies intense enough to resist being torn apart by the large-scale flow field. This hypothesis is currently being tested in fully nonlinear simulations, the results of which will be reported elsewhere.

The above considerations apply to the small-scale spectrum of two-dimensional turbulence. At large scales, two-dimensional turbulence exhibits a different universal spectrum, characterized by a $k^{-5/3}$ energy spectrum and a flux of energy from small scales to large scales. It is interesting to note that this spectrum is also compatible with a fractal vorticity structure. The energy spectrum implies a $k^{1/3}$ enstrophy spectrum, which yields a dimensionality $d = \frac{2}{3}$. Since $d = 0$ corresponds to "vortex clumps" and $d = 1$ corresponds to "vortex strands," the indicated geometry of the large-scale vorticity structures is "clumpy strands." The chaotic mixing results provide some dynamical understanding of the reasons for the small-scale vorticity geometry, but we do not yet have analogous explanation for the large-scale geometry.

5. The Predictability of Predictability: Fractal Structure of Lyapunov Exponents

We have seen that the trajectories associated with deterministic flow fields can exhibit chaos and loss of predictability. The mixing seen in Figure 10 is *prima facia* evidence that, given a small initial error in the position of a particle, after a sufficiently long period of time it will be more-or-less equally probable to find the particle anywhere within a broad two-dimensional region. Can we at least predict the rate at which predictability will be lost? If, for example, a radioactive cloud is released over Chernobyl, can we say with confidence how long its trajectory can be accurately predicted? In this section we take up this question, using the system described in §4 as an example.

The basis of this analysis is the computation of a variant of the *Lyapunov exponent.* This number measures the exponential growth rate of the separation of neighboring trajectories. To compute the Lyapunov exponent, we take two initial conditions separated by a distance δ and integrate the corresponding trajectories forward in time for a long period τ. The distance $d(t)$ between the trajectories is computed, with the (most unstable) Lyapunov exponent then being given by $\log d/\tau$. For a two-dimensional system, such as the Rossby wave example, there are only two Lyapunov exponents; one is positive and the other negative, their sum being zero, in accordance with the incompressibility of the velocity field. A technical detail of the calculation is that we actually use *three* trajectories, taking the maximum separation in order to ensure against the unlikely possibility of the initial condition projecting exactly onto the contracting direction rather than that of expansion.

The true Lyapunov exponent for a given trajectory is obtained in the joint limit as $\delta \to 0$ and $\tau \to \infty$. Of course, this is an impossible limiting operation to carry out numerically. Instead we deal with estimates of λ for finite times τ, which are in any case of greater physical interest since we're limited to only a finite time in which to observe a particular experiment anyway. It's generally assumed that any trajectory within the chaotic zone will yield the same Lyapunov exponent. But we take nothing for granted, especially since we are computing only a finite-time estimate. Hence, we cover the region of interest with a grid of initial conditions, and compute λ for trajectories starting from each point on the grid. In the subsequent discussion, we confine our attention to the finite-time results, and do not attempt to treat the formidable question of what happens as $\tau \to \infty$.

The spatial distribution of the finite-time estimates of the Lyapunov exponent are shown at three different scales in Figure 13,

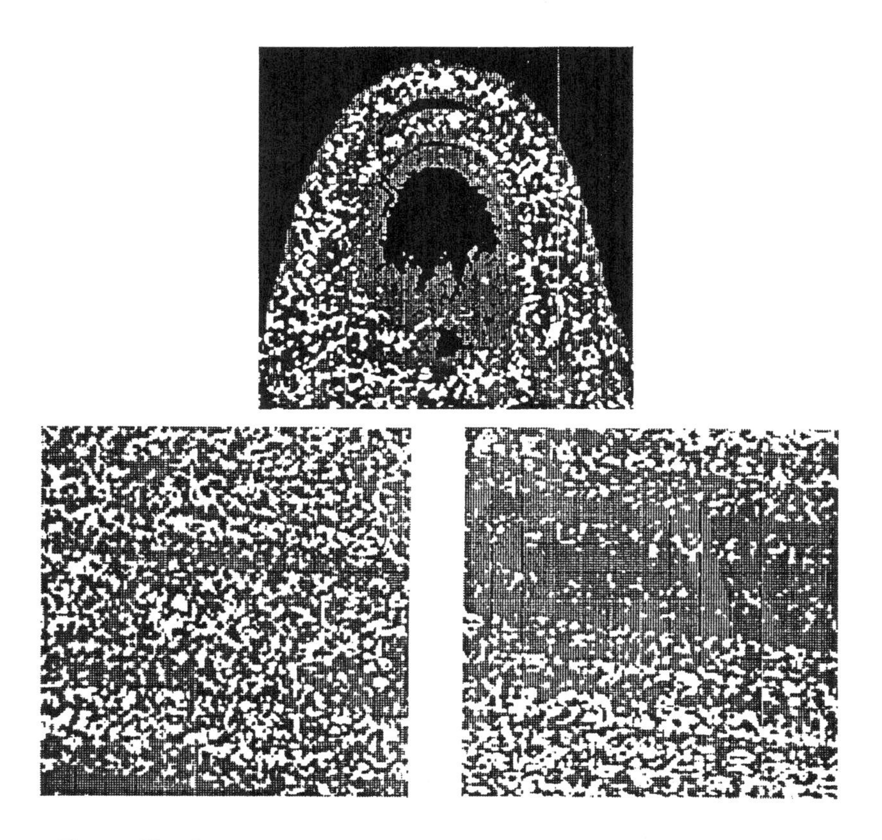

Figure 13. Spatial distribution of the Lyapunov exponents based on 50 Poincaré time periods. Black regions are nonchaotic ($\lambda = 0$), grey regions correspond to values of λ between 0 and the median of all positive values, and white regions correspond to above median values. The top panel shows the macroscopic structure of the chaotic zone, and covers the domain $x \in [0, \pi]$, $y \in [0, 2]$. The lower-left panel zooms in by roughly a factor of 10, covering the domain $x \in [1.520795, 1.620795]$, $y \in [0.2, 0.3]$. The lower-right panel zooms in by yet another factor of one thousand, covering the domain $x \in [1.570695, 1.570895]$, $y \in [0.2499, 0.2501]$.

computed for $\epsilon = 0.15$ and $\tau = 50T$. This analysis makes it easy to discern the macroscopic structure of the chaotic zone (see top panel), as non-chaotic regions are characterized by $\lambda = 0$. A striking feature of the distribution is that the boundary of the chaotic zone is not a smooth curve, but has the typical coastline-like appearance of a fractal. We leave it to the reader to decide whether or not this is surprising. Moreover, there is a great deal of fine-grained variability of λ within the chaotic zone. The lower panels of Figure 13 show that this variablity is present over at least four orders of magnitude of spatial scales.

From Figure 13, it's difficult to tell whether the positive values of λ are highly variable or sharply clustered about the median. Hence, in Figure 14 we show cross sections of actual pixel values of λ along a vertical centerline of each panel of Figure 13. It is clear that the variability of λ is substantial at all scales considered. Of course, for any fixed τ, it is a consequence of regularity theorems for differential equations that the estimate λ must be a smooth function of space. However, because of the exponential divergence of neighboring trajectories, the spatial scale at which one begins to see the smoothness decreases exponentially with increasing τ.

The physical implications of these results are twofold. The first concerns mixing times, a crucially important physical characteristic of the flow. Figure 11 showed that mixing proceeds in two stages, becoming diffusive only after the blob is stretched out over a distance comparable to the scale of the large-scale flow. The Lyapunov exponent approximates the time occupied by the first stage of mixing, our results suggesting a fractal pattern in the spatial distribution of Lyapunov exponents. Thus, in the limit of arbitrarily small initial blobs, the time it takes for the blob to become well mixed is arbitrarily sensitive to initial conditions. The second implication is that the rate of predictability loss is not very predictable. There are clearly macroscopic regions of tori in Figure 13, within which there is no predictability decay. However, within the fractal region of λ, the predictability of an given trajectory can differ wildly from that of its nearby neighbor.

6. Conclusions

We have given several examples here illustrating the power of the correlation dimension analysis as a tool for probing the patterns of variability in the atmosphere and in atmospheric models. Along the way, we have encountered a number of interesting dynamical systems, not least the atmosphere itself. The analysis picks up an expected one-dimensional component of atmospheric variability associated with the periodic sea-

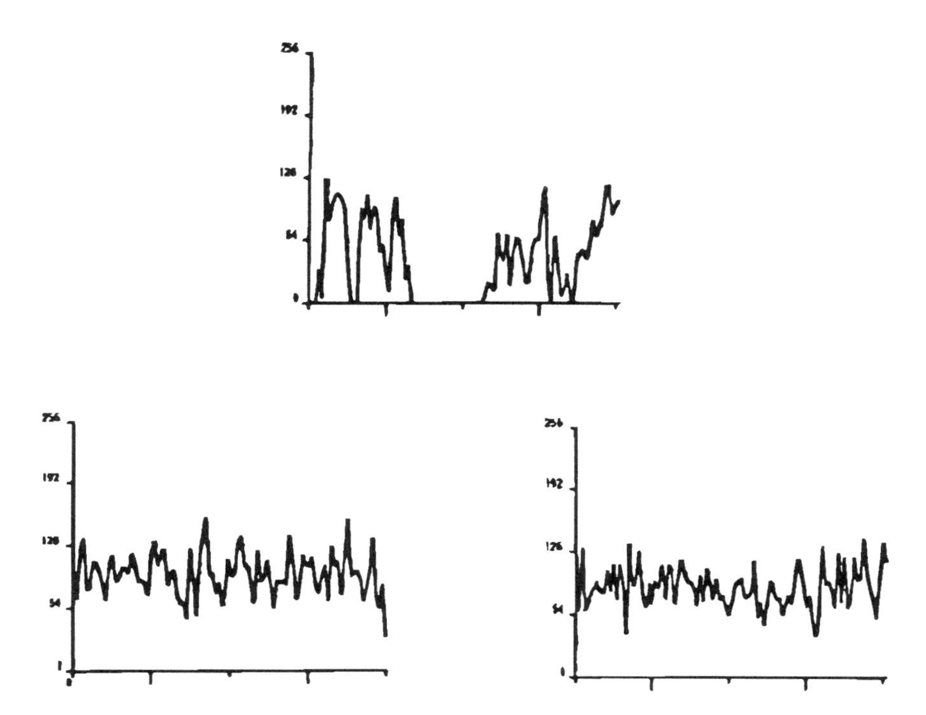

Figure 14. Cross section of values of λ along a vertical line at the centers of the three panels of Figure 13.

sonal cycle. For somewhat smaller amplitudes of fluctuation, it also reveals an additional two degrees of freedom linked to the seasonal cycle. The nature of this motion is obscure, and the next challenge will be to find some way to isolate the sequence of spatial patterns associated with it. For still smaller amplitudes of fluctuations, the indicated dimensionality is perhaps 20, and certainly less than 100. This regime is not tied to the seasonal cycle. A much longer record would be necessary to determine whether and how fast the dimensionality increases as the amplitude approaches zero. It would also be illuminating to carry out the analysis for higher-resolution data sets from computer models of the atmosphere's general circulation.

We are reluctant to conclude on the basis of these results that the atmospheric evolution could be reproduced with a model containing a few dozen variables (even assuming one had some way of identifying them.) Our analysis does not convincingly demonstrate that the embedding space we have dealt with is indeed a phase space. Considerations of vorticity versus streamfunction suggest that, on the microscopic level, it is indeed unlikely to be a phase space. The picture that emerges of the geometry of the atmospheric variability is the following: When we stand far back in the multimillion-dimensional embedding space of the atmospheric system, the trajectory appears to be confined to a hypersurface of rather low dimensionality (let us say it is 20). When we get closer to this hypersurface and take out our magnifying glass, we see that it is actually thick in several thousand orthogonal directions, and that the detailed course of the system trajectory depends not just on the projection onto the hypersurface, but also on knowlege of the value of these additional variables. This leads to several intriguing questions: Can one define a flow on the 20-dimensional hypersurface whose trajectories approximate those of the true system? As a function of time, what is the uncertainty in the future course of the true system, given knowlege only of the projection on the 20-dimensional hypersurface? And, finally, can one define a flow that at least reproduces the *climate,* i.e., the statistical behavior, of the true system projected on the hypersurface?

The correlation dimension analysis does, however, show that the state of the atmosphere can, in principal, be specified with considerable accuracy in terms of a few dozen variables. In order to carry out this data compression, one again comes up against the need to find a way to determine the *modes* of atmospheric variability, rather than just its dimensionality.

We have seen further that the rather low dimensionality characterizing the large-amplitude fluctuations of the wind fields is not incompatible with the observed spatial complexity of various conserved tracers in the atmosphere. An example was presented showing that even a precisely periodic large-scale flow field can produce spatial chaos in the tracer field it advects around. These tracer distributions on the plane were found to be characterized by a correlation dimension of very nearly 2. Further analysis revealed a simple connection between the correlation dimension of the spatial pattern and power spectrum of the concentration. This connection allows one to rephrase in geometric terms the search for universal spectra of concentration or vorticity in two-dimensional turbulence. It provides some insight into the prevalence of the classical spectra and the circumstances leading to deviations from the classical spectra.

Finally, there are some striking indications that the very rate at which trajectories lose their predictability is unpredictable, assuming there are even small errors in the initial point of the trajectory. The number characterizing predictability appears to be distributed in a spatially complex manner. We have made no attempt to formally characterize its variability. The fractal tools employed earlier in this essay were useful primarily for characterizing the geometry of objects. In the "predictability of predictability" problem we are dealing instead with a map assigning a real number to each point in space—or more precisely a measure assigning the average of the quantity to each small subset of the space—so the appropriate tool for effecting the characterization is multifractal analysis.

References

Essex, C., Lookman, T., and Nerenberg, M. A. H. 1987: "The Climate Attractor over Short Timescales," *Nature,* 326, 64–66.

Grassberger, P. 1986: "Do Climatic Attractors Exist?" *Nature,* 323, 609–612.

Grassberger, P. and Procaccia, I. 1983: "Measuring the Strangeness of Strange Attractors," *Physica D,* 9D, 189–208.

Knobloch, E. and Weiss, J. B. 1987: "Chaotic Advection by Modulated Traveling Waves," *Phys. Rev. A,* 36, 1522–1524.

Lorenz, E. N. 1969: "Atmospheric Predictability as Revealed by Naturally Occurring Analogues," *J. Atmos. Sci.,* 26, 636–646.

Maasch, K. A. 1989: "Calculating Climate Attractor Dimension from $\delta^{18}O$ Records by the Grassberger-Procaccia Algorithm," *Climate Dynamics,* 4, 45–55.

McIntyre, M. E. and Palmer, T. N. 1983: "Breaking Planetary Waves in the Stratosphere," *Nature,* 305, 593–600.

McWilliams, J. 1984: "The Emergence of Isolated Coherent Vortices in Turbulent Flow," *J. Fluid Mech.,* 146, 21–43.

Ottino, J. M., Leong, C. W., Rising, H. and Swanson, P. D. 1988: "Morphological Structures Produced by Mixing in Chaotic Flows," *Nature,* 333, 419–425.

Pedlosky, J. 1979: *Geophysical Fluid Dynamics,* Springer-Verlag, New York.

Polvani, L. and Pierrehumbert, R. T. 1989: "Characterization of Mixing by the Kida Vortex," Paper presented at the 1989 American Physical Society Symposium on Fluid Mechanics, Anaheim, CA.

Rhines, P. B. 1979: "Geostrophic Turbulence," *Ann. Rev. Fluid Mech.,* 11, 401–442.

Tsonis, A. A. and Elsner, J. B. 1988: "The Weather Attractor Over Very Short Timescales," *Nature,* 333, 545–547.

CHAPTER 7

Sir Isaac Newton and the Evolution of Clutch Size in Birds: A Defense of the Hypothetico-Deductive Method in Ecology and Evolutionary Biology

BERTRAM G. MURRAY, JR.

Abstract

In this chapter I explore the reasons why biologists have not been as successful as physicists in developing unifying, predictive, explanatory theory. Although most biologists claim that the failure results from the greater complexity of biological systems, I believe the failure is a result of most biologists being diversifiers, exploring the details of biology by observation and experiment, and of most theoretical biologists thinking inductively, whereas theoretical physicists often think deductively. After reviewing the diversifier, inductive unifier, and deductive unifier approaches to science, as exemplified by the development of our understanding of planetary motion, I compare two general explanation's accounting for the evolution of clutch size in birds—Lack's hypothesis and Murray's hypothetico-deductive theory.

1. Introduction

Biologists are essentially empirical scientists. We are trained to observe, to count, to measure, to design elaborate (sometimes elegant) experiments, and to perform sophisticated statistical tests on our data. We are not trained, however, in the techniques of developing and evaluating general theory. With regard to providing explanations for our data, we are on our own, more or less absorbing our epistemology from the customs of our teachers and our peers. As a result, biologists are overwhelmed with facts, and the journals are filled with descriptive studies. We are also overwhelmed with what purports to be theory. Indeed, there are probably two or more explanations for any particular observation. Multiple explanations for phenomena might be presumed to be a desirable state of affairs for scientists (Chamberlain 1890; Platt 1964), providing direction for further research, but often the multiple "theories" in biology are no more than untestable ad hoc hypotheses. There are few, if any, biological principles that have enjoyed the status among scientists as have Newton's Laws of Motion, Einstein's Special Relativity, or Heisenberg's Uncertainty Principle.

Much has been written regarding the methods of studying ecology and evolutionary biology (e.g., Hull 1973; Gould and Lewontin 1979; Strong 1980, 1983; Hilborn and Stearns 1982; Mayr 1983; Quinn and Dunham 1983; Roughgarden 1983; Simberloff 1983; Stenseth 1983; James and McCulloch 1985; Naylor and Handford 1985; Bartholomew 1986; Connor and Simberloff 1986; Murray 1986a, 1986b; Fagerström 1987; Loehle 1987, 1988; Haila 1988; Mentis 1988; Taylor 1989). These reviews and comments represent a diversity of opinion regarding the scientific method as it should be practiced by ecologists and evolutionary biologists. Unfortunately, these discussions are more often than not unaccompanied by specific examples of theories, hypotheses, models—or even speculations—from the scientific literature. Such examples would have served as illustrations and would have allowed readers to evaluate better the various methods being promoted or criticized.

In this paper, I will explore what I believe are the reasons for the biologists' general failure to develop universal statements and explanatory, predictive theories. In the remainder of this introduction, I discuss my biases regarding the nature of science. In Part 2, I review the development of our ideas concerning the motions of the sun, moon, and planets. I do this not because I have "physics envy" (Egler 1986), but because I believe physics offers unambiguous examples of ad hoc hypotheses, universal statements established by induction, and hypothetico-deductive theories. These examples serve as models against which we can evaluate biological ideas. In Part 3 I discuss a specific biological example, the evolution of clutch size in birds, with respect to the epistemological considerations of the first two parts.

• The Nature of Science •

I assume that the ultimate goal of science is "to discover unity in the wild variety of nature" (Bronowski 1965), to develop unifying *explanatory* theory (Popper 1972), and "to search out more general truths that particular facts illustrate and for which they are evidence" (Copi 1986). After making observations, counting, measuring, doing experiments, and applying statistical methods to the data, that is, after carefully describing "what is," scientists discover pattern. And, pattern requires explanation. Scientists explain how and why a particular pattern and not some other is what is observed. Initially, explanations take the form of ad hoc hypotheses (first meaning of Copi 1986, defined below). At this stage, observations seem chaotic, and ad hoc hypotheses are diverse. However, by exploring "hidden likenesses" (Bronowski 1965) among the observations, more general explanations are created. These more general, unifying explanations take the form of either inductive hypotheses or hypothetico-deductive theories. The best theories are those that make the fewest assumptions from which the widest range of predictions may be deduced, these predictions being corroborated by fact.

The goal of science and the process just described are not valued by every biologist or every philosopher of science. Among scientists, there are what Dyson (1988) has called "diversifiers" and "unifiers." Diversifiers are scientists who are largely experimentalists exploring the details of nature, whereas unifiers are "those people whose driving passion is to find general principles which will explain everything" and, in so doing, bring "more and more phenomena within the scope of a few fundamental principles." Both diversifiers and unifiers are essential for a healthy, active, growing science.

Dyson (1988) also points out that "Biology is the natural domain of diversifiers as physics is the natural domain of unifiers." I believe that Dyson has identified the primary reason why biologists have failed to produce unifying theory. Biologists are mainly diversifiers. They are interested in exploring the details of nature and are content with producing ad hoc hypotheses. They lack the "driving passion" of unifiers to discover general principles. Rosen (1985) has a similar view, suggesting that "the physicist ... has never doubted that he deals in the general and that biology concerns only the particular."

Of course, some biologists are unifiers, the most successful being Charles Darwin. This brings us to the second reason why biologists fail to produce unifying *predictive* theory. Unifying biologists think inductively, whereas unifying physicists think deductively. Charles Darwin, for example, thought inductively. As he wrote, "science consists in grouping facts so that general laws or conclusions may be drawn from them" (Barlow 1958, p. 70). However explanatory inductive hypotheses may be, they have limited predictive value. Even if Darwin's theory can be put in deductive form, its predictions are not specific (Ghiselin 1969; Rosenzweig 1974). In practice, biological observations are interpreted in terms of Darwin's theory rather than used in testing predictions of the theory (Gould and Lewontin 1979).

In contrast, unifying physicists think deductively. Their theories make specific predictions. Thus, having assumed that Newton's theory was correct, physicists expected to find unknown planet X at position Y at time Z. Or, having assumed that Einstein's theory was correct, physicists expected to see an apparent shift in the positions of the stars as their light passed the sun during a solar eclipse. In both cases, experimental physicists proceeded to corroborate the predictions empirically.

Briefly, then, I believe that the failure of biologists to develop unifying, explanatory, predictive theory is a result of (i) most biologists being diversifiers rather than unifiers and (ii) most unifying biologists thinking inductively rather than deductively. Both attitudes have greatly affected the development of theoretical biology.

• Some Terminology •

At the outset, perhaps we should distinguish between fact, hypothesis, and theory. I will use *fact* to refer to something that is known or knowable by direct observation. A *hypothesis,* however, is a statement whose truth or falsity is neither known nor knowable by direct observation. The truth or falsity of hypotheses is determined indirectly by testing the predictions deduced from them by direct observation. Although we may come to consider a particular hypothesis to be a *law of nature* because of the correspondence between its predictions and observed facts, the law remains a hypothesis about nature.

I will use the term *ad hoc hypothesis* to refer to a hypothesis "that was specially made up to account for some fact after that fact had been established" (first meaning of Copi 1986, p. 509). Copi (1986) continues, "In this sense, *all* hypotheses are ad hoc, since it makes no sense to speak of a hypothesis that was not devised to account for some antecedently established fact." I will use *ad hoc saving hypothesis* to refer to hypotheses that were invented for the purpose of explaining away discrepancies between the predictions of theory and the facts. Often, these account "*only* for the particular fact or facts [it was] invented to explain and [have] no other explanatory power, that is, no other testable consequences" (second meaning of Copi 1986, p. 509). These are unscientific. Nevertheless, some ad hoc saving hypotheses may suggest new lines of investigation. Copi (1986) proposes a third meaning for ad hoc hypothesis. These are statements of "mere descriptive generalization" of the facts. "Such a descriptive hypothesis will assert only that all facts of a particular sort occur in just some particular kinds of circumstances and will have no explanatory power or theoretical scope" (Copi 1986, p. 509). Eventually, all ad hoc hypotheses should be examined to determine if they are truly ad hoc or are of more general applicability and interest. And, eventually, good generalizations should be tested by using them as assumptions in hypothetico-deductive theory. (In all quotations from Copi, the italics are Copi's.)

A *theory* comprises a series of hypotheses and a series of predictions deduced from the hypotheses. Thus, I distinguish hypotheses from theories on structural grounds (a hypothesis is part of a theory), rather than on the often stated grounds of a difference in the quantity and quality of supporting evidence, theories being better supported by evidence than hypotheses (e.g., Copi 1986). Thus, Newton's Theory of Motion and Einstein's Theory of General Relativity were theories when originally proposed, simply because they comprised both hypotheses and deductions to be tested. Also, according to my definitions, "Darwin's theory," referred to above, becomes "Darwin's hypothesis," because there are no specific, testable predictions. In my terminology, *hypothetico-deductive theory* is redundant because all theories are hypothetico-deductive. Nevertheless, it seems a necessary redundancy.

Finally, for further clarification, biologists often confuse hypothesis with prediction. Biologists will state something like, "I hypothesize that the grass in Tibet is green. To test the hypothesis, one needs only to go to Tibet and look." In fact, the "hypothesis" is a prediction, which follows from the unstated induction, "all grass is green," which was based on the proposer's experience with grass in the United States, Australia, and Sweden. The simple rule is that predictions are tested directly and hypotheses are tested indirectly, that is, by testing the predictions that follow from them.

2. The Structure of Theory

I believe that there is a misunderstanding among biologists regarding the distinction between the inductive method and the hypothetico-deductive method. Induction is not simply the development of generalizations from specific facts, nor is deduction simply the prediction of specific facts from axiomatic general assumptions.

Induction and deduction are shorthand terms for profoundly different approaches to the exploration of nature. Normally, science begins with observations. Eventually, an explanation is proposed for the observations. When generalized (by induction), the hypothesis is often tested by searching for further examples that confirm the pattern.

Deduction also begins with observations, hypotheses, and inductions. In the hypothetico-deductive method, however, predictions are not deduced from a single statement, such as a hypothesis, which has been generalized by induction. A single statement, proposed as a universal generalization, cannot be tested by itself (Popper 1959, 1972; Copi 1986). In hypothetico-deductive theory, predictions are deduced from a series of hypotheses.

These distinctions should become clear (or, at least, clearer) after discussion of the examples in the remainder of this section.

• Scientific Questions •

I have proposed that there are only three functionally different questions that a scientist can ask with regard to explaining empirical facts (Murray 1986b).

(1) How can I explain my data in terms of theory X?

This is the question asked by most scientists, who are engaged in what Kuhn (1962) calls "normal science." They have adopted the currently acceptable or fashionable paradigm (i.e., a "universally recognized scientific [achievement] that for a time provide[s] model problems and solutions to a community of practitioners" [Kuhn 1962, p. x] and are filling in the gaps of knowledge exposed by that paradigm.

This is the question of the diversifiers. It results in a burgeoning collection of ad hoc hypotheses and ad hoc saving hypotheses. For example, Greek astronomers were limited to developing models of planetary motion based on Plato's dogma regarding the perfection of circular motion because they asked, "How could planetary motion be explained in terms of Plato's dogma?" As a result, Greek models of planetary motion involved circles in various combinations. The most successful was Ptolemy's models combining epicycles, eccentrics, and equants, a unique combination of which was required to describe the motions of each planet. Thus, each of Ptolemy's planetary models is an *ad hoc* hypothesis.

(2) How can I best explain my data?

This question is asked by theoretical scientists. From time to time, scientists recognize that their data are contrary to currently acceptable hypotheses or theories. Rather than create an ad hoc saving hypothesis, which makes their data consistent with conventional wisdom, they create a new but equally ad hoc hypothesis. This new ad hoc hypothesis provides a different explanation of the facts.

Once an ad hoc hypothesis has been proposed, scientists may generalize it by induction. For example, Kepler hypothesized that the orbit of Mars was elliptical in shape, based upon his study of detailed data on the positions of Mars in the sky. He proposed further (his first law of planetary motion) that "all planets have elliptical orbits." Thus, even if he were simply guessing about the orbits of other planets (Cohen 1980), Kepler generalized an ad hoc hypothesis into a universal law. Kepler was lucky. The paths of all known orbiting bodies approach being elliptical. We continue to accept Kepler's first law as an adequate description of the orbits of planets, even though we know that the orbits are not exactly elliptical.

We should concentrate on the process rather than on Kepler's luck. Kepler's first law is an induction, a generalization of his ad hoc hypothesis regarding the orbit of Mars. No matter how many orbits we examine and find to be elliptical in shape, the first law of planetary motion remains a hypothesis generalized by induction. We may consider it a "universal law," but logically it remains a hypothesis generalized by induction.

How do we test an induction? We could predict that the next body we discover in space will have an elliptical orbit. If it does, then the induction gains support; if it does not, the induction is refuted. But, no matter how many bodies with elliptical orbits we find, an induction remains an induction, and an induction is a *hypothesis.*

If inductions are falsifiable, can be tested, and can be rejected, we may wonder why Popper (1959, 1972) so severely criticizes the inductive method. I think there are several reasons. First, no matter how many orbits we find to be elliptical, the conclusion that **all** orbits in the universe are elliptical is unjustified. Second, I think Popper assumes that progress in science comes from the development of *explanatory* theory, and inductive hypotheses are often not explanatory. To say that planetary orbits are elliptical, even if universally accepted by scientists, does not explain why they are elliptical. Popper (1972) believes that explanatory theory cannot be achieved with the inductive method.

(3) What statements can I make from which I can deduce my data?

This is the question asked by deductive scientists. Rarely do biologists ask this question. At least, there are few examples one can point to in which a few explicitly stated universal assumptions lead to explicitly stated specific predictions. This is the question that physicists often ask themselves, or appear to ask themselves, even if this question is not explicitly formulated by them.

I think we biologists would appreciate physics more easily if we understood that theoretical physicists were often answering Question #3 in their research. Hypotheses may arise as inductions from observations (Newton 1729) or by inspiration (Bronowski 1965; Popper 1972). Thus, Newton (1729) hypothesized, as his first law of motion, that a body at rest will remain at rest, and a body in motion will remain in motion in a straight line at a constant speed, until acted upon by an outside force. This is not an empirical law. It is a hypothesis, which is not directly known or knowable from empirical facts. One cannot test it by searching the universe for objects at rest or objects moving in straight lines at constant speeds. Furthermore, we cannot deduce much. All we can say is that objects will be either at rest or in motion, and if in motion they will be moving in straight lines at constant speeds or slowing down, speeding up, or changing direction.

As Copi (1986, p. 508) pointed out, "where hypotheses of a fairly high level of abstractness or generality are involved, no observable or directly testable prediction can be deduced from just a single one of them." Thus, a group of hypotheses must be used together to make deductions that can be tested directly. Newton's hypothetico-deductive theory comprises his three laws of motion and his principle of gravitation. In addition to the assumption regarding inertial motion, he assumed that the force acting on a body is directly proportional to its acceleration, that for every action there is an equal and opposite reaction, and that the gravitational force acting between two bodies is directly proportional to the product of their masses and inversely proportional to the square of the distance between them. I repeat, each of these statements, which we call "laws," are *assumptions* or *hypotheses* about the world. They are unknowns in the sense that they cannot be known to be true or false by direct observation. These unknowns, however, allow us to predict what is empirically known or knowable. For example, from knowledge about the period and radius of the moon's orbit, we can predict with Newton's laws that an apple near the earth's surface should fall about 16 feet in the first second of free fall. This is easily tested. With Newton's laws we can predict that the earth's sphere is flattened at the poles. This is more difficult to test, but it is testable. And so on.

The corroboration of prediction (i.e., deduction) by fact gives us confidence in the hypotheses (i.e., the assumptions, premises, and unknowns) of the hypothetico-deductive theory. We should emphasize that hypotheses remain hypotheses, assumptions remain assumptions, premises remain premises, and unknowns remain unknowns, even if we call them laws, no matter how broadly applicable and how successful the hypothetico-deductive theory explains nature. Nevertheless, we believe we have an *explanation* for why an apple falls 16 feet in the first second of free fall, why the earth is flattened at the poles, why the tides rise and fall twice a day, and why the orbits of planets are elliptical in shape.

We would not be surprised if all the millions or billions of planets in the universe have essentially elliptical orbits. Our confidence that the orbits of undiscovered planets would be elliptical, however, is not based on Kepler's first law or on our knowledge of the few thousand elliptical orbits (including those of artificial satellites) observed in our planetary system. Rather, it is based upon the fact that elliptical orbits are deducible from Newton's universal laws (and, later, Einstein's theory of general relativity), whose many other disparate predictions have been corroborated by fact.

The Role of the Questions in Practice

In practice, no doubt, many scientists at one time or another engage in normal science, induction, and deduction, and ask Question #1, #2, or #3 when solving a problem. Perhaps, many biologists do not get beyond Question #1, offering ad hoc hypotheses or ad hoc saving hypotheses to explain their carefully obtained data. Certainly, inasmuch as biologists disagree on the mere existence of universal biological principles (cf., for example, Vandermeer 1972; Van Valen and Pitelka 1974; Murray 1979, 1986b; Roughgarden 1983; Loehle 1988), few biologists have answered Question #2 to the satisfaction of their colleagues. Few biologists have even asked Question #3.

Perhaps, biologists have become diversifiers and have not explored Questions#2 and #3 because they perceive biological systems to be too complex to allow generalization and unification. For them, biological complexity is to be described and understood in terms of ad hoc hypotheses. Bartholomew (1986) was probably expressing the view of most biologists when he suggested that "interactions [of biological entities at all levels of integration], like their future evolutionary consequences, are not amenable to detailed prediction." He added further that philosophers of science have misled biologists into believing that "we should use the same procedures that have worked so well in physics and chemistry," particularly, "deductive logic."

Of course, certain aspects of biology, as well as some of physics, such as atmospheric science, could be too complex to permit the kinds of analysis and prediction we biologists imagine to be typical of physical systems. But what those aspects are cannot be identified a priori. Unfortunately, biologists seem to have rejected hypothetico-deductive reasoning on a priori grounds. In contrast to the views of Egler (1986) and Bartholomew (1986), I believe that the failure of biologists to develop predictive and explanatory theory is a result of their *not* emulating physicists. Whereas physicists usually test hypotheses by first generalizing them and then combining the generalizations in order to deduce predictions, which can be compared with empirical facts, biologists usually test their inductions by searching for additional examples or by interpreting new discoveries with ad hoc hypotheses, which are consistent with the prevailing conventional wisdom.

The difference between physics and biology is not that one is inherently amenable to hypothetico-deductive thinking and the other not, or that one deals with simple systems and the other with complex systems, but that physicists rigorously test every speculation against empirically determined fact, discard those that fail the test, and move on. This, I think, distinguishes the science of unifiers from that of diversifiers. Biologists, being diversifiers, happily live with ad hoc hypotheses, ad hoc saving hypotheses, unsubstantiated speculations, and even hypotheses contrary to fact and general principles (Fagerström 1987; Taylor 1989). Mathematical modelling in biology, especially, often bears no relationship to reality (Bartlett 1973; Pielou 1977).

In order to evaluate my view that complex biological systems are amenable to hypothetico-deductive reasoning, let us compare two proposed explanations for the evolution of clutch size in birds.

3. The Evolution of Clutch Size

• Variations in Clutch Size in Birds •

The clutch size of birds varies not only among species but geographically, annually, seasonally, and with the age of the female parent (Lack 1947, 1948, 1954, 1968; Cody 1966, 1971; Klomp 1970; von Haartman 1971).

• *Specific differences*—Clutch size varies from one egg to more than fifteen among all species of birds. Species with one-egg clutches are among the longest lived (e.g., albatrosses and eagles), whereas tits and phasianids with clutch sizes of up to fifteen or more eggs are among the shortest lived. One-egg clutches are invariable in size, whereas the clutches of species with larger average sizes often vary individually, annually, and seasonally.

• *Geographic*—The best known and perhaps the most interesting variation is the increase in clutch size with increasing latitude. This occurs in single-brooded species (one brood reared per year) as well as in multibrooded species (more than one brood reared per year).

There is a suggestion of longitudinal variation of clutch size in Europe with the clutch size increasing from west to east, that is, from maritime climates to continental climates.

Finally, populations on islands often have smaller clutches than those on the adjacent mainland.

Some groups of birds, however, have no geographic variation in clutch size. Albatrosses and their relatives invariably have a single-egg clutch, whether breeding in the tropics or at high latitudes. Among other, geographically widespread groups (e.g., nightjars, pigeons, doves, and hummingbirds), clutches are of either one or two eggs. Furthermore, the latitudinal trend is not always uniform.

- *Annual and seasonal*—The clutch size of individuals may vary annually and seasonally. Annual variations are widespread among birds, even in those species with average clutches of two or three eggs (e.g., gulls) but are best known in raptors (i.e., hawks and owls) and among birds with larger clutches (e.g., tits). In some species, females often lay larger clutches early in the season; in other species, they may lay larger clutches in mid-season.

This is the minimum set of observations that a general theory of clutch size must account for. We must, however, be careful to distinguish between clutch size differences among populations or species and clutch size differences among and within individuals. The annual and seasonal variations of individuals may reflect adaptations that allow females to adjust their clutch size and reproductive effort to short-term fluctuations in the local environment (e.g., food availability, population density, weather), whereas differences between populations or species may be adaptations to the longer term conditions under which the populations or species have evolved (e.g., long-term average probabilities of surviving and of breeding successfully).

• An Inductive Hypothesis •

The great diversity of clutch size variations in birds lends itself to a variety of ad hoc hypotheses. These are not of great interest to a unifier in search of general theory. For example, several ornithologists have proposed that the clutch size of tropical species may be limited by selection for small nests (B. K. Snow 1970; Lill 1974; D. W. Snow 1978), but many tropical species have small clutch sizes even though they place their eggs on the ground or on large platforms. Selection for nest size is not a likely general explanation for the clutch size variations in birds. A second example of an ad hoc hypothesis, that the small clutch size of pigeons and doves is limited by the ability of the parents to produce crop milk (Blockstein 1989), cannot be of general interest because no other birds produce crop milk. In the search for general theory, a unifier must simply ignore ad hoc hypotheses that cannot readily be generalized by induction.

Theoretical ornithologists usually think inductively, asking themselves, in effect, Question #2, "How can I best explain my data?" The only general hypothesis regarding the evolution of clutch size is Lack's hypothesis (Lack 1947, 1948, 1954, 1968), which has sometimes been called the *food limitation hypothesis.* Although Cody's (1966) general model might be construed to be hypothetico-deductive, it makes no explicit, testable predictions. In my view, his model only formalizes Lack's hypothesis, suggesting that energy from food resources be optimally allocated among efforts to defend one's mate, progeny, and self against predators, to contend with competitors, and to reproduce. An alternative hypothesis, the *nest predation hypothesis,* originally proposed to explain the small clutch size of tropical birds (Skutch 1949), has been proposed as a possible explanation for clutch size variations in passerine birds (Slagsvold 1982). Let us examine the inductive thinking of ornithologists by considering, even if too briefly, the historical development of Lack's hypothesis.

Lack (1947, 1948) reviewed the clutch size variations in birds. Focusing attention on latitudinal variations in clutch size, he proposed that parents at higher latitudes could rear more young than those at lower latitudes because the longer days at high latitudes allowed parents to find and deliver more food to their young. This is an ad hoc hypothesis, which Lack (1947, 1948, 1954, 1966, 1968) extended by induction to all birds. He stated, in one way or another in his various publications, that the "number of eggs in the clutch has been evolved to correspond with that from which, on the average, the most young are raised. In nidicolous species, the limit is set by the amount of food which the parents can bring for their young, and in nidifugous species by the average amount of food available for the laying female, modified by the size of the egg" (Lack 1968, p. 5).

Subsequent research by him and others was interpreted in terms of this hypothesis. Investigators were often answering Question #1 ("How do I explain my data in terms of Lack's hypothesis?"). For example, "Clutch size tends to be larger in tropical savanna than tropical evergreen forest, and larger on the European Continent than in England at the same latitude, *presumably* due to differences in the availability of food for the young. Clutch size does not vary significantly with the type of pair-bond, *presumably* because polygyny or promiscuity tend[s] to be evolved only where food is so easy to obtain that one parent can feed a brood unaided" (Lack 1968, p. 285, italics added). I agree with Cody (1969, p. 1186) that, "Unfortunately, the wide applicability of the theory depends on the existence of largely undemonstrated food variations which must *ex hypothesi* follow observed clutch size trends [and, which] at times stretch our credulity," a statement that is as true today as it was in 1969.

Lack's hypothesis was put to the test in other ways. It was reasoned that parents should not be able to rear on average more than the average size brood. This led to what has been called the twinning experiment, in which one or more eggs were added to a clutch, or one or more chicks were added to a brood. The results of these experiments have been mixed (Murray 1979; Lessells 1986). In about one-third of the experiments, parents failed to rear larger broods, whereas in two-thirds the parents succeeded with little apparent cost (Lessells 1986).

Twinning experiments showing that parents could raise larger broods than they normally did were often explained away by ad hoc saving hypotheses. For example, in an early experiment, North Atlantic Gannets *Morus bassanus,* which normally lay only one egg and rear at most one young during a breeding season, were able to hatch two eggs or rear two young of nearly normal weight (Nelson 1964). Do these data constitute evidence sufficient to refute Lack's hypothesis? No. Lack (1966, p. 248) proposed, "The most probable explanation of these successful twinning experiments is that the food situation is unusually favourable for Gannets on the Bass Rock at the present time, as suggested by their recent increase in numbers, and that the present clutch of one was evolved under more rigorous conditions in the past in which only one chick could normally be reared." This, clearly, is an ad hoc saving hypothesis. Furthermore, it is untestable. We cannot recreate or even know what conditions prevailed for the Gannet during the evolution of its clutch size.

A second problem for Lack's hypothesis is the fact that nocturnal species, such as owls, have an increasing clutch size with increasing latitude (von Haartman 1971). Presumably, nocturnally foraging species should have a decreasing clutch size because of the shorter nights at higher latitudes. Do these data constitute evidence sufficient to refute Lack's hypothesis? No. To account for this discrepancy, Owen (1977) proposed that because prey species are more diverse and less abundant in the tropics than at higher latitudes, both diurnal and nocturnal tropical birds should have a more difficult time in forming a proper search image of their prey and, thus, in obtaining food for their young. Although this is an ad hoc saving hypothesis, it does seem potentially testable, although with great difficulty.

In another examination of Lack's hypothesis, Ashmole (1961) proposed that the available food resources of temperate populations fluctuated widely between breeding and nonbreeding seasons, whereas those of tropical populations did not. Because temperate populations would suffer greater mortality during the nonbreeding season than tropical populations, the availability of resources during the breeding season would be greater for temperate populations. Thus, temperate populations would have disproportionately more food available to them than would be available to populations in tropical regions, and parents would be able to rear disproportionately (with respect to differences in day length) larger broods in temperate regions.

In testing this interpretation, Lack and Moreau (1965) assumed that savanna and evergreen forest habitats in equatorial Africa simulated temperate and tropical conditions, respectively, savannas being seasonal and evergreen forest being relatively aseasonal. They compared the average clutch sizes of birds within the families (Lack and Moreau 1965) or subfamilies (Lack 1968) found in these habitats. Such a comparison of species between seasonal and aseasonal habitats within the tropics eliminates the effect of day length on clutch size and focuses attention on seasonal differences in resource availability. Presumably, seasonal and aseasonal habitats in the tropics would differ in the relative amount of food available to parents for feeding young. In most comparisons the average clutch size of evergreen forest species was smaller than the average clutch size of savanna species. The results were interpreted as support for Ashmole's modification of Lack's hypothesis.

Later, Ricklefs (1980) tested Ashmole's hypothesis by comparing average clutch sizes of birds with seasonal variation in actual evapotranspiration (AE), which was presumed to be proportional to primary production, in a variety of geographical regions. He found that clutch size was directly proportional to the ratio between summer AE and winter AE and that clutch size was inversely proportional to winter AE and independent of summer AE. Ricklefs (1980) suggested that the geographical trends in clutch size were caused primarily by factors that limit populations during the nonbreeding season, which is consistent with the implications of Ashmole's modification of Lack's hypothesis.

Hussell (1985) corrected the equation of Ricklefs (1980), showing that clutch size was correlated with winter AE and with the breeding season surplus in resources (measured by the differences between summer AE and winter AE), although it was not possible to distinguish the effects of winter AE and breeding season surplus because they were correlated with each other. Nevertheless, he added, "We must not be seduced by an attractive hypothesis, however, into believing that correlation of climatic or other environmental variables with clutch size provides more than meager support for any hypothesis [because many], perhaps most, environmental and biological variables are correlated with latitude and also will be correlated with clutch size in those avian species and groups within which clutch size varies with latitude" (Hussell 1985, p. 634).

• *General Criticism*—The inductive hypotheses of Lack, Cody, Ashmole, Ricklefs, and others paint "predictions" with broad strokes. They propose what are essentially correlations between clutch size and some environmental variable. As with most correlations, there are exceptions. Yet, evolution proceeds within populations, requiring that the clutch size of each population, including the exceptions, be accounted for by a general theory. Thus, even if 9 of 10 populations fit an expected trend, a general theory should account specifically for the exceptional population.

This, of course, is a philosophical point. I think it must be a difference between inductive thinkers and deductive thinkers. For inductive thinkers, among biologists anyway, correlations, and exceptions to them, seem acceptable. A deductive thinker is, or should be, dissatisfied with a hypothetico-deductive theory that could not account for exceptions to general trends.

I believe Lack's hypothesis, in its various modified forms, is an inadequate explanation of the facts of clutch size variations in birds. There are too many discrepancies, too many ad hoc saving hypotheses, and, even more importantly, too few specific predictions. The many ad hoc hypotheses, apparently being proposed with increasing frequency (reviewed by Itô and Iwasa 1981; Winkler and Walters 1983; Murphy and Haukioja 1986), are, by definition, inadequate general explanations for clutch size variation in birds and cannot replace Lack's hypothesis as a unifying explanation. The only other attempt to explain the full range of clutch size variations in birds is the hypothetico-deductive theory of Murray (1979, 1985a), which we shall now consider.

• A Hypothetico-deductive Theory •

In trying to explain the evolution of clutch size in birds (Murray 1979, 1985a), I ask Question #3, "What statements can I make from which I can deduce my observations (i.e., data available in the literature)?" Those statements comprise the assumptions of the theory. The deductions derived from these hypotheses are the predictions, which must be sufficiently specific in content to be readily testable. If the theory is any good, the predictions must be confirmed by empirical evidence.

Assumptions

(1) The measure of fitness of females of a given phenotype is the Malthusian parameter ρ, as determined from the equation,

$$1 = \sum \lambda_x \mu_x e^{-\rho x}, \quad (1)$$

where λ_x is the probability of a female's surviving from birth to age class x, μ_x is the average number of female eggs produced by a female while a member of age class x, e is the base of the natural logarithms, ρ is the annual rate of increase, and x is the age class (Murray and Gårding 1984).

(2) The probability of survival of females, of their progeny, or of both to later ages decreases as eggs are added to the clutch (Figure 1A).

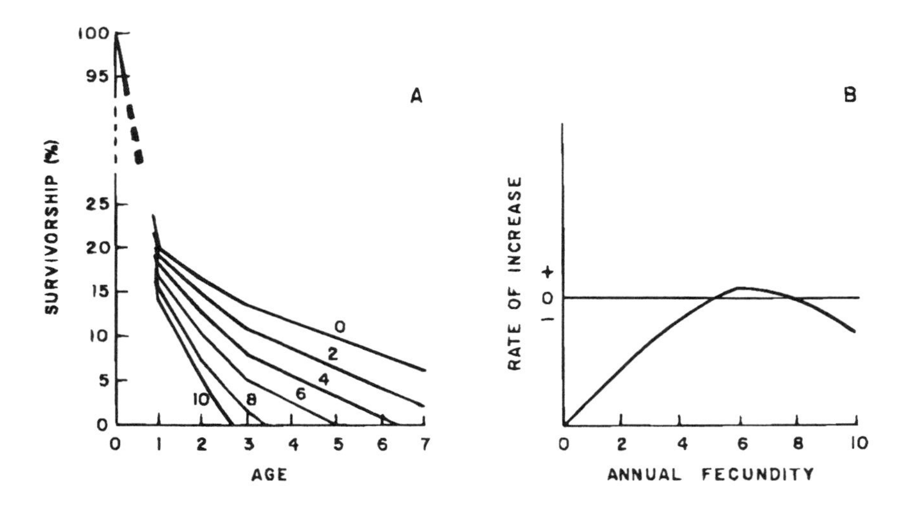

Figure 1. A graphical model of the evolution of annual fecundity. (A) Partial survivorship curves of birds with different gene-determined differences in annual fecundity: 0, 2, 4, 6, 8, 10 (odd number of eggs are not included in order to reduce clutter). Curves for annual fecundity of 0 and 2 (and 1) actually extend beyond age 7. These curves show the costs in survivorship of differences in annual fecundity. (B) Annual rate of increase (ρ) of birds with different annual fecundities. The shape of the curve may be somewhat different but always increases to a maximum at the smallest annual fecundity that results in replacement. See text for fuller explanation. (From Murray 1985a).

(3) Selection favors those females that lay as few eggs as are necessary for replacement because they have the highest probability of surviving to breed again, their young have the highest probability of surviving to breed, or both (Figure 1B).

(4) Females divide their annual fecundity between two or more clutches if they are able to rear two or more broods during a breeding season.

(5) Survivorship (λ_x) and fecundity (μ_x) schedules represent the average annual probabilities of surviving and breeding.

These five assumptions clearly constitute unknowns—as universal assumptions their truth or falsity cannot be determined directly by reference to empirical data. With regard to assumption (1), the Malthusian parameter, as a measure of fitness, is a subject of current debate (Charlesworth 1980; Stenseth 1983; Nur 1984, 1987; Sober 1984; Murray 1985b, 1988, in press.). Even if there were universal agreement among evolutionary biologists that the Malthusian parameter was *the* measure of fitness, the presumption that it is would remain an assumption.

With regard to assumption (2), several evolutionary biologists have proposed a tradeoff between survivorship and fecundity (e.g., Lack 1954; Cody 1966; Williams 1966; Hussell 1972; Charnov and Krebs 1974; Stearns 1976; Ricklefs 1977, 1983; Pianka 1978; and Haukioja and Hakala 1979). There is some indication of such a trade-off from empirical studies (Murray 1979; Reznick 1985), but several investigators claim to have failed to find support for an increasing mortality rate with increasing clutch size in birds (Högstedt 1980, 1981; Smith 1981; Pettifor et al. 1988). Empirical studies, however, seem to have woefully small sample sizes (see Lewontin 1974, p. 240). Furthermore, empirical studies have focused on parental mortality. The survivorship schedule of a genotype, however, begins with the egg stage (Murray, in press). Thus, the parental mortality is only one of several potential "costs" of reproduction. Larger clutches may instead result in reduced survivorship of the young. Furthermore, if one assumes that populations cannot grow indefinitely in a finite environment, there must be some kind of "tradeoff" between survivorship and fecundity.

Assumptions (3) and (4) are without question contrary to conventional wisdom. Only Moreau (1944) had suggested that natural selection may favor birds with small broods because their reduced effort with a small brood would increase their probability of success with later broods during the same season or their probability of surviving to breed at a later season. Spencer and Steinhoff (1968) proposed a similar explanation for latitudinal variation in the litter sizes of small mammals. Recently, experimenters (Slagsvold 1984; Lindén 1988) have shown that female birds with smaller first clutches have a higher probability of producing a second brood.

Assumption (5) is clearly hypothetical.

We test the whole group of assumptions at once by comparing predictions deduced from them. Several qualitative predictions have been made (Murray 1979, 1985a), and these will be treated below. The assumptions also suggested quantitative relationships between population parameters. Subsequently, Murray and Nolan (1989) developed an equation with which the clutch size of a population could be predicted from other demographic parameters. From the assumptions of the theory, they proposed that clutch size (CS) should be,

$$CS = \frac{\mu_s}{P_1 + P_2 \ldots} \tag{2}$$

where P_1 and P_2 are the probabilities of a female's rearing a first and second brood, respectively, to independence during a breeding season, and μ_s is the total number of eggs a female should lay in all clutches that produce one or more young to independence during a breeding season, if she is to replace herself. The replacement number of eggs is given by

$$\mu_s = \frac{a+1}{\sum_{\alpha}^{\omega} \lambda_x} \tag{3}$$

where a is the primary sex ratio (i.e., at time eggs are laid), α is the average age class of first breeding, and ω is the age class of last breeding.

Equation (2) represents the quantitative relationships described by assumptions (1) to (5) and, thus, summarizes the hypotheses in such a way that specific predictions can be deduced and tested by comparison with empirical facts.

Murray and Nolan (1989) used data on the Prairie Warbler *Dendroica discolor* in Indiana to illustrate the calculation of clutch size with the equation. The equation predicted a clutch size of 3.49, and the actual average clutch size of the Prairie Warbler was 3.89 (Nolan 1978). We believed this to be encouraging support for Murray's theory. Because data on the Prairie Warbler were used in developing the final formulation of the equation, the correspondence between prediction and empirical fact in this case does not constitute a rigorous test.

Later, Murray et al. (1989) used the equation to predict the average clutch size of the Florida Scrub Jay *Aphelocoma c. coerulescens* at the Archbold Biological Station in central Florida. This population's biology has been thoroughly studied (Woolfenden and Fitzpatrick 1984). Since 1969, each banded individual has been censused monthly. Because emigration and immigration were rare, the time of death could be determined almost certainly for each individual of the population. Furthermore, every nest has been found, and thus the reproductive success of each female could be determined with great accuracy. The relevant parameters were calculated and substituted into the equation, which predicted a clutch size of 3.43. This predicted value was extraordinarily close to the actual average clutch size of 3.33.

The success of the equation in predicting the clutch size of the Florida Scrub Jay gives greater confidence in the equation. The first importance of the equation is that it identifies the parameters presumably affecting clutch size in birds. These are survivorship of individuals reared from clutches that produce at least one young to independence, age of first breeding, primary sex ratio, and the probability of a female's rearing one, two, or more broods during a breeding season.

Predictions

In addition to predicting the average clutch size from knowledge of the other demographic characteristics of a population, the equation allows one to infer several specific predictions regarding clutch size variation. Although discussed in terms of birds, the predictions should hold for many other groups of organisms.

- *Between population variation*

(1) Breeding seasons vary in length. If breeding seasons are short relative to the time required to rear a brood to independence, then females could not be expected to rear more than a single brood. Also, with a relatively short breeding season, females would have limited opportunities to lay replacement clutches following losses to predators or other causes. In contrast, if breeding seasons are long relative to the time required to rear a brood to independence, females would not only have more opportunities to rear a first brood, increasing the probability of being successful, but could rear two or more broods.

Thus, in regions where breeding seasons are long, P_1 would be expected to be larger than in regions where breeding seasons are short. Also, P_2, P_3, and so on would be expected to be greater where breeding seasons are long than where they are short. For these two reasons, we should expect that clutch size should be smaller in regions with longer breeding seasons.

Breeding seasons are on average longer in tropical regions than at higher latitudes. Thus, we should expect that clutch size should increase with increasing latitude because females at higher latitudes should have fewer chances to rear a first brood by laying replacement clutches (lower P_1) and fewer chances to rear second, third, or more broods (lower P_2, $P_3, \ldots$) than do females at lower latitudes.

What has attracted attention in tropical species is the high rate of loss of eggs or young in nests to predators (Marchant 1960; Willis 1961, 1974; Snow and Snow 1964; Skutch 1966, 1985; Ricklefs 1969; Fogden 1972; but see Oniki 1979). The implication of Eq. (2) is that in tropical species, even though the probability of successfully rearing young from a particular clutch is low because of high predation, the probability of successfully rearing one or more broods during a breeding season is high. According to the equation, what matters in the evolution of clutch size is the probability of rearing young during a breeding season rather than the probability of rearing young from a particular clutch.

Unfortunately, there are no quality data available on the frequency of clutch starts or on the production of multiple successful broods in either tropical or temperate species (Cody 1971; von Haartman 1971; Ricklefs 1973), except for those on the Prairie Warbler and Florida Scrub Jay. What skimpy data exist are suggestive. The largest number of broods reared during a breeding season occurs in tropical species (Ricklefs 1973; Skutch 1976). For example, the Southern House Wren *Troglodytes aedon musculus,* with a clutch size of 4, rears as many as four broods in Costa Rica (Skutch 1953, 1976), whereas the Northern House Wren *Troglodytes a. aedon,* with a clutch size of 6, usually rears only two broods in Illinois (Kendeigh et al. 1956). Interestingly, the Mourning Dove *Zenaida macroura,* a widespread species in temperate North America, not only has a "tropical" clutch size (i.e., 2) but a "tropical" breeding season (i.e., long—8–10 months), compared with other North American birds. In Iowa, the average pair of Mourning Doves laid six clutches and reared three broods (McClure 1942). Some Mourning Doves rear as many as five or six broods in a year (cf. Nice 1957). By comparison, the average female Prairie Warbler, with clutch size 3.89, laid 3.1 clutches and reared 0.77 broods (Nolan 1978; Murray and Nolan 1989). The average female Florida Scrub Jay, with clutch size 3.33, laid 1.37 clutches and reared 0.74 broods (Murray et al. 1989).

By analogy, we might expect that in milder climates, such as in maritime regions, the breeding seasons should be longer than farther inland where the climate is more severe. Thus, P_1 for all birds and P_2, P_3, etc. for multibrooded species might be expected to be greater in maritime regions, resulting in smaller clutches there.

(2) Habitats or situations differ with respect to the degree of predation on eggs or broods in nests. Considering species living in areas where breeding seasons are of about equal length, in those areas where predation rates are low, we should expect P_1 for all species and P_2, P_3, etc. for multibrooded species to be high compared with where predation rates are high. Thus, clutch sizes should be smaller where predation rates are low compared with clutch sizes where predation rates are high (contrary to the expectations of the nest predation hypothesis [Perrins 1977; Slagsvold 1982, 1984; Martin 1988]).

Again, data are scarce. In a study of New World warblers (Emberizidae: Parulinae), which are usually single-brooded, species that nest on the ground suffer higher losses from predators and have larger clutch sizes than do those species in the same region that breed above ground level (Martin 1988), consistent with the present theory but inconsistent with the nest predation hypothesis (Martin 1988).

If predation rates are lower on islands than on adjacent mainland areas, as may be the case (Carlquist 1965; Lack 1968), then we should expect lower clutch sizes on islands, which would be consistent with observations.

(3) If the environments of diurnal and nocturnal animals vary in the same way with latitude (e.g., length of breeding season and predation rates), both groups should have a higher probability of rearing a first brood (greater P_1) and higher probabilities of rearing two (greater P_2), three (greater P_3), or more broods at lower latitudes. Thus, we should expect that clutch size of nocturnal species, as well as of diurnal species, should increase with increasing latitude.

This, in fact, is the case for owls (von Haartman 1971) and small mammals (Lord 1960).

(4) If larger species have longer maximum life expectancies (Lindstedt and Calder 1976) and greater juvenile and adult annual survivorship (Saether 1989) than smaller species, we should expect that larger species would have longer breeding lifetimes. If they do, then their $\sum_{\alpha}^{\omega} \lambda_x$ should be larger than those of smaller species, resulting in smaller μ_s and smaller clutch sizes.

In general, this seems to be the case. The smallest clutch sizes (one egg) occur in large species, such as albatrosses, eagles, and vultures, and the largest clutch sizes (> 10 eggs) in small species (for instance, tits). There is, however, considerable variation, some largish species (e.g., ratites, galliformes) having large clutches and smallish birds (e.g., hummingbirds, swifts, storm-petrels) having small clutches. Consistent with prediction (2), however, the clutches of the ground-nesting ratites and galliformes are subject to high predation rates, whereas the clutches of the cavity-nesting swifts and storm-petrels are subject to low predation rates. Hummingbirds, instead, rear several broods in rapid succession, consistent with prediction (1).

(5) If species of similar body size have similar maximum life expectancies (Lindstedt and Calder 1976) and similar juvenile and adult annual survivorship (Saether 1989), then they should have similar values for $\sum_{\alpha}^{\omega} \lambda_x$. If they do, then we should expect species of the same size living at the same latitudes to have similar clutch sizes regardless of the number of parents providing food.

We have little information on this point, but Skutch (1949, 1976, 1985) has presented data on selected pairs of species that differed in parental care (uniparental vs. biparental care) but nevertheless had similar clutch sizes. Skutch argued that if Lack's hypothesis were correct, then the species with biparental care should have larger clutches than species with uniparental care. A comparison of European passerines with uniparental and biparental care had a similar result—no difference in clutch size (von Haartman 1955).

(6) If birds nesting in cavities or other seemingly protected sites were in fact protected from predation (this may, however, be an artifact of studies of birds breeding in man-made nest-boxes [Nilsson 1984a, 1984b]), we should expect a lower μ_s and a higher P_1, both resulting in smaller clutch sizes.

This prediction is not entirely borne out by the evidence. Although cavity-nesting storm-petrels and swifts do have small clutch sizes, hole-nesting species typically have larger clutches than open-nesting species (Lack 1947, 1948, 1954; von Haartman 1957; Nice 1957; Ricklefs 1969; Klomp 1970; Cody 1971).

On the other hand, if suitable holes were limited in number, all females may not be able to breed, postponing the average age of first breeding, decreasing $\sum_{\alpha}^{\omega} \lambda_x$. This would result in increasing μ_s, which would result in greater clutch size.

Unfortunately, data on age of first breeding, survivorship, and probabilities of success of hole-nesting species are poorly known. Nevertheless, the placement of nest boxes in habitats having few natural holes and low populations of hole-nesting species often results in an in-

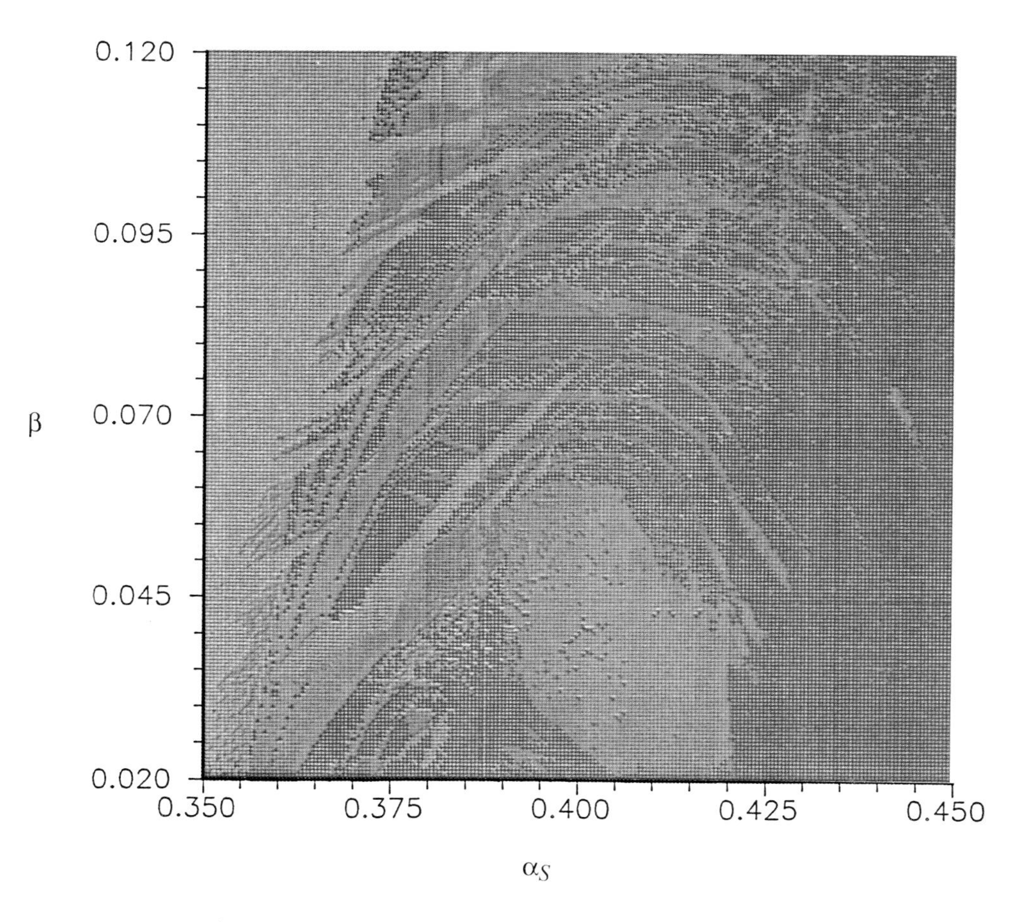

Chapter 9, Figure 11. Magnification of the region in policy space $0.35 \leq \alpha_S \leq 0.45$, $0.02 \leq \beta \leq 0.12$. Note how fingers of stable behavior penetrate into the region of unstable behavior. Also note the complex distribution of periodic and aperiodic modes. In certain regions, a stable solution may be surrounded by chaotic solutions on all sides. Here a slight change in α_S or in β will produce completely different solutions.

crease in numbers (von Haartman 1971; Brawn and Balda 1988). Most spectacularly, in southern Finland, Pied Flycatchers *Ficedula hypoleuca* increased from 3 to 137 nesting pairs in one year after experimenters put up nest boxes (von Haartman 1971). These results indicate that at least some hole-nesting birds are limited by the availability of holes and that, therefore, some females are forced to postpone breeding. This should result in larger clutch sizes.

• *Individual variation*

(1) If selection favors females that minimize the costs of their reproductive effort, then we should expect females to vary their clutch size with changing environmental conditions. When conditions for laying eggs and rearing young are poor, females should reduce their clutch size, and when conditions are especially good, they should increase their clutch size, relative to their average clutch size. The alternative is to produce clutches of the same size under all conditions (Figure 2). The cost of maintaining the clutch size under adverse conditions is often permanent (i.e., death). Thus, there would be no possibility of making up the loss in later breeding seasons. In contrast, the cost of reducing clutch size under adverse conditions can be made up in later breeding seasons by increasing the clutch size when conditions are better. Thus, seemingly, females that vary their clutch size with environmental conditions, whether between years or within a breeding season, should be favored over those that do not.

There is abundant evidence that clutch sizes and other measures of effort vary with proximate conditions (e.g., Lack 1954, 1968; Klomp 1970; Högstedt 1980; Hussell and Quinney 1987). This has usually been interpreted to mean that females are laying as many eggs or rearing as many young as conditions allow. The interpretation here is that proximate conditions may cause the clutch size to be smaller or larger than the average, but the average clutch size is the "replacement" clutch size, which is ultimately determined by the female's probabilities of rearing young successfully, her expectation of further life, and her age of first breeding.

(2) If selection favors reducing reproductive effort to a minimum (i.e., no more eggs than required for replacement), then we should expect younger individuals to invest less effort in reproduction (i.e., smaller clutch size, fewer broods) than older individuals. The extreme in reduction is not breeding at all (i.e., postponing the age of first breeding), which occurs in long-lived birds.

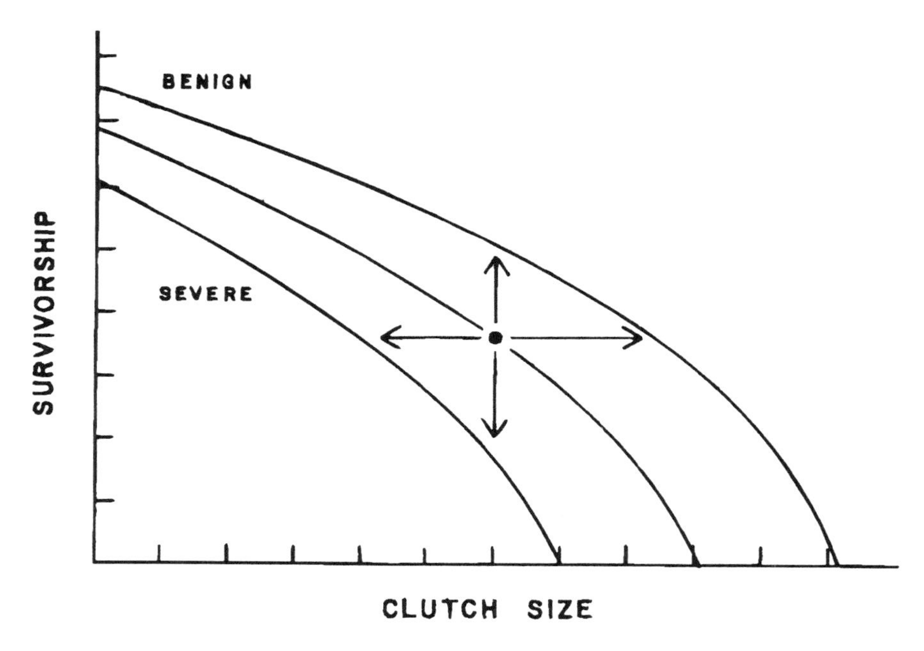

Figure 2. Relationship between clutch size and annual survivorship. As clutch size increases, the probability of surviving to the next breeding season declines, regardless of environmental conditions. A particular clutch size, say six, becomes more costly as conditions worsen. In a fluctuating environment, a female can either lay the same number of eggs with varying cost in survivorship (vertical rows) or lay a different number of eggs with no change in cost in survivorship (horizontal arrows). (From Murray 1979).

The evidence indicates that this is usually so (Lack 1966; Skutch 1976; Curio 1983). Although this has often been attributed to maturation effects, Curio (1983) proposed the "restraint hypothesis," suggesting that younger birds could increase their lifetime reproductive success by reducing annual effort. Curio's restraint hypothesis is a predictable consequence of the theory proposed here. Dying at an early age has a much higher cost than dying at an older age. A young animal with a high $\sum_x^\omega \lambda_x$ and high residual reproductive value (e.g., Williams 1966) should not jeopardize itself by exerting an effort in reproduction that increased its probability of dying. The cost of the extra effort exerts a great effect on its lifetime reproductive success. An older female with a low $\sum_x^\omega \lambda_x$ and low residual reproductive value might benefit from increasing its reproductive effort. Although increasing its probability of dying or failing, the extra cost exerts only a small effect on its lifetime reproductive success.

• *Non-avian groups*

The preceding predictions refer to birds. I believe that many of them are applicable to other groups of organisms. A completely unexpected prediction of the equations concerns groups with parthenogenetic reproduction. Birds are not parthenogenetic, but individuals of some other taxa are. From Eq. (2), all other things being equal, we expect that clutch sizes in parthenogenetic populations, in which unfertilized eggs produce only female offspring ($a = 0$), should be smaller than in populations of amphimictic populations in which eggs fertilized by sperm produce both male and female offspring (often $a = 1$).

The evidence is dreadfully limited (Bell 1982). The parthenogenetic fish *Rivulus marmoratus* produces fewer eggs than does the related bisexually reproducing *Oryzias latipes* (Harrington 1971). In the gynogenetic salamander, *Ambystoma jeffersonianum*, females are about two-thirds as fecund as the bisexually reproducing females of the same population (Uzzell 1964). The parthenogenetic species of *Cnemidophorus* lizards, however, seem no less fecund than the amphimictic species (Schall 1978).

• Discussion •

The hypothetico-deductive theory proposed here to account for clutch size variations in birds has the advantage of being at once both specific and general in its predictions. For example, all other things being equal, a population with a higher P_1 than another should have a smaller clutch size. But, differences in P_1 could result from differences in predation on eggs or young in the nest, from differences in the parents' ability to find and deliver food, from differences in the age of the parents, and so on. Thus, we should not expect to find any single factor to determine or to be tightly correlated with clutch size. Neither the food limitation hypothesis nor a properly-formulated nest predation hypothesis should be expected to account fully for clutch size variations in birds.

The same is true for the other parameters of the equation, survivorship of adults, age of first breeding, and probabilities of rearing two or more broods in a breeding season. Many environmental factors could affect each of these parameters. As the diversifiers explore the intricacies of nature, they should find that in one species some factor may be of importance and another factor of no importance, whereas in another species the relative importance of the two factors could be reversed. The hypothetico-deductive theory proposed here accounts for the fact that factors could have different effects in different populations.

Furthermore, each prediction is based upon the notion, all other things being equal. But, all other things are not always equal. For example, within the procellariiforms, variation in size is great from the small storm-petrels to the large albatrosses. Prediction (4) could suggest that the smaller storm-petrel should have a larger clutch than the albatross, but both groups have a clutch size of one. If the storm-petrels do have lower annual survivorship and shorter maximum lifetimes than albatrosses, a clutch size of one for each group could result from differences in P_1 or from differences in age of first breeding. Indeed, albatrosses have a later age of first breeding than do the smaller procellariiforms (Lack 1968). Because differences in clutch size can result from differences in survivorship, age of first breeding, and the probabilities of rearing one, two, or more broods, as indicated in Eq. (2), there is no reason to believe that we should find a close correlation between clutch size and any other variable.

There is much work to be done. The hypotheses of the theory lead to specific, testable predictions. The data required are accurate survivorship schedules, ages of first breeding, and the probabilities that a female is successful in rearing one, two, or more broods during a breeding season. Although some data on survivorship and age of first breeding are available, these are usually inadequate for calculating clutch size with Eq. (2). Data on the probabilities of success in rearing broods are completely unavailable, except for the Prairie Warbler (Murray and Nolan 1989) and Florida Scrub Jay (Murray et al. 1989) because ornithologists have focused their attention on the proportion of clutches rather than on the proportion of females that are successful in hatching eggs or producing chicks.

Some Implications for General Evolutionary Theory

Eq. (1) represents the neo-Darwinian hypothesis of natural selection written in mathematical form. In states simply that the trait with the greatest ρ, given its possessors' probabilities of surviving (λ_x) and reproducing (μ_x) under prevailing conditions, should increase more rapidly than other traits. A trait (or genotype) that increases survivorship or fecundity, relative to those of alternative traits (or genotypes), should be selected for and increase relative to the alternatives. Because no population can increase indefinitely, as individuals with the favored trait come to predominate, their survivorship or fecundity or both must decrease, resulting in a "steady-state" (i.e., the population neither increasing nor decreasing). I have placed steady-state within quotes because no population is in a steady-state (Murray 1979). Rather, populations fluctuate about some average size (often called the carrying capacity of the environment) as a result of short-term environmental changes affecting annual survivorship, fecundity, or both.

With regard to the evolution of clutch size, I have proposed that the favored trait (or genotype) is what may be construed to be the "replacement" clutch size, given the probabilities of rearing, one, two, or more broods. But, the replacement clutch size is defined as the minimum number of eggs that a female can lay in a clutch and still be able to replace herself. A replacement clutch size does not mean that $\rho = 0$. For example, suppose the replacement clutch size is 3.7. A genotype that produces a variable clutch size, averaging 3.7 eggs, should be favored over genotypes producing an average clutch size of 3 or even 4 eggs. Clutches of 3 eggs clearly result in extinction. Genotypes producing 3 or 4 eggs (under poor or good conditions, respectively), however, should prevail over genotypes producing 4 or more eggs because the latter would be subjected to greater mortality when conditions for laying eggs or rearing young were poor (Figure 2). When conditions are good the genotype for 3.7 increases in absolute numbers. When conditions are poor the genotype may decrease in numbers, but it increases relatively with respect to other genotypes. As a result the genotype comes to prevail in the population.

In populations with an invariable clutch size (as occurs in many shorebird species, for example), a genotype producing 4 eggs should be favored, if the replacement clutch size is 3.7. As the population increases and becomes crowded at high density, survivorship decreases, such that the predicted replacement clutch size becomes 4, or reproduction decreases (i.e., some older females do not breed, some younger females postpone first breeding, some breeding females produce fewer broods) such that the average clutch size becomes 3.7, or both survivorship and reproduction decrease. Survivorship and fecundity are brought into balance, and the population achieves a "steady-state" (Murray 1979).

The theory predicts that there is only one clutch size (or average clutch size for populations with a variable clutch size) that should be selected for because it produces a steady-state population under specific conditions. The selected average clutch size is not an "optimum," "maximum," or "minimum"; it is the **only** clutch size that can produce a steady-state population. Theoretically, two phenotypes (or genotypes) with different survivorship and fecundity schedules and, therefore, different clutch size could coexist in a balanced polymorphism, but I consider this an unlikely occurrence.

What I am proposing, essentially, is that natural selection tends to result in what may be called an *evolutionary "steady-state"* (E"SS"). An E"SS" superficially resembles what Maynard Smith (1982) has called an evolutionary stable strategy (ESS). According to Maynard Smith (1982), a strategy is "a specification of what an individual will do in any situation in which it may find itself," and an ESS is "a strategy such that, if all the members of a population adopt it, no mutant strategy can invade." Both the E"SS" and ESS seem to suggest that under specific conditions only one phenotype (or genotype) could be selected for and prevail. Beyond that, there is no similarity between the two conceptions.

I would not want to describe the evolutionary "steady-state" as a synonym for evolutionary stable strategy for three reasons. First, the ESS is a conception derived from game theory (Maynard Smith 1982), whereas E"SS" is a conception derived from a consideration of the population dynamics of evolutionary change (Murray 1979). Populations are dynamic systems approaching a steady-state.

Second, evolutionary stable strategy is a misleading phrase because the standard definitions of "stable" and "strategy" do not seem applicable to evolving systems. Populations are not really stable, and natural selection does not produce strategies. Nevertheless, an author may define terms as he or she wishes. Maynard Smith (1982) and others may choose to use stable and strategy in their writing and thinking. I prefer not to.

Finally, game theory models are ad hoc hypotheses providing rationales for the occurrence of specific phenotypes. Game theory does not seem to make specific predictions from general hypotheses. My approach to understanding the action of natural selection is to produce hypothetico-deductive theories (Murray 1971, 1979, 1981, 1984, 1986b; Jehl and Murray 1986) and to think in terms of natural selection producing phenotypes (and genotypes) that result in evolutionary "steady-states." The game theory approach seems limited, and the phrase "evolutionary stable strategy" seems misleading. The two approaches to understanding evolution should not be confused by equating E"SS" with ESS.

4. Concluding Remarks

The message of this paper is that biologists are ignoring or incorrectly applying a method of investigation that has proved so successful in the physical sciences, that is, the hypothetico-deductive method. I believe that, by avoiding, if not deprecating, the hypothetico-deductive method in biological research, biologists are missing important and interesting insights into biological problems.

In supporting the usefulness of the hypothetico-deductive method, I emphasize that it is only one tool that biologists can use in their research. Furthermore, I emphasize that not all scientific problems are necessarily amenable to the hypothetico-deductive method. Not every fact in biology can be explicitly predicted from unifying principles any more than every physical fact (e.g., tides, weather) can be deduced from unifying principles. In order to predict the tides, the weather, or the consequences of harvesting a fish population at a given rate may require elaborate ad hoc models combining principles and conditional statements.

Without question, there is much room for diversifiers exploring the details of biological systems, whether by uncontrolled field research or more elaborate experimental methods. Good theoretical research begins by explaining the facts as they are known, and they cannot be known without accurate description. Initial explanations are *always* in the form of ad hoc hypotheses, and theoretically-oriented biologists may or may not generalize a particular ad hoc hypothesis by induction. Inductions, however, are limited in applicability and in explanatory power. In order to test *any* general hypothesis, whether an inference from observations or an inspired guess, we must combine several into a hypothetico-deductive theory and compare the deductions with empirical evidence (Popper 1959, 1972; Copi 1986).

Advances in biological knowledge, great as they have been, have suffered because biologists have ignored, rejected, or misapplied the hypothetico-deductive method. I hope that this may change.

Acknowledgments

I thank Charlotte Avers, Kees Hulsman, David J. T. Hussell, Joseph R. Jehl, Jr., Craig Loehle, William M. Shields, and Daniel Simberloff for reading and commenting on at least portions of earlier versions of the manuscript, and Brian Goodwin for a most interesting discussion of evolution.

References

Ashmole, N. P. 1961. The biology of certain terns. Unpubl. Ph.D. dissert., Oxford University, Oxford, England.

Barlow, N. 1958. *The Autobiography of Charles Darwin,* Norton, New York.

Bartholomew, G. A. 1986. The role of natural history in contemporary biology. *BioScience,* 36: 324–329.

Bartlett, M. S. 1973. Equations and models of population change. Pp. 5–21 in *The Mathematical Theory of the Dynamics of Biological Populations,* M. S. Bartlett and R. W. Hiorns, eds., Academic Press, New York.

Bell, G. 1982. *The Masterpiece of Nature: The Evolution and Genetics of Sexuality,* Univ. California Press, Berkeley.

Blockstein, D. E. 1989. Crop milk and clutch size in Mourning Doves. *Wilson Bull.,* 101: 11–25.

Brawn, J.D., and R.P. Balda. 1988. Population biology of cavity nesters in Northern Arizona: Do nest sites limit breeding densities? *Condor,* 90: 61–71.

Bronowski, J. 1965. *Science and Human Values, Rev. ed.,* Harper and Row, New York.

Cajori, F. 1946. *Newton's Principia* (Motte's translation, revised), Univ. Calif. Press, Berkeley, California.

Carlquist, S. 1965. *Island Life,* Natural History Press, Garden City, New York.

Chamberlin, T. C. 1890. The method of multiple working hypotheses. (Reprinted in 1965, *Science,* 148: 754–759).

Charlesworth, B. 1980. *Evolution in Age-structured Populations,* Cambridge Univ. Press, Cambridge.

Charnov, E. L., and J. R. Krebs. 1974. On clutch-size and fitness. *Ibis,* 116: 217–219.

Cody, M. L. 1966. A general theory of clutch size. *Evolution,* 20: 174–184.

Cody, M. L. 1969. The evolution of reproductive rates. *Science,* 163: 1185–1187.

Cody, M. L. 1971. Ecological aspects of reproduction. Pp. 461–512 in *Avian Biology, Vol. 1,* D. S. Farner and J. R. King, eds., Academic Press, New York.

Cohen, I. B. 1980. *The Newtonian Revolution,* Cambridge Univ. Press, Cambridge.

Connor, E. F., and D. Simberloff. 1986. Competition, scientific method, and null models in ecology. *Amer. Sci.,* 74: 155–162.

Copi, I. M. 1986. *Introduction to Logic, 7th ed.,* Macmillan, New York.

Curio, E. 1983. Why do young birds reproduce less well? *Ibis,* 125: 400–404.

Dyson, F. J. 1988. *Infinite in All Directions,* Harper and Row, New York.

Egler, F. E. 1986. "Physics envy" in ecology. *Bull. Ecol. Soc. Amer.,* 67: 233–235.

Fagerström, T. 1987. On theory, data and mathematics in ecology. *Oikos,* 50: 258–261.

Fogden, M. P. L. 1972. The seasonality and population dynamics of equatorial forest birds in Sarawak. *Ibis,* 114: 307–343.

Ghiselin, M. T. 1969. *The Triumph of the Darwinian Method,* Univ. Chicago Press, Chicago.

Gould, S. J., and R. C. Lewontin. 1979. The spandrels of San Marco and the Panglossian paradigm: a critique of the adaptationist programme. *Proc. Roy. Soc. Lond. B,* 205: 581–598.

Haartman, L. von. 1955. Clutch size in polygamous species. *Proc. XI Internatl. Ornithol. Congr.,* 450–453.

Haartman, L. von. 1957. Adaptation in hole-nesting birds. *Evolution,* 11: 339–347.

Haartman, L. von. 1971. Population dynamics. Pp. 391–459 in *Avian Biology, Vol. 1,* D. S. Farner and J. R. King, eds., Academic Press, New York.

Haila, Y. 1988. The multiple faces of ecological theory and data. *Oikos,* 53: 408–411.

Harrington, R. W. 1971. How ecological and genetic factors interact to determine when self-fertilizing hermaphrodites of *Rivulus marmoratus* change into functional secondary males, with a reappraisal of the modes of intersexuality among fishes. *Copeia,* 1971: 389–432.

Haukioja, E., and T. Hakala. 1979. On the relationship between avian clutch size and life span. *Ornis Fenn.,* 56: 45–55.

Hilborn, R., and S. C. Stearns. 1982. On inference in ecology and evolutionary biology: the problem of multiple causes. *Acta Biotheor.,* 31: 145–164.

Högstedt, G. 1980. Evolution of clutch size in birds: adaptive variation in relation to territory quality. *Science,* 210: 1148–1150.

Högstedt, G. 1981. Should there be a positive or negative correlation between survival of adults in a bird population and their clutch size? *Amer. Nat.,* 118: 568–571.

Hull, D. 1973. *Darwin and His Critics,* Univ. Chicago Press, Chicago.

Hussell, D. J. T. 1972. Factors affecting clutch size in arctic passerines. *Ecol. Monogr.,* 42: 317–364.

Hussell, D. J. T. 1985. Clutch size, daylength, and seasonality of resources: comments on Ashmole's hypothesis. *Auk,* 102: 632–634.

Hussell, D. J. T., and T. E. Quinney. 1987. Food abundance and clutch size of tree swallows *Tachycineta bicolor. Ibis,* 129: 243–258.

Itô, Y., and Y. Iwasa. 1981. Evolution of litter size. I. Conceptual reexamination. *Res. Popul. Ecol.,* 23: 344–359.

James, F. C., and C. E. McCulloch. 1985. Data analysis and the design of experiments in ornithology. Pp. 1–63 in *Current Ornithology, Vol. 2,* R. F. Johnston, ed., Plenum, New York.

Jehl, J. R., Jr., and B. G. Murray, Jr. 1986. The evolution of normal and reverse sexual size dimorphism in shorebirds and other birds. Pp. 1–86 in *Current Ornithology, Vol. 3,* R. F. Johnston, ed., Plenum, New York.

Kendeigh, S. C., T. C. Kramer, and F. Hammerstrom. 1956. Variations in egg characteristics of the House Wren. *Auk,* 73: 42–65.

Klomp, H. 1970. The determination of clutch-size in birds. A review. *Ardea,* 58: 1–124.

Kuhn, T. S. 1962. *The Structure of Scientific Revolutions,* Univ. Chicago Press, Chicago.

Lack, D. 1947. The significance of clutch-size. Parts I and II. *Ibis,* 89: 302–352.

Lack, D. 1948. The significance of clutch-size. Part III. *Ibis,* 90: 25–45.

Lack, D. 1954. *The Natural Regulation of Animal Numbers,* Oxford Univ. Press, Oxford.

Lack, D. 1966. *Population Studies of Birds,* Clarendon Press, Oxford.

Lack, D. 1968. *Ecological Adaptations for Breeding in Birds,* Methuen, London.

Lack, D., and R. E. Moreau. 1965. Clutch-size in tropical passerine birds of forest and savanna. *Oiseau,* 35 (specl. no.): 76–89.

Lessells, C. M. 1986. Brood size in Canada Geese: a manipulation experiment. *J. Anim. Ecol.,* 55: 669–689.

Lewontin, R. C. 1974. *The Genetic Basis of Evolutionary Change,* Columbia Univ. Press, New York.

Lill, A. 1974. The evolution of clutch size and male "chauvinism" in the White-bearded Manakin. *Living Bird,* 13: 211–231.

Lindén, M. 1988. Reproductive trade-off between first and second clutches in the Great Tit *Parus major:* an experimental study. *Oikos,* 51: 285–290.

Linstedt, S. L., and W. A. Calder. 1976. Body size and longevity in birds. *Condor,* 78: 91–94.

Loehle, C. 1987. Hypothesis testing in ecology: psychological aspects and the importance of theory maturation. *Quart. Rev. Biol.,* 62: 397–409.

Loehle, C. 1988. Philosophical tools: potential contributions to ecology. *Oikos,* 51: 97–104.

Lord, R. D., Jr. 1960. Litter size and latitude in North American mammals. *Amer. Midl. Nat.,* 64: 488–499.

Marchant, S. 1960. The breeding of some S.W. Ecuadorian birds. *Ibis,* 102: 349–382, 584–599.

Martin, T. E. 1988. Nest placement: implications for selected life-history traits, with special reference to clutch size. *Am. Nat.,* 132: 900–910.

Maynard Smith, J. 1982. *Evolution and the Theory of Games,* Cambridge Univ. Press, Cambridge.

Mayr, E. 1983. How to carry out the adaptationist program? *Am. Nat.,* 121: 324–334.

McClure, H. E. 1942. Mourning Dove production in southwestern Iowa. *Auk,* 59: 64–75.

Mentis, M. T. 1988. Hypothetico-deductive and inductive approaches in ecology. *Funct. Ecol.,* 2: 5–14.

Moreau, R. E. 1944. Clutch-size: a comparative study, with special reference to African birds. *Ibis,* 86: 286–347.

Murphy, E. C., and E. Haukioja. 1986. Clutch size in nidicolous birds. Pp. 141–180 in *Current Ornithology, Vol. 4,* R. F. Johnston, ed., Plenum, New York.

Murray, B. G., Jr. 1971. The ecological consequences of interspecific territorial behavior in birds. *Ecology,* 52: 414–423.

Murray, B. G., Jr. 1979. *Population Dynamics: Alternative Models,* Academic Press, New York.

Murray, B. G., Jr. 1981. The origins of adaptive interspecific territorialism. *Biol. Rev.,* 56: 1–22.

Murray, B. G., Jr. 1984. A demographic theory on the evolution of mating systems as exemplified by birds. Pp. 71–140 in *Evolutionary Biology, Vol. 18,* M. K. Hecht, B. Wallace, and G. T. Prance, eds., Plenum, New York.

Murray, B. G., Jr. 1985a. Evolution of clutch size in tropical species of birds. Pp. 505–519 in *Neotropical Ornithology,* P. A. Buckley, M. S. Foster, E. S. Morton, R. S. Ridgely, and F. G. Buckley, eds., Ornithol. Monogr. 36. Lawrence, Kansas.

Murray, B. G., Jr. 1985b. Population growth rate as a measure of individual fitness. *Oikos,* 44: 509–511.

Murray, B. G., Jr. 1986a. Natural history and deductive logic. *BioScience,* 36: 513–514.

Murray, B. G., Jr. 1986b. The structure of theory, and the role of competition in community dynamics. *Oikos,* 46: 145–158.

Murray, B. G., Jr. 1988. On measuring individual fitness: a reply to Nur. *Oikos,* 51: 249–250.

Murray, B. G., Jr. in press. Population dynamics, genetic change, and the measurement of fitness. *Oikos.*

Murray, B. G., Jr., and L. Gårding. 1984. On the meaning of parameter x of Lotka's discrete equations. *Oikos,* 42: 323–326.

Murray, B. G., Jr., and V. Nolan Jr. 1989. Evolution of clutch size. I. An equation for predicting clutch size. *Evolution,* 43: 1699–1705.

Murray, B. G., Jr., J. W. Fitzpatrick, and G. E. Woolfenden. 1989. Evolution of clutch size. II. A test of the Murray-Nolan equation. *Evolution,* 43: 1706–1711.

Naylor, B. G., and P. Handford. 1985. In defense of Darwin's theory. *BioScience,* 35: 478–484.

Nelson, J. 1964. Factors influencing clutch-size and chick growth in the North Atlantic Gannet *Sula bassana*. *Ibis*, 106: 63–77.

Newton, I. 1729. [See Cajori 1964.]

Nice, M. M. 1957. Nesting success in altricial birds. *Auk*, 74: 305–321.

Nilsson, S. G. 1984a. Clutch size and breeding success of the Pied Flycatcher *Ficedula hypoleuca* in natural tree-holes. *Ibis*, 126: 407–410.

Nilsson, S. G. 1984b. The evolution of nest-site selection among hole-nesting birds: the importance of nest predation and competition. *Ornis Scand.*, 15: 167–175.

Nolan, V., Jr. 1978. The ecology and behavior of the Prairie Warbler *Dendroica discolor*. Ornithol. Monogr. 26. Allen Press, Lawrence, Kansas.

Nur, N. 1984. Fitness, population growth rate and natural selection. *Oikos*, 42: 413–414.

Nur, N. 1987. Population growth rate and the measurement of fitness: a critical reflection. *Oikos*, 48: 338–341.

Oniki, Y. 1979. Is nesting success of birds low in the tropics? *Biotropica*, 11: 60–69.

Owen, D. F. 1977. Latitudinal gradients in clutch size: an extension of David Lack's theory. Pp. 171–179 in *Evolutionary Ecology*, B. Stonehouse and C. Perrins, eds., University Park Press, Baltimore, Maryland.

Perrins, C. M. 1977. The role of predation in the evolution of clutch size. Pp. 181–191 in *Evolutionary Ecology*, B. Stonehouse and C. Perrins, eds., University Park Press, Baltimore, Maryland.

Pettifor, R. A., C. M. Perrins, and R. H. McCleery. 1988. Individual optimization of clutch size in Great Tits. *Nature*, 336: 160–162.

Pianka, E. R. 1978. *Evolutionary Ecology, 2nd ed.*, Harper and Row, New York.

Pielou, E. C. 1977. *Mathematical Ecology*, Wiley, New York.

Platt, J. R. 1964. Strong inference. *Science*, 146: 347–353.

Popper, K. R. 1959. *The Logic of Scientific Discovery*, Hutchinson, London.

Popper, K. R. 1972. *Objective Knowledge: An Evolutionary Approach*, Oxford Univ. Press, Oxford.

Quinn, J. F., and A. E. Dunham. 1983. On hypothesis testing in ecology and evolution. *Am. Nat.*, 122: 602–617.

Reznick, D. 1985. Costs of reproduction: an evaluation of the empirical evidence. *Oikos*, 44: 257–267.

Ricklefs, R. E. 1969. An analysis of nesting mortality in birds. *Smithsonian Contrib. Zoology*, 9: 1–48.

Ricklefs, R. E. 1973. Fecundity, mortality, and avian demography. Pp. 366–435 in *Breeding Biology of Birds*, D. S. Farner, ed., National Academy of Sciences, Washington, D.C.

Ricklefs, R. E. 1977. On the evolution of reproductive strategies in birds: reproductive effort. *Am. Nat.*, 111: 453–478.

Ricklefs, R. E. 1980. Geographical variation in clutch size among passerine birds: Ashmole's hypothesis. *Auk*, 97: 38–49.

Ricklefs, R. E. 1983. Comparative avian demography. Pp. 1–32 in *Current Ornithology, Vol. 1*, R. F. Johnston, ed., Plenum, New York.

Rosen, R. 1985. Organisms as causal systems which are not mechanisms: an essay into the nature of complexity. Pp. 165–203 in *Theoretical Biology and Complexity: Three Essays on the Natural Philosophy of Complex Systems*, R.Rosen, ed., Academic Press, Orlando, Florida.

Rosenzweig, M. L. 1974. *And Replenish the Earth: The Evolution, Consequences, and Prevention of Overpopulation*, Harper and Row, New York.

Roughgarden, J. 1983. Competition and theory in community ecology. *Am. Nat.*, 122: 583–601.

Saether, B.-E. 1989. Survival rates in relation to body weight in European birds. *Ornis Scand.*, 20: 13–21.

Schall, J. J. 1978. Reproductive strategies in sympatric whiptail lizards *(Cnemidophorus)*. *Copeia*, 1978: 108–116.

Simberloff, D. 1983. Competition theory, hypothesis-testing, and other community ecological buzzwords. *Am. Nat.*, 122: 626–635.

Skutch, A. F. 1949. Do tropical birds rear as many birds as they can nourish? *Ibis*, 109: 579–599.

Skutch, A. F. 1953. Life history of the Southern House Wren. *Condor*, 55: 121–149.

Skutch, A. F. 1966. A breeding bird census and nesting success in Central America. *Ibis*, 108: 1–16.

Skutch, A. F. 1976. *Parent Birds and Their Young*, Univ. Texas Press, Austin, Texas.

Skutch, A. F. 1985. Clutch size, nesting success, and predation on nests of neotropical birds, reviewed. Pp. 575–594 in *Neotropical Ornithology,* P. A. Buckley, M. S. Foster, E. S. Morton, R. S. Ridgely, and F. G. Buckley, eds., Ornithol. Monogr. 36. Allen Press, Lawrence, Kansas.

Slagsvold, T. 1982. Clutch size variation in passerine birds: the nest predation hypothesis. *Oecologia,* 54: 159–169.

Slagsvold, T. 1984. Clutch size variation of birds in relation to nest predation: on the cost of reproduction. *J. Anim. Ecol.,* 53: 945–953.

Smith, J. N. M. 1981. Does high fecundity reduce survival in Song Sparrows? *Evolution,* 35: 1142–1148.

Snow, B. K. 1970. A field study of the Bearded Bellbird in Trinidad. *Ibis,* 112: 299–329.

Snow, D. W. 1978. The nest as a factor determining clutch-size in tropical birds. *J. Ornithol.,* 119: 227–230.

Snow, D. W., and B. K. Snow. 1964. Breeding seasons and annual cycles of Trinidad land birds. *Zoologica,* 49: 1–39.

Sober, E. 1984. *The Nature of Selection,* MIT Press, Cambridge, Massachusetts.

Spencer, A. W., and H. W. Steinhoff. 1968. An explanation of geographic variation in litter size. *J. Mammal.,* 49: 281–286.

Stearns, S. C. 1976. Life-history tactics: a review of the ideas. *Quart. Rev. Biol.,* 51: 3–47.

Stenseth, N. C. 1983. Grasses, grazers, mutualism and coevolution: a comment about handwaving in ecology. *Oikos,* 41: 152–153.

Strong, D. R., Jr. 1980. Null hypotheses in ecology. *Synthese,* 43: 271–285.

Strong, D. R., Jr. 1983. Natural variability and the manifold mechanisms of ecological communities. *Am. Nat.,* 122: 636–660.

Taylor, P. 1989. Revising models and generating theory. *Oikos,* 54: 121–126.

Uzzell, T. M. 1964. Relations of the diploid and triploid species of the *Ambystoma jeffersonianum* complex (Amphibia, Caudata). *Copeia,* 1964: 257–300.

Vandermeer, J. H. 1972. Niche theory. *Ann. Rev. Ecol. Syst.,* 3: 107–132.

Van Valen, L., and F. A. Pitelka. 1974. Commentary—Intellectual censorship in ecology. *Ecology,* 55: 925–926.

Williams, G. C. 1966. Natural selection, the costs of reproduction, and a refinement of Lack's principle. *Am. Nat.*, 100: 687–690.

Willis, E. O. 1961. A study of nesting ant-tanagers in British Honduras. *Condor,* 63: 479–503.

Willis, E. O. 1974. Populations and local extinctions of birds on Barro Colorado Island, Panama. *Ecol. Monogr.,* 44: 153–169.

Winkler, D. W., and J. R. Walters. 1983. The determination of clutch size in precocial birds. Pp. 33–68 in *Current Ornithology, Vol. 2,* R. F. Johnston, ed., Plenum, New York.

Woolfenden, G. E., and J. W. Fitzpatrick. 1984. *The Florida Scrub Jay: Demography of a Cooperative-breeding Bird,* Princeton University Press, Princeton.

CHAPTER 8

The Generic Properties of Morphogenetic Fields

BRIAN C. GOODWIN

Abstract

One of the fundamental issues in biology is whether organismic morphology can be understood as the result of a generative process with generic properties. If so, then biological form is intelligible in dynamical terms, and biology is the realm of expression of a characteristic type of order, making it a rational science. This contrasts with the dominant view that biology is an historical science, species being indviduals, not types, that result exclusively from the contingencies of survival. This chapter explores these issues in relation to a model of morphogenesis of a particular species, which suggests that the forms generated may be dynamic attractors in morphogenetic space, so that biological forms may be natural kinds.

1. Introduction

Although it may sound rather ambitious it is nevertheless true that the objective of biological study is to make the realm of living beings intelligible. There are essentially two different ways of doing this, both of which are reflected in distinct traditions of biological thought. The one that happens to be dominant at the moment is based upon the proposition that whatever order there is over the biological realm as a whole is a direct result of the history of organismic adaptation to changing environments. Thus the way to make biological phenomena intelligible is to reconstruct their history in as much detail as possible in terms of the postulates of random genetic variation and natural selection. In this view, which comes to us essentially from Darwin, there are no intrinsic principles of dynamic order in organisms and so there are no universal properties of life other than reproduction, variation, and adaptation by natural selection.

The alternate view was expressed by the founder of systematic biology, Linnaeus. He assumed that the regularities in the structural relationships among organisms that make possible a hierarchical taxonomy of species reflect intrinsic principles of order, which could eventually be expressed in terms of laws of form. This view informed the work of the rational morphologists of the 18th and early 19th centuries such as Cuvier, Geoffroy St. Hilaire and Reichert (see Cassirer, 1950). Although Goethe articulated a highly dynamic, transformational concept of biological form, the tradition of rational morphology suffered from too static a conception of structure and was unable to respond to the two great biological developments of the 19th century, evolution and embryology. As a result, the problem of biological form went into eclipse as evolution and genetics transformed biology into an historical science preoccupied with particulars rather than universals. However, what was left out of the modern synthesis, embryology (or more generally, developmental biology) has now matured to the point where a dynamic solution of Linnaeus' proposition becomes a real possibility. The consequences are potentially momentous. If biological intelligibility resides in the generative principles of development, then the historical explanations of neo-Darwinism turn into a description of necessary conditions for life cycles, while developmental dynamics assumes the primary explanatory burden of biological form and relational order. This essay explores these issues in terms of a narrowly focused enquiry from which some broad propositions are developed.

2. Evolution as Genealogy

The goal of neo-Darwinism is to give an account of evolution in terms of the historical relationship of species, and in so doing to construct a taxonomy that is basically a genealogy, an account of actual historical descent. This, according to Darwin (1859), would reveal the plan of creation. But genealogy is essentially the historical pattern of inheritance and this is now interpreted as the transmission of hereditary information, primarily in the form of the genetic material, DNA. Hence neo-Darwinian evolution is essentially an account of historical patterns of DNA sequence transmission. In this account, organisms have disappeared and have been replaced by genotypes, with phenotypes as their causal effects. Development is an irrelevant epiphenomenon in this account of evolution, despite the fact that natural selection is considered to act on phenotypes. It is sufficient to represent phenotypic characters by selection coefficients that weight the contribution to overall fitness of different genes involved in character generation. It then emerges that the theory of natural selection is a theory of dynamic stability of genotypes in particular environments, represented by particular selection coefficients.

What destroys the organism in this description of evolution is an hereditarian essentialism in which a distinct part, the hereditary material, is taken to embody the essence of the living process through its powers of self-reproduction. This then results in a sharp focus on inheritance as the primary problem of biology, with the subject taking on the character of an historical science whose major objective is the tracing of hereditary lineages. This has led to the immensely successful development of genetics and the commonly-held belief that all the important properties of organisms are encoded in their genomes. The reasons why this belief is mistaken are presented in detail elsewhere (Goodwin 1985, 1989), amounting to a refutation of the proposition that organisms can be reduced to their hereditary essences. The theory of evolution by natural selection thus turns out not to be a theory about the generative origins of organisms, but about the dynamics of populations of abstract self-reproducing entities (genes or replicators). So let us turn to a perspective in which organisms exist as real entities, the generators of biological order.

3. Evolution as Generation

Figure 1 shows the life cycle of the marine alga, *Acetabularia acetabulum*. This species combines the features of a basically simple reproductive cycle with a distinctive morphology that reveals just how large and complex single cells can become as a result of an interesting generative process. Other species of plant or animal may have simpler or more complex life cycles, but the basic characteristics are always the same: a simple initial form (bud or zygote) develops into an organism of specific form, a part of which is then capable of reinitiating the process. This cycle must be dynamically stable in a particular range of environments if that species is to persist. Also, there is a hereditary aspect to the process that limits the range of forms generated in these environments, resulting in a distinctive set of characters that identify species morphology.

These properties describe the conditions for survival and the essential quality of heredity in relation to organismic form. One can then raise the question whether an understanding of the dynamics of this type of generative process can give an insight into the distinctive order that is revealed in the biological realm through the process of evolution. This order is recognized through the existence of a taxonomy of species in which relationships of similarity and difference of morphology, together with other characteristics, are used to produce a logically ordered array of species.

In a genealogical approach to taxonomy, such as Darwin's, it is assumed that relationships of similarity and difference are dependent upon patterns of historical descent. The alternative proposition, which

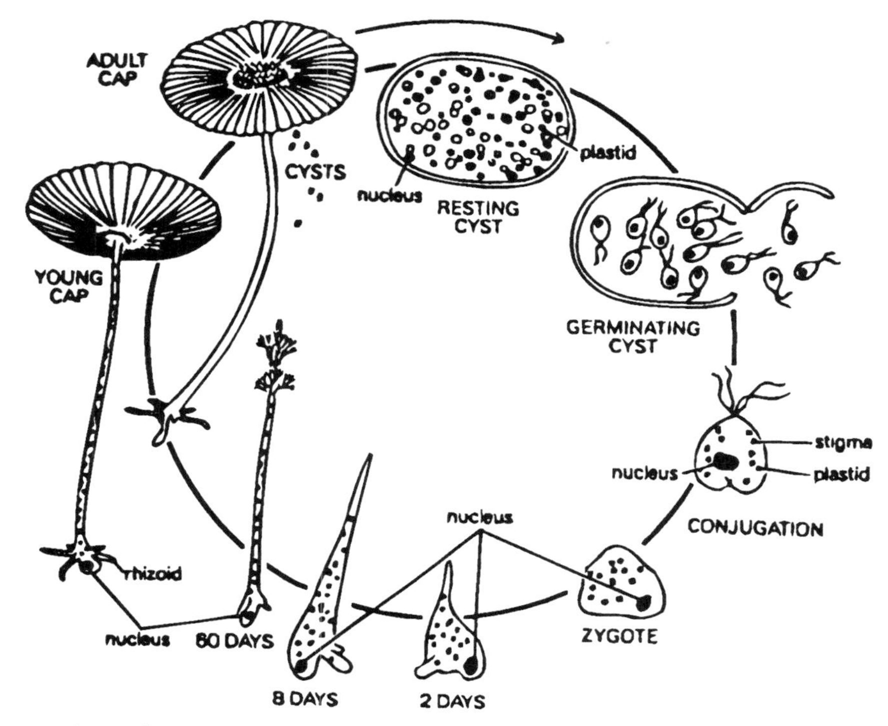

Figure 1. The life cycle of *Acetabularia* acetabulum.

was held by Linnaeus and the whole pre-Darwinian school of rational morphology, is that biological taxonomy reveals an order intrinsic to the distinctive qualities of the living state, i.e., of organisms. In 20th-century biology, this becomes the proposition that taxonomic order is a result of the dynamic properties of the process by which organisms are generated. The actual historical sequence of species (their genealogy) reflects the contingencies of time, place, and circumstance that operate in evolution as a historical process; but the logical order of the biological realm is to be found in the distinctive dynamics of its fundamental generative process, organismic life cycles.

4. The Generation of Form

The study of morphogenesis is still in a primitive state of development and the analysis to be presented here is a mere scratching at the surface of a potentially significant research program. The procedure will be to use *Acetabularia* to illustrate the principles of the approach being used, and then to suggest how these may be extended to other illustrative examples.

Referring to the life cycle depicted in Figure 1, the process that needs to be described and understood is the growth of the alga and the formation of the characteristic structures at the growing tip, culminating in a cap. The cycle is completed by the formation of haploid gametes in the cap, their subsequent release and fusion to form a diploid zygote and the reinitiation of the growth process.

If the cap of a mature *Acetabularia* is cut off, a new one is regenerated by the same morphogenetic sequence as the normal development of an adult from a zygote: generation and regeneration proceed by the same route. Certain experimental studies are more conveniently carried out on regenerating plants; and the modelling exercise is based on these conditions. However, the model applies equally well to development from a sphere rather than a cylinder. Figure 2 shows schematically the sequence of shapes that the tip assumes during regeneration to the first whorl. Only the wall is shown, but inside this is a thin layer of cytoplasm, with the usual eukaryotic cytoskeleton, while the central space is occupied by a vacuole.

When the cap is cut off, a new membrane rapidly forms around the exposed cytoplasm and within 4 hours a new cell wall has formed. By 24 hours this is more or less hemispherical, osmotic pressure exerted by the vacuole maintaining the shape. By about 48 hours after the cut a small tip is produced, followed by growth and extension. The growing tip then flattens at about 72 hours after cutting and a crown of small bumps, the primordia of leaf-like structures called hairs or verticils, is produced. These grow and bifurcate. The whole crown-like structure is called a whorl. From the centre of the whorl a new tip is initiated and the process is repeated. This sequence recurs a variable number of times at roughly 24–48 hour intervals until a cap primordium is produced in place of a whorl (Figure 3). It grows laterally into the delicately sculpted form of the adult cap and the growth ceases. The hairs fall off, and the adult morphology (Fig. 4) is stable for several months. The single nucleus remains throughout this developmental process in a branch of the root-like base, the rhizoid. The organism depends for its energy on photosynthesis, exchanges gases and ions with its environment (sea-water), and maintains its shape by the mechanical properties of the cell wall and the turgor pressure of the large central vacuole, whose osmotic pressure is controlled by the regulation of ions and organic acids.

An extensive program of experimental study on *Acetabularia* has revealed the importance of the calcium ion in morphogenesis (see Goodwin et al, 1983; Harrison and Hillier, 1985; Harrison et al, 1988;

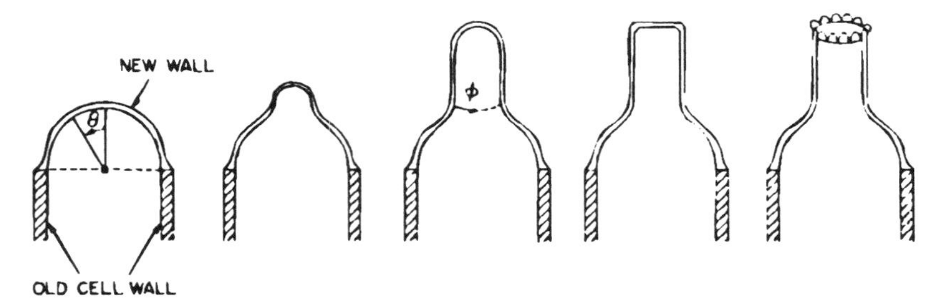

Figure 2. Schematic sequence of shapes during regeneration in *Acetabularia,* up to first whorl formation.

Figure 3. Formation of a cap primordium, after three whorls.

Cotton and Vanden Driessche, 1987). It is also evident that the cytoskeleton and the cell wall must be intimately involved in the processes of growth and shape change at the growing tip. These components have been put together into a model based upon equations describing the dynamics of the cytoskeleton-calcium interactions derived by Goodwin and Trainor (1985), incorporating also a description of the relationship between the cytoskeleton and the cell wall (Brière and Goodwin, 1988). This describes a morphogenetic field defined on the domain of regeneration, which starts as a shell of cytoplasm within a cell wall whose shape is determined by the osmotic pressure exerted by the vacuole and by its elastic properties. These vary with the strain (stretching or compression) in the cytoskeleton, which is itself dependent upon interaction with cytosolic free calcium. The equations describing the cytoplasmic regulation of calcium also depend upon cytoskeletal strain, so that there is a reciprocal coupling between these variables in the cytoplasm. It is this coupling that gives to the cytoplasm the properties of an excitable

Figure 4. Morphology of the mature alga, showing rhizoid, stalk and cap. The whorls have dropped off.

medium: it can spontaneously break symmetry (spatial homogeneity) to generate spatially ordered patterns (Goodwin and Trainor, 1985). This is the basis of the whole generative process. Changes of shape in this system depend upon local changes in the elastic properties of the cell wall, and upon a growth process which is based upon experimentally observed properties: when wall domains are stretched beyond a critical amount, new material is added to the wall and the deformation becomes plastic. Below this critical value, deformations are elastic. The full model is simulated in three dimensions using a finite-element analysis. The cytoplasm is then described as a mesh of filaments that obey the morphogenetic field equations, as shown in Figure 5. A similar treatment is used for the cell wall. A detailed description of the equations and the analytical procedure is given in Brière and Goodwin (1988), and an abbreviated version in Appendix I.

5. Generic Dynamic Properties of Morphogenesis

The objective of morphogenetic modelling is two-fold: to gain insights into the particular process under investigation, such as *Acetabularia*

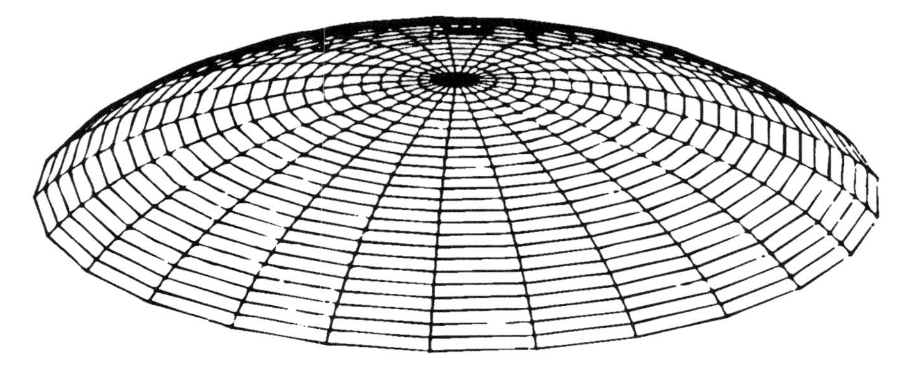

Figure 5. Computer simulation of regeneration in which the cytoplasm and the cell wall are described as shells made up of finite elements which obey equations describing their dynamics as mechanochemical or elastic media, respectively.

morphogenesis, and to explore the possibility that there are general morphogenetic principles characteristic of whole groups of organisms. *Acetabularia acetabulum* belongs to a group called the *Dasycladaceae,* all of which share basic morphological features. They have a primary apico-basal growth axis with secondary rings of hairs or verticils, but not all produce a terminal cap. In these latter species the hairs grow into structures within which cysts and gametes are produced, thus serving a reproductive role. In *Acetabularia acetabulum* it is possible to 'phenocopy' such morphologies simply by reducing the concentration of calcium in the medium from the normal value of 10mM to 4mM. Then whorls are formed but no caps are produced on most algae (Goodwin et al, 1983). However, these capless algae have not been observed to form gametes, so they are unable to complete the life cycle: the phenocopy is not reproductively stable.

The results of the morphogenetic simulation of *Acetabularia* are of interest in relation to both objectives of the modelling exercise. The rules of the game are that parameters are set initially and the system then does its own thing. All that the model has available to it to generate form is a dynamic that can spontaneously produce spatially inhomogenous patterns (technically, the parameters of the field equations are in a range that permits bifurcations to occur) and a defined growth algorithm. It was found that there is a significant range of parameter values in which the following behavior occurs. First the cytoskeleton-calcium field generates a gradient of elevated Ca^{2+} and strain at the pole of the regenerating region, resulting in wall softening and the formation of a tip (Figure 6). Axial growth then follows. Gradually, as growth proceeds, the gradient in Ca^{2+} with a maximum changes into an annulus, with the maximum occurring as a ring. This results in an annular region of wall softening and maximum curvature, resulting in tip flattening. The annulus of Ca^{2+} continues to increase as growth continues. These changes of shape and field state are shown in Figures 7–9, in which shape is represented by a section through the growth domain from base to tip, the base having radius 1.

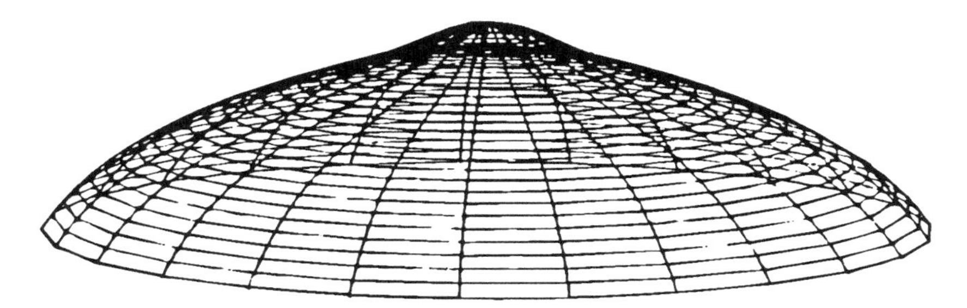

Figure 6. Tip initiation in the model, resulting from spontaneous symmetry-breaking of the cytogel-calcium dynamic, gradient formation with maxima of strain and calcium at the tip, and wall softening as a result of interaction between cytoplasm and the wall.

The level of the calcium annulus at the stage of growth shown in Fig. 9 is such that a second bifurcation can occur. Under random perturbation, the ring of elevated calcium can break up into a ring of calcium peaks with the same basic structure as a whorl. When this was tested the result was as anticipated, the ring of calcium peaks arising in the region of maximum curvature of the tip where it bulges slightly before going flat towards the pole. This is precisely where the ring of hair primordia form, as shown in Fig. 2. The calcium peaks coincide with regions of maximal strain in the cytogel and this results in wall softening. So these are the regions where the wall bulges out, corresponding to hair primordia. This sequence of events is something of a revelation, because it provides for the first time a possible explanation for the tip flattening that was always observed prior to whorl formation but never understood. It also gives a clear prediction about the expected behaviour of free calcium: it should form an annulus prior to whorl formation, and break up into a series of peaks just before hair primordia appear. There is some interesting evidence for this (Harrison et al, 1988), and a more detailed imaging study of cytosolic calcium changes is currently in progress.

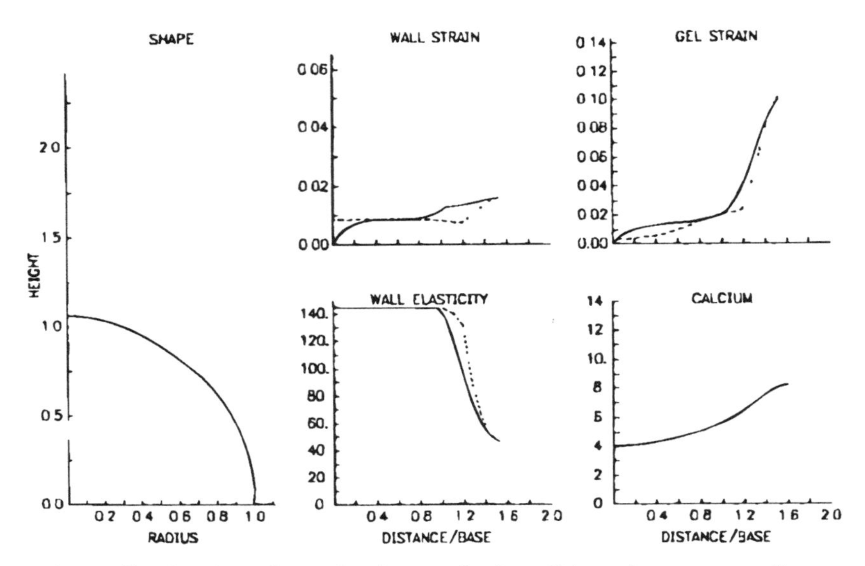

Figure 7. Section through the mesh describing the regenerating tip, showing shape, wall strain and elastic modulus, cytogel strain and free calcium concentration as a function of distance from base (origin O, on the abscissa of the graphed variables, radius 1 in the curve showing shape). A gradient in calcium forms spontaneously, with a maximum at the tip.

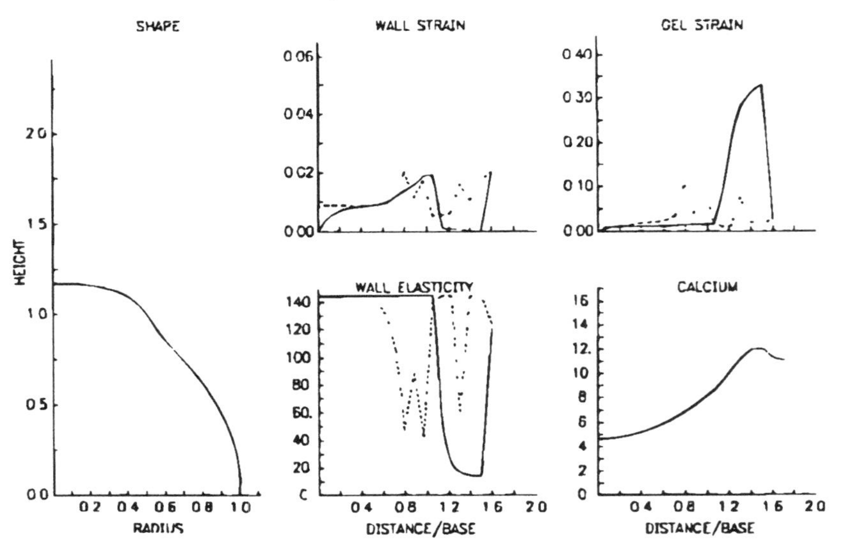

Figure 8. Later stage of regeneration: the maximum of cytosolic free calcium is now displaced from the tip, to the region of maximum curvature of the wall shape where the wall elastic modulus is reduced. In three dimensions, this defines an annulus.

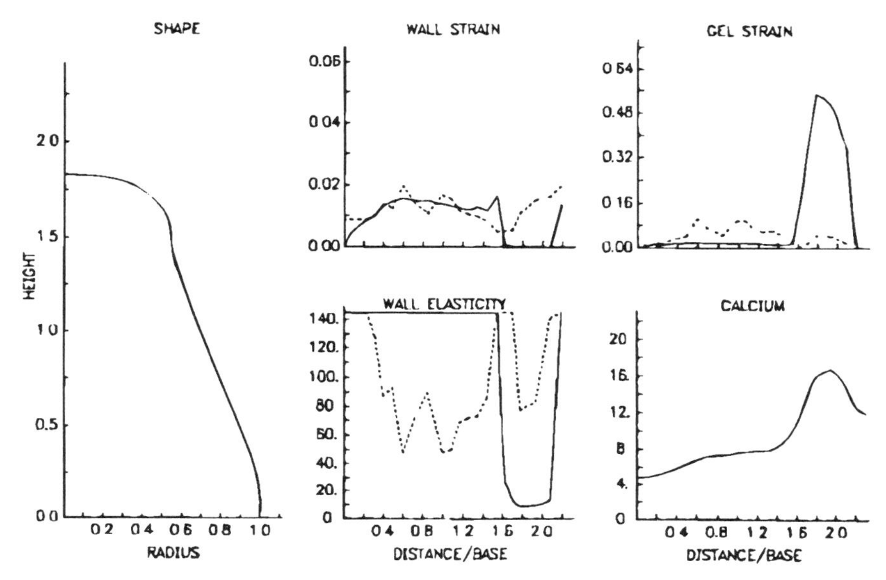

Figure 9. The pattern in the simulation corresponding to the state of tip flattening described in Figure 2, with a well-defined annulus of calcium at the region of maximum curvature in the shape. This annulus then spontaneously breaks into a series of peaks, assumed to initiate the whorl pattern shown in Figure 2, last drawing.

The model is not yet able to generate and grow a ring of hairs, for technical reasons (the size of the mesh required for this is too small and the computing time needed for such fine detail is too large for the VAX cluster used, requiring a supercomputer). However, the simulation has illuminated a further step in the morphogenetic sequence. The calcium annulus of Fig. 9 is not stable, but increases and decreases with an irregular rhythm as growth proceeds. This may account for the periodic formation of whorls, which is also quite irregular. However, after a variable number of waxings and wanings of the annulus, a new pattern arises. The system spontaneously enters a phase of growth in which the whole tip expands, rather like a cap but less flattened, giving a terminal structure at the end of its axial growth.

What emerges from these simulation studies is very suggestive. It appears that there is an interaction between the geometry of the growing form and the intrinsic dynamics of the morphogenetic field that results in a parametrically robust generic sequence of events with a qualitative similarity to those observed in the developing organism. These are: tip initiation, growth, annulus formation and tip flattening, bifurcation to a whorl prepattern, a sequence which can then be repeated a number of times, followed finally by the formation of a terminal structure. These processes generate the general characteristics of *Dasyclad* morphology, all without anything that would normally be regarded as a detailed genetic program. The suggestion that thus emerges is that a morphogenetic system based upon essential dynamic and mechanical properties of 'plant' cells will, if it does anything interesting, generate a structure with the basic characteristics of the living system. This could be regarded as the generic basis of the whole group. The differences between members of the group could then be explained by differences in parameter values such as the elastic modulus of the wall, binding constants of calcium regulatory proteins, cytoskeletal parameters, etc., all of which are under genetic control. What the genotype then does is determine the precise region of parameter space in which a particular species lives, giving it a particular morphogenetic sequence and a specific adult morphology. The set of possible morphologies within a group is then defined by the basic characteristics of the morphogenetic process of the group and the regions of parameter space that reliably give particular forms.

6. Rational Morphology

Despite the preliminary nature of this analysis, the surprising result emerges that, when growth occurs in the model, it follows a pattern that is similar in basic features to that observed in *Acetabularia,* generating forms with properties characteristic of the *Dasycladaceae.* None of the details were programmed into the model. The morphogenetic field embodies properties common to all eukaryotes and so can be used to describe morphogenesis in either plants (with cell walls and vacuoles) or animals (without these structures and just the cytogel, together with the extracellular matrix and cell-cell interactions in the metazoa). The first model of this type of morphogenetic field was actually applied to gastrulation in animal embryos (Odell et al, 1981) and a variety of other applications to animal morphogenesis have since been presented (Murray and Oster, 1984; Oster et al, 1988). The possibility is that the calcium cytoskeletal field is a eukaryotic morphogenetic universal that is differently contextualized in different types of organism, resulting in characteristically different types of morphological structure.

For example, the distinctive sequences of bone patterns in tetrapod limbs that constitute one of the classic examples of morphological homology can be understood in terms of a basic generative process that naturally produces sequences of elements of the type observed in these structures. This is just an ontogenetic version of the rational morphology of Cuvier, Geoffrey and Owen (see, e.g., Webster and Goodwin,

1982). The different types of limb, from amphibian to horse and human, can be understood as genetic stabilizations (canalizations) of particular epigenetic sequences, each of which is one of the possible forms that the generative process can produce (Shubin and Alberch, 1986). The explanation of form in the case of the giant algae or the tetrapod limb is then clearly not in terms of genealogy, the historical sequence in which the forms are generated. It is in terms of what forms are possible in relation to the morphogenetic process under consideration, and in relation to the stability of the overall organismic life cycle within which these forms play a role. The first defines the generative process that makes the forms possible in the first place (a structuralist analysis); the second is a dynamic stability analysis of the life cycle of the organism with the structure under consideration (a functionalist analysis). The first is necessary to explain existence; the second to explain persistence. History as genealogy explains neither; it simply describes the sequence in which particular forms are realized. Rational taxonomic orderings of species based upon morphologies depend upon the intrinsic order of the generative process, using a distance measure that defines neighborhood relationships between the morphogenetic trajectories leading to the possible forms. Taxonomy as genealogy describes the historical sequence in which particular forms are realized. This need bear no relationship to the rational taxonomy of organismic morphology, though in general there are likely to be some correlations.

7. Conclusions

The partial differential equations and functional relationships in Appendix I define a morphogenetic field whose solutions describe the geometry and the space- and time-dependent variables that identify field states. This is a complex moving-boundary problem in which dynamics and geometry interact to produce characteristic forms. By varying parameters it is possible to generate a variety of shapes, and our research program has barely begun to explore systematically the set of possible forms. We have found that a range of experimentally observed variants of normal morphology can be generated, such as algae with bulbous tips that result from reduced calcium levels in the medium (Goodwin and Pateromichelakis, 1979). These fail to undergo normal morphogenesis. Such shapes can be simulated simply by reducing the elastic modulus of the cell wall, which is a consequence of reduced calcium. Furthermore, the model simulates the initial phase of morphogenesis from the zygote (Fig. 1) when spherical boundary conditions are used: symmetry-breaking occurs on a sphere, and a tip is produced that then grows and elongates as in initial axis formation. What the model so far fails to do is stop growing. Clearly some equilibrium between the forces involved arises and results in stable geometry not yet realized in the simulation. And we have little idea how to complete the life cycle, i.e., to simulate gamete and cyst formation.

Nevertheless the limited success of the model encourages the view that moving boundary problems of the type described may have dynamic attractors whose behaviour is suggestive of generic properties of morphogenetic processes. The production of whorls of bracts or leaves during plant growth is a very common generative process over a great range of species. A crucial question arises, however, in relation to multicellularity. Can the morphogenetic field model used to describe tip growth in a unicellular organism be applied to the morphogenetic dynamics of the multicellular meristem of the higher plant? What modifications are required? Will they substantially change morphogenetic patterns? These questions are currently being investigated in collaboration with Paul Green, whose strain field descriptions of leaf and floral meristem dynamics are fully compatible with the concepts developed in the tip growth model (Green, 1987). We have already observed the initiation of patterns in our fields that are similar to leaf primordium sequences. If the geometry is held constant, spiral waves propagate around the field, which is one of the dynamical modes of excitable media. We assume that, with the right conditions for growth and leaf initiation, these will generate the spiral phyllotaxis patterns observed in higher plants.

The research program is to explore the conjecture that there are universal generative principles of development over a great diversity of species, plant and animal. The potential value of this enterprise is very great. For the demonstration and classification of generic properties of morphogenetic fields could provide a basis for a rational taxonomy of biological forms as natural kinds. And this would transform the study of morphological taxonomy from an historical into an exact science.

Appendix I

(A) Viscoelastic Properties

The first equation describes the displacements of the cytogel (cortical cytoplasm containing the cytoskeleton) near an equilibrium state. The variables to consider are the displacement ξ from this equilibrium and the calcium concentration χ (in this theory the cytogel is represented as a continuum). From the displacement field, we can calculate at any point the strain tensor ϵ and the strain rate tensor $\dot{\epsilon}$, which account for the elastic deformation and for the viscous motion, respectively.

The stress tensor σ, whose elements are the components of the forces per unit area exerted in the three principal space directions, depends on the strain and strain-rate tensors and on the mechanical properties of the cytogel. Since these properties depend on the local calcium concentration, we can write at any point M:

$$\sigma(M) = \sigma(\epsilon(M), \dot{\epsilon}(M), \chi(M))$$

A linear expansion of σ about the equilibrium stress tensor σ_0 gives:

$$\sigma = \sigma_0 + S \cdot \epsilon + A \cdot \dot{\epsilon} \tag{1}$$

In general, A and S are fourth-order tensors (calcium dependent). But for an isotropic material (which we assume for the cytogel), the 81 components of S can be expressed in terms of just two elastic moduli, the Lamé coefficients λ and μ; similarly, in the theory of simple liquids, the 81 components of A can be expressed in terms of two coefficients, the shear viscosity ζ and the bulk viscosity η.

Consider now a very small unit element of cortical material, with a volume density ρ; it accelerates according to Newton's second law in response to the various elastic, viscous and mechanical forces acting on it such that

$$\text{div } \sigma + f = \rho \frac{\mathrm{d}^2 \xi}{\mathrm{d}t^2} \tag{2}$$

where div σ accounts for the elastic and viscous forces and f describes the external forces.

From (1) we have

$$\text{div } \sigma = \text{div } \sigma_0 + \text{div } (S \cdot \epsilon) + \text{div } (A \cdot \dot{\epsilon}) \tag{3}$$

But, since σ_0 depends on the calcium concentration

$$\text{div } \sigma_0 = \frac{\partial \sigma_0}{\partial \chi} \cdot \nabla \chi$$

then

$$\text{div } \sigma = S \cdot \nabla \epsilon + A \cdot \nabla \dot{\epsilon} + \frac{\partial \sigma_0}{\partial \chi} \cdot \chi + \left(\frac{\partial S}{\partial \chi} \cdot \epsilon + \frac{\partial A}{\partial \chi} \cdot \dot{\epsilon} \right) \cdot \nabla \chi \tag{4}$$

Replacing S and A by the elastic moduli and viscosity coefficients and expressing ϵ and $\dot{\epsilon}$ in terms of the displacement and the velocity vectors ξ and $\dot{\xi}$, we finally get the Goodwin and Trainor equation for viscoelasticity

$$\rho\frac{\mathrm{d}^2\xi}{\mathrm{d}t^2} = \mu\nabla^2\xi + (\lambda+\mu)\nabla(\nabla\xi) + \eta\nabla^2\dot{\xi} + (\zeta+\eta/3)\nabla(\nabla\cdot\dot{\xi}) - F\cdot\nabla\chi - R\dot{\xi} + \text{second-order terms} \tag{5}$$

where

$$F = -\frac{\partial\sigma_0}{\partial\chi} \tag{5.1}$$

F is a calcium-dependent second-order tensor and the external force term $R\xi$ stands, in a linear approximation, for the restoring forces due to structural components (e.g. microtubules) that resist local displacement of the gel. It can be assumed (Goodwin and Trainor, 1985) that the contribution of the acceleration term is negligible, so that the left-hand side of (5) is zero.

(B) Calcium Kinetics

The Goodwin and Trainor calcium equation describes the simplest aspect of calcium kinetics. These authors assume a reaction with a stoichiometry n between calcium ions and a macromolecule C:

$$C + n\mathrm{Ca}^{2+} \underset{k_{-1}}{\overset{k_1}{\rightleftharpoons}} C^*$$

where C^* represents the complex of n calcium ions bound to the macromolecule C.

Assuming that the total concentrations of the binding macromolecules and of calcium are constant, a straightforward derivation leads to the kinetic equation.

$$\frac{\mathrm{d}\chi}{\mathrm{d}t} = k_{-1}(K-\chi) - k_1(\beta+\chi)\chi^n \tag{6}$$

where χ represents the concentration of free calcium, and β, K are constants (NB: $\beta + K = nC$, the total concentration of calcium-binding macromolecules). In order to take account of stretching or compression effects on calcium release or calcium binding, the rate constant k_{-1} is assumed to be a function of strain. Expanding this function to first order gives

$$k_{-1} = a + a_{ij}\frac{\partial \xi_i}{\partial \chi_j} \tag{6.1}$$

Finally, the Goodwin and Trainor equation for calcium kinetics is:

$$\frac{\partial \chi}{\partial t} = \left(a + a_{ij}\frac{\partial \xi_i}{\partial \chi_j}\right)(K - \chi) - k_1(\beta + \chi)\chi^n + D\nabla^2\chi \tag{7}$$

where a term for the diffusion of calcium, with diffusion coefficient D, has been added.

References

Brière, C. and Goodwin, B. C. (1988). Geometry and dynamics of tip morphogenesis in *Acetabularia*. *J. Theoret. Biol.*, 131, 461–475.

Cassirer, E. (1950). *The Problem of Knowledge.* New Haven: Yale University Press.

Cotton, G. and Vanden Driessche, T. (1987). Identification of calmodulin in *Acetabularia:* its distribution and physiological significance. *J. Cell Science*, 87, 337–347.

Darwin, C. (1859). *The Origin of Species, 1st edition,* Penguin: Harmandsworth, UK.

Dawkins, R. (1986). *The Blind Watchmaker.* Longmans, UK.

Goodwin, B. C. (1985). What are the causes of morphogenesis? *BioEssays,* 3, 32–35.

Goodwin, B. C. (1989). Morphogenesis: Gene Action Within the Context of Cytoplasmic Order. In *Cytoplasmic Organization in Development,* G. M. Malacinski, ed., in press.

Goodwin, B. C. and Brière, C. (1987). The concept of the morphogenetic field in plants. In *Le Développement des Végétaux,* H. Le Guyader, ed., Paris: Masson, pp. 329–337.

Goodwin, B. C., and Pateromichelakis, S. (1979). The role of electrical fields, ions and the cortex in the morphogenesis of *Acetabularia. Planta,* 145, 427–435.

Goodwin, B. C., Skelton, J. C., and Kirk-Bell, S. M. (1983). Control of regeneration and morphogenesis by divalent cations in *Acetabularia mediterranea. Planta,* 157, 1–7.

Goodwin, B. C. and Trainor, L. E. H. (1985). Tip and whorl morphogenesis in *Acetabularia* by calcium regulated strain fields. *J. Theoret. Biol.,* 117, 79–106.

Green, P. B. (1987). Inheritance of pattern: analysis from phenotype to gene. *Amer. Zool.,* 27, 657–673.

Harrison, L. G. and Hillier, N. A. (1985). Quantitative control of *Acetabularia* morphogenesis by extracellular calcium: a test of kinetic theory. *J. Theoret. Biol.,* 114, 177–192.

Harrison, L. G., Graham, K. T., and Lakowski, B. C. (1988). Calcium localization during *Acetabularia* whorl formation: evidence supporting a two-stage hierarchical mechanism. *Development,* 104, 255–262.

Murray, J. D. and Oster, G. (1984). Generation of biological pattern and form. *IMA J. Math. in Med. and Biol.,* 1, 51–75.

Odell, G., Oster, G. F., Burnside, B. and Alberch, P. (1981). The mechanical basis of morphogenesis. *Devel. Biol.,* 85, 446–462.

Oster, G. F., Shubin, N., Murray, J. D. and Alberch, P. (1988). Evolution and morphogenetic rules: the shape of the vertebrate limb in ontogeny and phylogeny. *Evolution,* 42, 862–884.

Shubin, N. H. and Alberch, P. (1986). A morphogenetic approach to the origin and basic organization of the tetrapod limb. *Evolutionary Biology,* 20, 319–387.

Webster, G. C. and Goodwin, B. C. (1982). The origin of species: a structuralist approach. *J. Soc. Biol. Struct.,* 5, 15–47.

CHAPTER 9

Coping With Complexity: Deterministic Chaos in Human Decisionmaking Behavior

ERIK MOSEKILDE, ERIK LARSEN, AND JOHN STERMAN

Abstract

This chapter describes an experiment with human decisionmaking behavior in simulated microeconomic environments. Participants were asked to operate a simplified production- distribution chain to minimize costs. Performance was systematically suboptimal, however, and in many cases the subjects were unable to secure the stable operation of the system. As a result, large-scale oscillations and various forms of highly nonlinear dynamic phenomena was observed.

A model of the applied ordering policy is proposed. Econometric estimates show that the model is an excellent representation of the actual decisions. With different parameters, computer simulations of the estimated order policy produce a great variety of complex dynamic behaviors. Analyses of the parameter space reveal an extremely complex structure having a fractal boundary between the stable and unstable solutions, and with fingers of periodic solutions penetrating deeply into regions representing quasiperiodic and chaotic solutions. In certain parts of the parameter space, any neighborhood of a given solution contains a qualitatively different solution. Thus, changes on the margin can produce a completely different system behavior.

Our results provide direct experimental evidence that chaos can be produced by the decisionmaking behavior of real people in simple managerial systems. The consequent implications for the ability of human subjects to cope with complex dynamical systems are explored.

1. Introduction

Since the days of Newton the underlying presumption of most scientific activity has been that the behavior of a system can be precisely predicted if the laws of motion are known sufficiently accurately. This mechanistic world view implies a conceptual stance in which empirically observable phenomena follow well-defined trajectories, and in which unpredictability and randomness can be ascribed to incomplete knowledge of the relevant equations of motion or to unforeseeable exogeneous events. Thus, much economic theory has been characterized by attempts to explain human systems deterministically in terms of assumed preferences and rational behavior. Likewise, it has been assumed that the result of a given decision can be precisely evaluated, at least in principle, and that this outcome will vary in a continuous manner with the parameters characterizing the decision.

With the discovery of deterministic chaos [1-3], it has become clear that systems that are controlled by purely deterministic equations of motion can behave in an irregular and completely unpredictable manner. Classical examples are the Lorenz and Rössler models [4, 5], which were originally derived as simplified descriptions of atmospheric turbulence. Another example is Duffing's equation, which was applied by Ueda [6] to study nonlinear electronic circuits. During the past few years, similar unpredictable motion has been observed in a great variety of systems, including enzymatic reactions [7], kidney pressure and flow regulation [8], stimulated heat cells [9], interacting nerve cells [10], and periodically-driven microwave diodes [11]. For living systems, it looks almost as if chaotic behavior is the rule rather than the exception.

These discoveries have naturally prompted speculation that similar dynamics might also occur in economic, managerial, and other human systems. Obviously, actual economic time series are seldom characterized by the regular dynamics typical of linear deterministic systems. Instead, various types of irregularities in the form of large-scale fluctuations over a broad spectrum of different frequencies are observed. The question therefore arises as to what degree these fluctuations are generated internally within the system by some kind of chaotic dynamics, and to what degree they are the result of random exogeneous events. Indeed, by allowing for the influence of environmental constraints on per-capita income in a neoclassical growth model, Day [12] has shown how irregular behavior can emerge from simple macroeconomic interactions. Lorenz [13, 14] has illustrated the occurrence of chaos in a multisector business cycle model, and van der Ploeg [15] has investigated complex dynamics in a financial markets model. Furthermore, Chiarella [16] has demonstrated the emergence of unpredictability in the classical cobweb model of market equilibration, while Rasmussen, et al. [17] have studied chaotic phenomena in a simple model of the economic long wave. Turning to the microeconomic realm, Rasmussen and Mosekilde [18] have shown how allocation of resources between two departments of a corporation can lead to deterministic chaos through a cascade of period-doubling bifurcations, while similar behavior has been found by Hald, et al. [19] in a generic commodity market model.

The lack of predictability for chaotic systems is related to their sensitivity to the initial conditions. For dissipative systems, this sensitivity arises from instabilities that cause small random fluctuations to grow exponentially until they become strong enough to influence the future course of the system. This often requires many decades of amplification. By comparison with most systems considered in the natural sciences, economic systems display exceptional behavior precisely because of the prevalence of instability generating, positive feedback loops. Well-known examples include the accelerator-multiplier loops of ordinary Keynesian business cycle theories. Other loops work through self-enhancement of growth expectations, amplification of capital requirements due to capital/labor substitution, and acceleration of orders in the face of increasing delivery delays. Combined with nonlinear limitations associated, for instance, with constraints on capacity utilization or on labor availability, each of these loops is a candidate for a chaos-producing mechanism. In this connection it is of interest to note that application of conventional regression techniques to estimate the parameters of a linearly-unstable nonlinear system in terms of a linear model is likely to underestimate the gain significantly. The estimation normally produces parameter values suggesting that the linear model is stable. Moreover, the statistics of such an experiment may not provide grounds to reject the assumed linear model structure [20].

While the above analyses provide interesting insights into the behavior of particular models, the significance of the results to economic theory hinges on whether these models represent real-world phenomena. Perhaps the chaotic regimes lie in inaccessible regions of parameter space, or the damping and stabilization mechanisms, which exist in all economic systems, have been misspecified and underestimated. It has turned out to be extremely difficult to resolve these problems by appeal to aggregate economic data. Despite intriguing efforts involving the development of a whole new set of statistical techniques [21, 22], economic time series are usually too short and influenced too much by external events to allow us to identify a strange attractor with any degree of confidence. From their very thorough investigation of US series for the unemployment rate, employment, real GNP, gross private domestic investment, and industrial production, Brock and Sayers [23] thus conclude that convincing evidence of deterministic chaos cannot be produced.

As a complementary approach, we report here the results of an experiment designed to explore the nature of human decisionmaking behavior in simulated corporate environments. This approach has allowed us to demonstrate how deterministic chaos can arise in extremely simple economic structures by means of the decision heuristics applied by real people.

Management students at MIT and experienced managers from major US companies were asked to operate a four-stage production-distribution chain consisting of a factory, a distributor, a wholesaler, and a retailer. By providing a decentralized network of inventories to buffer unanticipated variations in demand, such systems are meant to dampen the required adjustments in production. The objective of the individual participant was to minimize the costs of maintaining inventories, while at the same time avoiding out-of-stock conditions. Performance, however, was systematically suboptimal. By virtue of the built-in delays and nonlinear constraints, many players found that they were unable to achieve the stable operation of the system and, consequently, that large-scale fluctuations developed.

Assuming that the participants applied anchoring and adjustment heuristics to determine their orders [24], we propose a model for the subjects' decision rule. With parameter values that vary from participant to participant, econometric estimates show that the model can satisfactorily reproduce the actual decisions. Computer simulations of the production-distribution system with the estimated order policies show a great variety of highly nonlinear dynamic modes of behavior. More detailed analyses of the parameter space reveal an extremely complex mode distribution with a fractal boundary between stable and unstable solutions, and with fingers of periodic solutions penetrating deeply into regions of quasiperiodic and chaotic solutions.

In certain regions of parameter space, any neighborhood of a given solution appears to contain a qualitatively different solution. Thus, marginal changes of the ordering policy can completely change the behavior of the system. In other regions, chaotic transients are observed. With this type of behavior, the system moves in a highly irregular and unpredictable manner until, after a randomly distributed time, it suddenly latches on to a stable solution.

This evidence showing that deterministic chaos can be produced by actual managerial decisions in simplified corporate systems raises a number of issues for social scientists. Policy interventions often imply changes in the parameters of a decision rule. But how can policy analysis be conducted if changes on the margin can produce qualitatively new modes of behavior? To what degree does the complexity in behavior slow down the discovery of cause-and-effect relationships by the economic agents? How can experience be transferred between circumstances that differ only slightly? Much further development of theory and experiment is required to answer these questions and, hence, to assess the significance of chaos in economic systems.

2. The Beer Game

In order to reach a widespread market, it is customary for breweries (and other industries as well) to utilize a hierarchical distribution system with dealers at several different levels: a distributor receives the beer from the factory and ships it to the main markets; regional wholesalers receive the beer from the distributor and allocate it to local outlets such as liquor stores and bars; these retailers then finally disperse the products for consumption.

To guard itself against irregularities in demand and supply, each level in the distribution system maintains a suitable set of inventories. Besides securing availability of beer at the level of individual customers, the cascaded distribution system is meant to facilitate swift replacements if a dealer runs low in inventory. The chain should also function as a filter to protect the production line from fluctuations in final consumption. Seasonal and other low frequency components of the demand variations should propagate towards the factory in a damped fashion.

Figure 1 shows the basic structure of the simplified distribution system considered in the present experiment. To allow hand simulations to be performed, there is only one inventory at each level. Orders for beer propagate from right to left, and products are shipped from stage to stage in the opposite direction. Developed some 30 years ago at The Sloan School of Management to illustrate Industrial Dynamics [25, 26], the game has been played by thousands of people. Ranging from high school students to chief executive officers and government officials, participants from many different countries have experienced both the fun and the aggravation of trying to control the extremely complicated behavior of this simple system.

The hand simulations are conducted using a board on which the main structural elements of the distribution chain are outlined. Cases of beer are represented by chips that are physically manipulated by the players, with orders written on slips of paper that are sent from sector to sector. The passing of orders and the production and shipment of beer involve time delays. It is assumed that there is a mailing delay of 1 week (1 time period) from one stage to the next, and in the same way it takes 1 week to ship beer between two sectors. The production time is taken to be 3 weeks, and it is assumed that the production capacity of the brewery is unlimited.

Each week customers order beer from the retailer, who ships the requested quantity out of inventory. Customer demand is exogeneous.

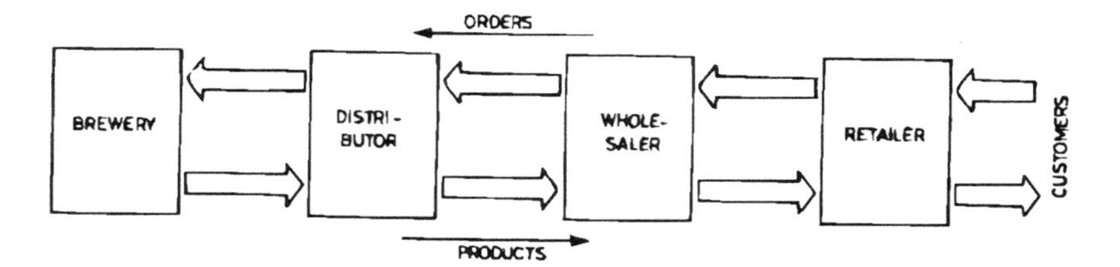

Figure 1. Basic structure of the production-distribution system considered in the Beer Game. Orders for beer propagate from left to right, and brewery products are shipped in the opposite direction. Designed originally as an educational tool, the parameters of the game are chosen to facilitate hand simulation more than to match real world conditions. All by itself, however, the game represents a managerial system.

In all the simulations reported here, customer demand consists of a constant level of 4 cases of beer per week until week 5, at which time it is increased to 8 cases of beer per week. The demand is then maintained at this level for the rest of the game. Each hand simulation may thus be seen as a particular example of the response of the chain to a step increase from 4 to 8 cases per week in customer demand. The game is always initialized with 12 cases of beer in each inventory.

In response to variation in customer demand and to other pressures, the retailer adjusts the order for beer placed with the wholesaler. As long as the inventory is sufficient, the wholesaler ships the beer requested. Orders that cannot be met are kept in backlog until delivery can take place. Similarly, the wholesaler orders and receives beer from the distributor, who in turn orders and receives from the brewery. According to the rules of the game, orders must always be filled if they are covered by the available inventory. Orders that have already been placed cannot be cancelled, and deliveries cannot be returned.

The objective for the participants is to minimize cumulative costs over the length of the game (40 weeks). Because of the costs associated with holding inventory, stocks should be kept as small as possible. On the other hand, failure to deliver on request may force customers to seek alternative suppliers. For this reason, there are also costs associated with having backlogs of unfilled orders. Therefore, each stock manager must attempt to keep the inventory at the lowest possible level while at the same time avoiding out-of-stock conditions. If the inventory begins to fall below the desired level, extra beer must be ordered to rebuild inventory. If stocks begin to accumulate because of a slackening in demand, the order rate must be reduced. In the present case, the inventory holding costs are taken to be \$0.50 per case per week, and the costs of having backlogs are set at \$2.00 per case per week.

For all sectors the decision variable in each round is the amount of beer to be ordered from the immediate supplier. The participants can base their ordering decisions on all information locally available to them, i.e., the current value of their inventory/backlog, previous values of these variables, expected orders, anticipated deliveries, and so forth. In addition, the participants may utilize their overall conception of the way in which the distribution chain functions. To assist the players in acquiring the necessary information, in each round they are required to plot the value of the orders placed and their current inventory/backlog on graph paper. This requirement allows us to analyze the course of the game and afterward to estimate the applied decision rules.

Figure 2(a) shows a typical result of a hand simulation. Here, we have plotted the variation in the effective inventory for the different sectors. The effective inventory is inventory minus backlog. A negative effective inventory represents a backlog of unfilled orders. Of course, the simulations develop differently in detail from trial to trial. Figure 2(b) shows the result of a hand simulation in which the participants have been particularly successful in avoiding large backlogs. Qualitatively, however, the results exhibit a number of significant regularities, suggesting that the players apply relatively consistent heuristics in determining their orders.

The hand simulations are characterized by large-scale oscillations that grow in amplitude from retailer to wholesaler and from wholesaler to distributor. Thus, by the time the original stepwise increase in customer orders reaches the factory, it typically leads to an expansion of production by a factor of more than 6. A second characteristic feature is the increase in orders, which propagates in a wavelike fashion down the chain, depleting the inventories one by one until it is finally reflected at the factory when the large surplus of orders placed during the out-of-stock period is eventually produced. These features clearly indicate the existence of an amplification mechanism in the system. At the same time, the behavior is restricted by various nonlinearities associated, for instance, with the nonnegativity of orders and shipments. Together with the relatively high number of state variables, this allows for an extraordinary variety of complex dynamic behaviors.

The amplification phenomenon observed in the beer distribution chain is connected with the built-in delays. Assume that a particular sector suddenly experiences a significant increase in demand. To find out if the change in demand is of a more permanent character, the

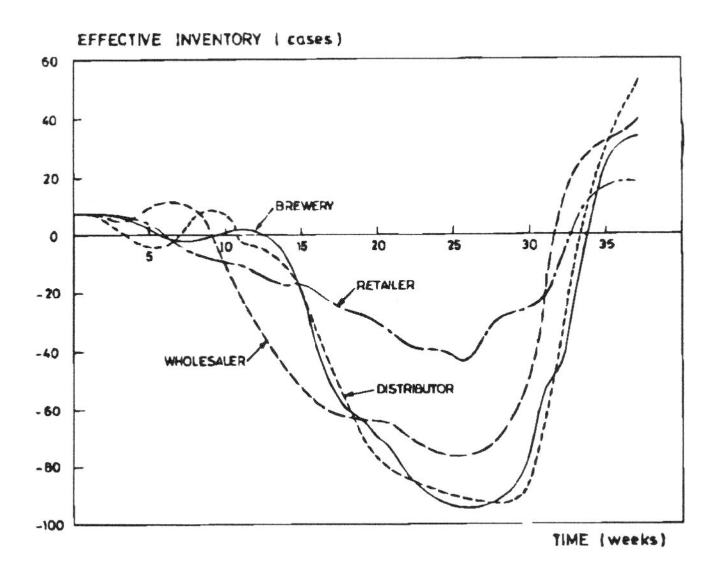

Figure 2(a). Typical experimental results obtained in a hand simulation of the Beer Game. These results represent the response of the distribution system to a stepwise increase in customer orders from 4 to 8 cases/week. The curves show the variation in the effective inventory for each of the four sectors.

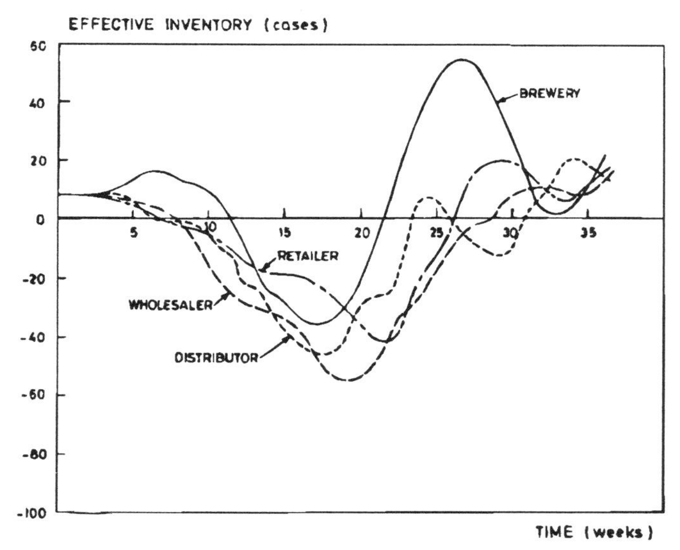

Figure 2(b). Results of a hand simulation in which the subjects have been particularly successful in avoiding large backlogs. Note the significant amplification which occurs in the system.

player usually hesitates a little before adjusting his own orders by a similar amount, since the very purpose of an inventory is to absorb high frequency components in the demand fluctuations. However, because of this hesitation, and by virtue of the built-in mailing and shipping delays, demand will exceed inventory replacements for several weeks. So during this period the inventory will decrease. To bring the inventory back to its desired level, the player must therefore increase the orders placed with his supplier beyond incoming orders. As the player realizes that the increase in demand is of a more lasting character, he usually increases his orders even further with an eye toward expanding the inventory.

The amplification phenomenon is a natural consequence of the structure of the chain in much the same way as the structure of the economy amplifies small fluctuations in the demand for consumer goods into significant variations in the demand for capital goods via the well-known accelerator mechanism. On the other hand, it is important to realize that the beer distribution chain can be operated in a stable manner. In fact, our experience indicates that many players are capable of doing so, and that only about 25% of the participants produce ordering policies leading to deterministic chaos. Large-scale oscillations are always observed in the transient behavior, however, and in all cases the ordering decisions are highly suboptimal, producing costs exceeding the minimal possible costs in the game by a factor of more than 5.

3. The Stock Management Problem

The decision task in the beer game is an example of a more general stock management problem in which managers seek to maintain a quantity at a target level or at least within an acceptable range. For the individual firm, managers must order parts and raw materials to maintain inventories sufficient for production to continue without unnecessary interruptions. Other managers must hire and lay off personnel to maintain an adequate workforce, and they must order and discard capital units to adjust the production capacity in accordance with the needs of the company. At the macroeconomic level, national banks seek to manage the stock of money to provide sufficient credit for economic growth without fueling inflation.

A stock cannot be controlled directly; acquisitions must be ordered from the supplier, and the time lag involved in receiving the ordered products is a potential variable. In the beer game, for instance, the retailer will receive the beer requested after 2–3 weeks only if the wholesaler carries a sufficient inventory. If the wholesaler has run out of stock, the retailer must wait until additional cases have been supplied from the distributor. If the distributor is also out of stock, the retailer must await shipments from the brewery to the distributor before the distributor can deliver to the wholesaler who, in turn, can then deliver to the retailer.

The stock management problem may be illustrated in a generic form by the flow diagram in Figure 3. Here, the upper part portrays the stock and flow structure associated with ordering, acquisition, and shipment of stock units, while the lower part represents the decision rule applied by the manager. In addition to the inventory, the supply line of orders is a relevant stock variable. In the same way that the inventory accumulates differences between acquisitions and shipments, the supply line of orders accumulates differences between ordering and acquisition. Thus, the supply line represents orders that have been placed, but for which products have not yet been received. Shipments may depend on various endogeneous and exogeneous variables. Usually they are also controlled by the stock itself. The acquisition rate is determined by the supply line and by AL, the average acquisition lag. In general, AL will depend on the supply line itself as well as on a number of other variables.

For the beer distribution chain, acquisition and shipment follow automatically from the rules of the game. The interesting part of the problem focuses on the ordering policy. Consistent with behavioral decision theory [24, 27] and the theory of bounded rationality [28, 29], we propose an ordering policy that utilizes information available locally to the decisionmaker only. Although the participants may have a general conception of the structure of the chain, it is not assumed that they have a global knowledge of the overall state of the system.

Our generic decision rule recognizes three motives for ordering: provision for expected demand, adjustment of inventory, and adjustment of supply line. The first motive is straightforward. In equilibrium, when inventory and supply line are both at their desired levels, the stock manager must continue to order enough to replace ongoing shipments. Failure to replace losses would cause the inventory to fall, reducing the ability of the sector to cope with future demands. Instead of determining replacement orders from expected demand, actual shipments could form the basis for the orders. This would create an additional destabilizing feedback loop. When the inventory of a given sector has been depleted, the shipments vanish even though orders continue to arrive. Cutting orders under these conditions would delay the acquisition of adequate inventory, preventing a recovery of the sector.

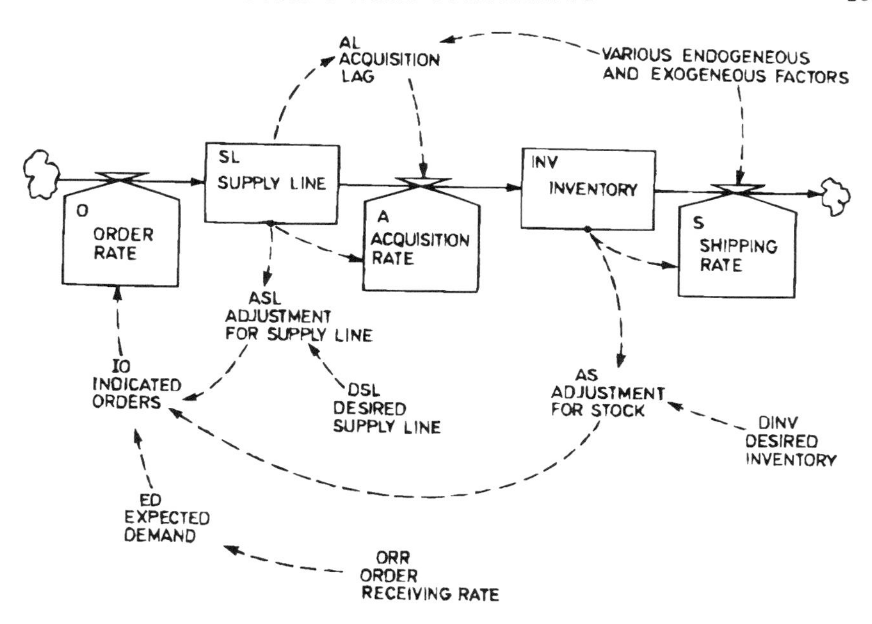

Figure 3. Flow-diagram illustrating the generic stock management problem envisaged by each of the participants in the hand simulation experiment. The proposed decision rule recognizes three motives for ordering: provision for expected demand, adjustment of inventory, and adjustment of supply line.

Conversely, when the sector does acquire enough inventory it will begin to ship the current customer orders plus the backlog. Shipments will therefore exceed customer orders, boosting the ordering rate above equilibrium and causing excess inventory accumulation.

Neither replacement of shipments nor provision for expected demand are sufficient to capture the applied ordering policy. Errors in forecasting demand and irregularities in the supply of beer can cause the inventory to wander away from its desired level. Faced with the increasing costs of such a development, the stock manager must adjust the orders above or below replacements so as to bring the inventory back into line. This will introduce a negative feedback loop regulating the stock. Thus, a principal policy question is how aggressively this policy should be carried out, i.e., how fast should inventory discrepancies be corrected?

Adding stock adjustments to replacements is still not sufficient as a general ordering policy. In fact, with such a policy managers would place orders to rectify a lack of inventory, immediately forget that this beer has been ordered, and then reorder it in the next round. The problem with such a policy is that it neglects orders that have already been placed but for which products have not yet been received. This can generate enormous oscillations in the system as the excess beer requested week after week during the out-of-stock period is finally produced and delivered to the various inventories. Experience shows that many participants take notice of their supply line and try to maintain it at a relatively constant level. Therefore, a second important policy question is what weight supply line adjustments should be given relative to stock adjustments in the ordering policy?

The above description is in accordance with anchoring and adjustment heuristics. Anchoring and adjustment is a common judgmental strategy in which an unknown quantity is estimated by first recalling a known reference point (the anchor) and then adjusting for the effects of other factors that may be less salient or whose influences are obscure. This heuristic has been shown to apply to a wide variety of decision-making tasks [30, 31]. In the present case the unknown quantity is the ordering rate, the anchor is the expected demand ED, and adjustments are made in response to discrepancies between desired and actual stock and between the desired and actual supply line.

The following equations formalize the proposed heuristics: Firstly, orders must be non-negative, i.e.,

$$O_t = \max(0, IO_t), \tag{1}$$

where IO_t denotes the indicated order rate. The subscript t indicates that the value of the variable is considered at time t.

The expected demand may be formulated in various ways. Common assumptions in economics and management science include static expectations, regressive expectations, adaptive expectations, and extrapolative expectations. Adaptive expectations are widely assumed in simulation studies of corporate systems. They are often a good model of the development of expectations in the aggregate, and they represent one of the simplest formulations flexible enough to be usable under nonstationary conditions. Postulating adaptive expectations in the beer game, we express the expected demand as the participant's forecast of incoming orders:

$$ED_t = \theta \cdot ORR_{t-1} + (1 - \theta) \cdot ED_{t-1}. \tag{2}$$

ED_t and ED_{t-1} are here the expected demand at time t and time $t-1$, respectively, ORR is the order receiving rate, and $\theta (0 \leq \theta \leq 1)$ is a parameter controlling the rate at which expectations are updated.

Stock adjustments are assumed to be linear in the discrepancy between the desired and the actual inventory:

$$AS_t = \alpha_S(DINV - INV_t), \tag{3}$$

where the stock adjustment parameter α_S is the fraction of the discrepancy ordered in each round. Because the participants lack the time and information to determine the optimal inventory level, the desired inventory $DINV$ is taken to be constant throughout the course of a simulation, although $DINV$ may vary from participant to participant. Similarly, supply line adjustments are expressed as:

$$ASL_t = \alpha_{SL}(DSL - SL_t), \tag{4}$$

where DSL is the desired supply line, and α_{SL} is the fractional adjustment rate.

Defining $\beta = \alpha_{SL}/\alpha_S$ and $Q = DINV + \beta \cdot DSL$, the generic expression for the indicated order rate becomes

$$IO_t = ED_t + \alpha_S(Q - INV_t - \beta \cdot SL_t). \tag{5}$$

Since $DINV, DSL$, and β are all non-negative, $Q \geq 0$. Further, it is unlikely that participants place more emphasis on the supply line than on the inventory itself. The supply line does not directly affect the costs, nor is it as salient as the inventory. Therefore, $\alpha_{SL} \leq \alpha_S$ and $\beta \leq 1$. β may be interpreted as the fraction of the supply line taken into account by the participants. If $\beta = 1$, the subjects fully recognize the supply line and do not double order. If $\beta = 0$, then the orders placed are forgotten until the beer arrives.

4. *Experimental Results*

The results reported here were drawn from 48 games, involving a total of 192 participants. The games were performed over a period of 4 years, following precisely the same protocol in each trial. Participants were undergraduate and graduate students at The Sloan School of Management at MIT and senior executives from major US firms. Inventories and orders were continuously recorded during the game, and a computer model was used to test the records for consistency. Trials in which there were accounting errors of more than a few cases per week for more than a few weeks were discarded from further consideration, leaving 11 trials (44 participants) for analysis. This subset generated slightly lower costs than the full sample, indicating that these participants understood and performed best in the experiment.

The results show the behavior of the subjects to be far from optimal. The average team costs were more than 10 times greater than the minimum costs produced by simulation using the proposed decision rule for orders. Even though the beer distribution system is extremely simple by comparison with real managerial systems, the system is apparently complex enough to prevent the participants from discovering a policy that is even close to being optimal. For a system as simple as the beer distribution chain, learning may improve this situation. It is likely, however, that the complexity of real managerial systems is so much higher, and the conditions under which decisions must be made so variable, that learning by experience is insufficient in many cases for optimal policies to be attained.

For each of the participants we have estimated the parameters θ, α_S, β, and Q of the proposed ordering policy. Nonlinear least squares estimation was applied, subject to the conditions that $0 \leq \theta,\ \alpha_S,\ \beta \leq 1$, and $Q \geq 0$. A large majority of the estimated parameters are significant. The mean R^2 is 0.71, and only for 6 out of 44 subjects is $R^2 \leq 0.50$. The proposed ordering heuristic therefore seems to be a good representation of actual decisions.

Distributions of several of the obtained parameter values are shown in Figure 4. Averaged over the 44 participants, $\theta = 0.36$ with a standard deviation of 0.35. On the average, then, the subjects update their demand expectations over a period of about 3 weeks. However, a significant fraction of the participants have relatively low θ values, indicating that either they apply almost static demand expectations or that the replacement motive is weak in their ordering policy.

The average value of α_S is found to be 0.26 with a standard deviation of 0.18. This implies that the participants typically attempt to rectify about one-quarter of their stock discrepancy in each round. A significant maximum in the distribution is found in the range $0.3 \leq \alpha_S \leq 0.4$. Our experience shows that this value is too high to ensure stable operation of the system. By directing the players' attention towards inventory discrepancies, the cost function may indirectly cause the costs to increase.

The estimated value of β averages at 0.34 with a standard deviation of 0.31. Thus, supply line adjustments are weighted with a typical factor of $\frac{1}{3}$ relative to inventory adjustments. However, a considerable number of participants give very little attention to their supply line. This is a major cause of poor performance in the game. For Q, the

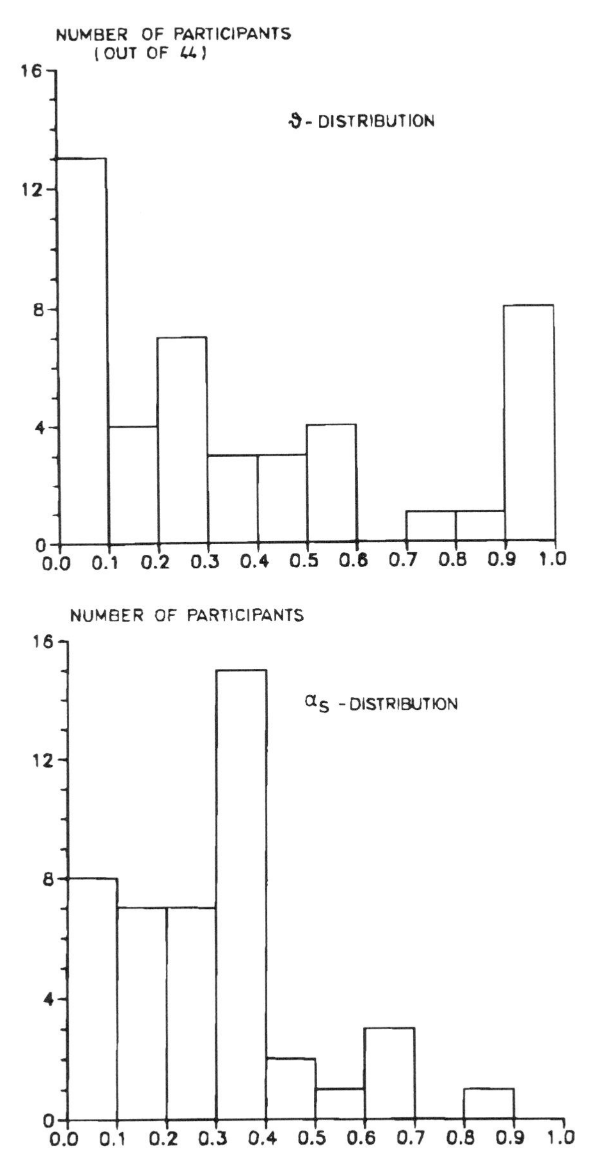

Figure 4. Distribution of estimated order parameters for 44 participants. θ measures the rate at which expectations are updated, α_S measures the fractional rate at which inventory discrepancies are removed, β measures the weight asigned to supply line adjustments relative to inventory adjustments, and Q is the sum of the desired inventory and β times the desired supply line. Low values of α_S combined with high values for β and Q produce stable behavior.

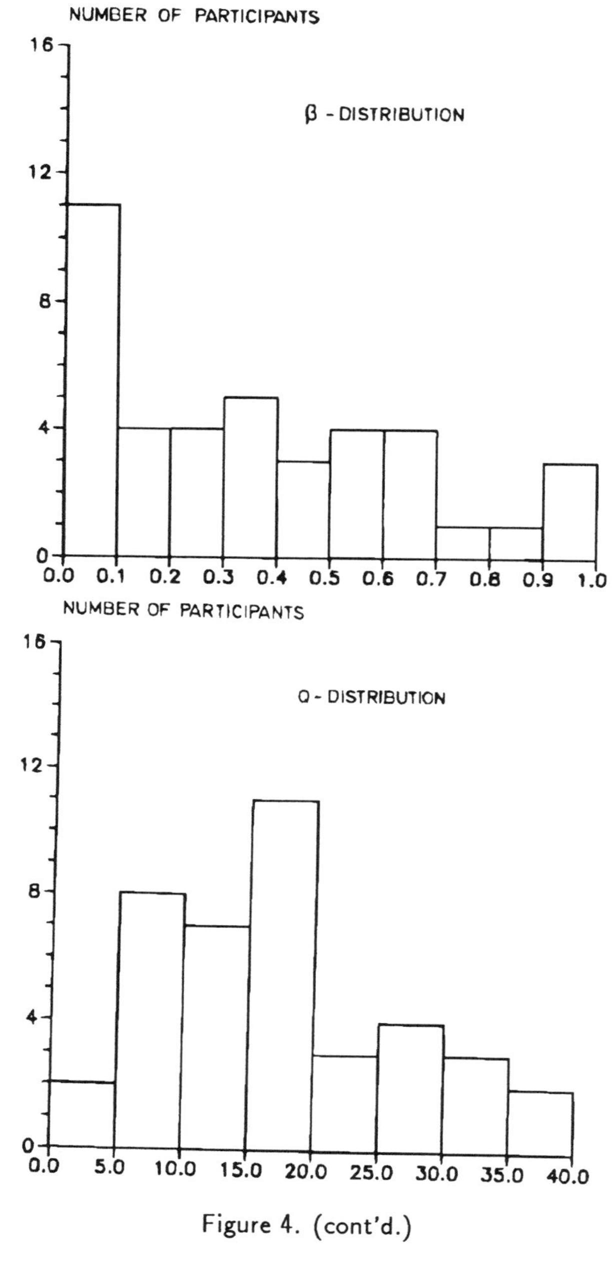

Figure 4. (cont'd.)

estimated value is 17 with a standard deviation of 9. Q is the parameter whose distribution is closest to normal. Presumably, this indicates that the players have maintained a relatively clear appreciation of their desired inventory.

Next, the beer game has been simulated on a computer using the estimated ordering policies. Each hand simulation involves four subjects, each of whom can apply his own policy. It appears that there is little correlation between the participants' position in the chain (retailer, wholesaler, distributor or factory) and the parameters of the applied policy. To simplify the analysis, the computer simulations have been performed with the same parameter values in the ordering policy for all four sectors.

Out of the 44 parameter sets, 31 produce stable behavior. Three produce limit cycles of period-1, period-4, and period-12, respectively, and 10 (23%) yield chaotic or quasiperiodic behavior. Among those simulations producing stable behavior, the majority have very long transients, and in several cases chaotic transients are observed. With this type of behavior, the system moves in an unpredictable and highly irregular manner until, after a randomly distributed time, it suddenly locks on to a stable solution. Chaotic transients to periodic solutions are also observed.

Figure 5 shows a typical example of a simulation of the beer game. In this case $\theta = 0.00$, $\alpha_S = 0.30$, $\beta = 0.15$, and $Q = 17$. Figure 5(a) shows the variation of the effective inventory for each of the four sectors over the first 60 weeks, while Figure 5(b) shows the corresponding variation in order rates. The computer simulation reproduces the amplification and phase shifts characteristic of the hand simulations.

Figure 6 illustrates the results of extending the computer simulation to a time horizon of 1000 weeks while maintaining the same values of the policy parameters. The model behavior is clearly aperiodic: the system never repeats itself, but continues to explore new regions in phase space. Calculations show that the largest Lyapunov exponent is positive. Therefore, the system is sensitive to the initial conditions and the behavior is chaotic [32].

Figure 7 shows a similar simulation for $\theta = 0.05$, $\alpha_S = 0.50$, $\beta = 0.00$, and $Q = 15$. In this case the behavior is hyperchaotic: it has now lost completely all traces of regularity, and the three largest Lyapunov exponents are positive. Hyperchaos has previously been studied by Rössler [33]. Similar behavior is obtained for parameter sets such as $\theta = 0.00$, $\alpha_S = 0.40$, $\beta = 0.10$, and $Q = 7$, or $\theta = 0.55$, $\alpha_S = 0.65$, $\beta = 0.00$, and $Q = 9$. The chaotic and hyperchaotic decision rules are characterized by relatively low values for β and Q and relatively high values for α_S, i.e., this is an aggressive stock adjustment policy, which operates with a low desired inventory, tending to neglect supply line adjustments.

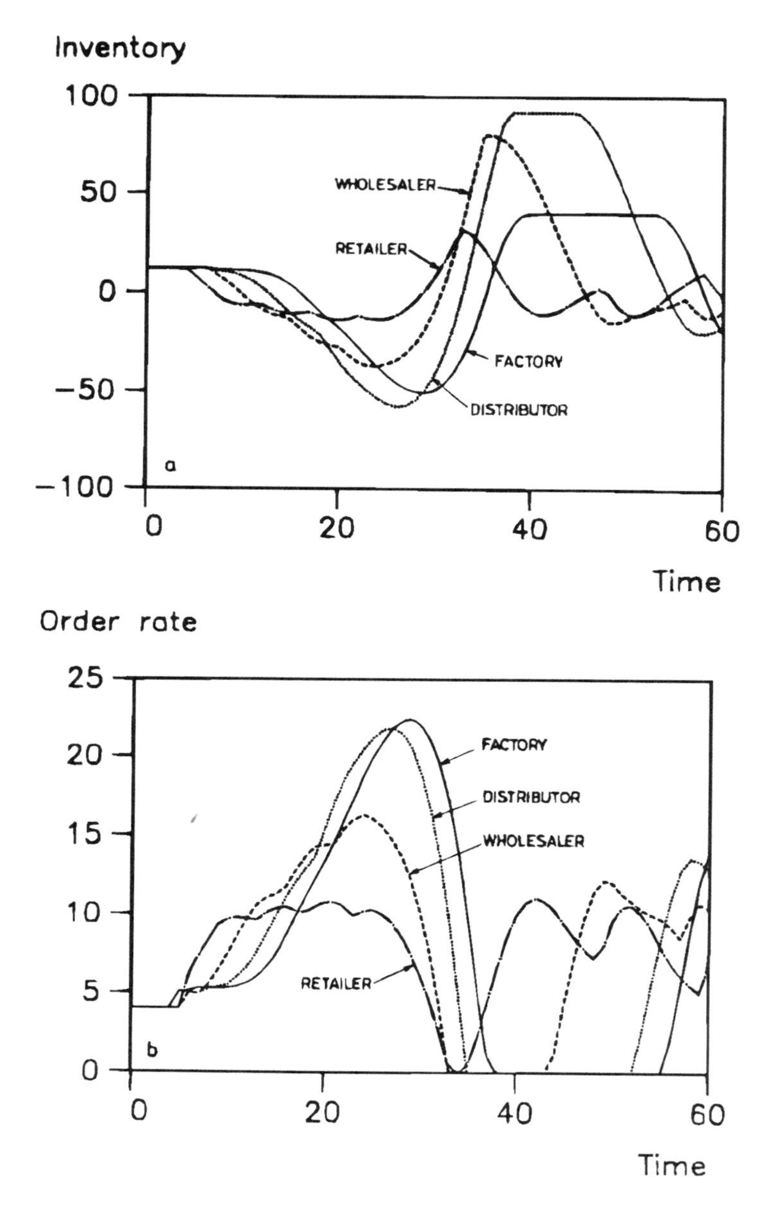

Figure 5. Computer simulation of the variation in (a) the effective inventory and (b) the corresponding order rate for each of the four sectors in the beer distribution chain. The calculations were performed for $\theta = 0.00, \alpha_S = 0.30, \beta = 0.15$, and $Q = 17$. The numerical simulation reproduces the amplification and phase shifts characteristic of the hand simulations.

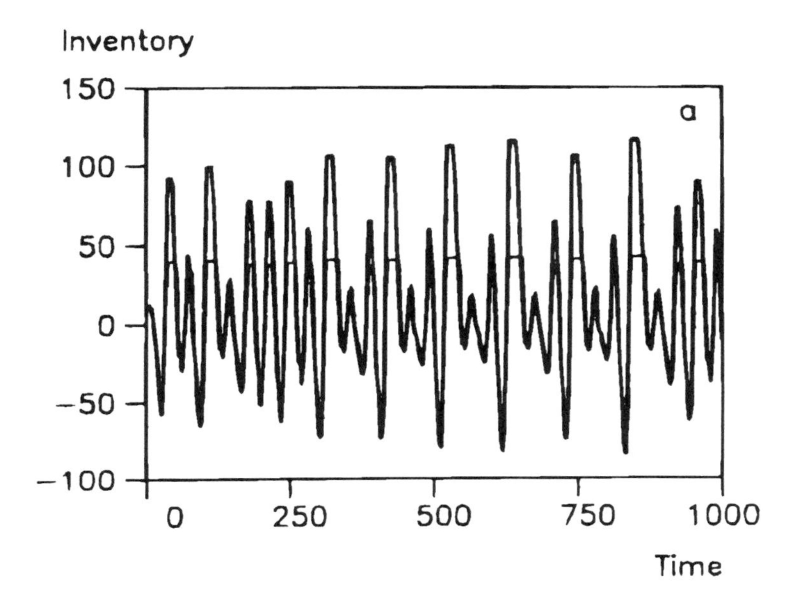

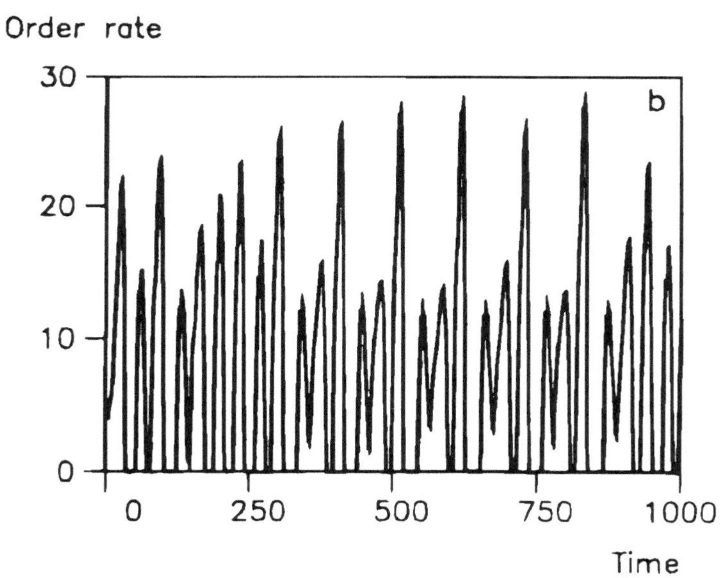

Figure 6. Extended simulation of the beer distribution chain with the same parameters as in Figure 5. The model behavior is clearly aperiodic: the system never repeats itself, but continues to find new ways in phase space. From the positive value of the largest Lyapunov exponent, this type of behavior is characterized as chaotic.

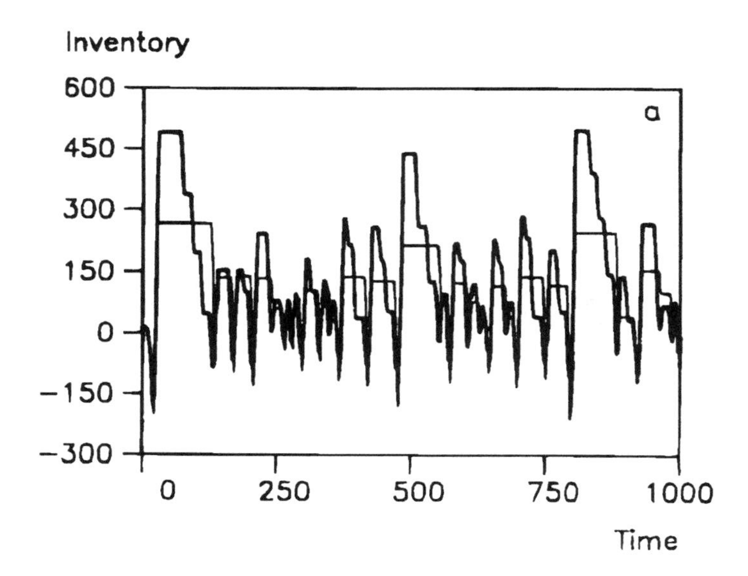

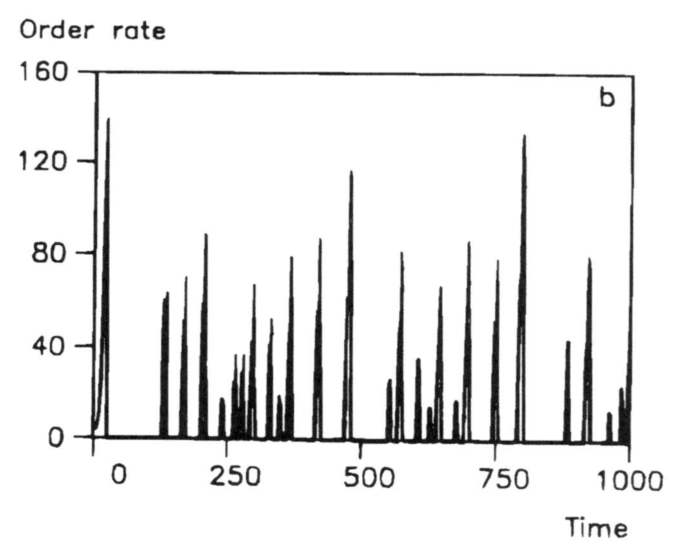

Figure 7. Variation of the effective inventory (a) and the corresponding order rate (b) for $\theta = 0.05, \alpha_S = 0.50, \beta = 0.00$, and $Q = 15$. Thin line for brewery, and heavy line for distributor. In this case, the variation is hyperchaotic, i.e., more than one Lyapunov exponent is positive.

Figure 8 shows an example of a periodic trajectory obtained for $\theta = 0.25$, $\alpha_S = 0.30$, $\beta = 0.65$, and $Q = 12$. Here the transient is a regular damped approach to the limit cycle attractor.

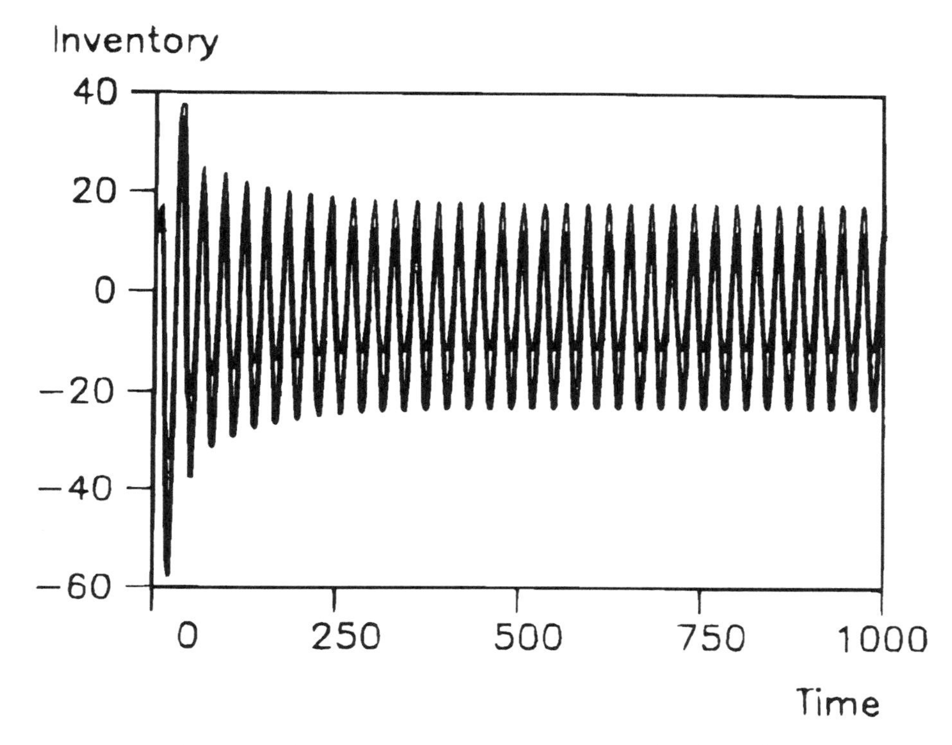

Figure 8. Example of a periodic trajectory obtained for $\theta = 0.25$, $\alpha_S = 0.30, \beta = 0.65$, and $Q = 12$. The thin line is for the brewery, while the heavy line is for the distributor. The transient is a regular damped approach to the limit cycle attractor.

By contrast, Figure 9 shows an example of a chaotic approach to a period-4 oscillation. In this simulation $\theta = 0.25, \alpha_S = 0.425$, $\beta = 0.09$, and $Q = 17$. To clearly illustrate the periodic nature of the stationary solution obtained with these parameter values, Figure 9(b) shows a phase plot of this behavior obtained by plotting simultaneous values of the distributor and factory inventories over many oscillation periods. The sharp corners on the phase-space projection reflect the discrete-time approach used for the model. To obtain a periodic trajectory, the internally generated oscillation must adjust its period to be commensurate with the time period of the model (1 week). This entrainment process gives rise to characteristic Arnol'd tongues of periodic solutions penetrating into the region of quasiperiodic and chaotic solutions. This will be illustrated in more detail in the next section.

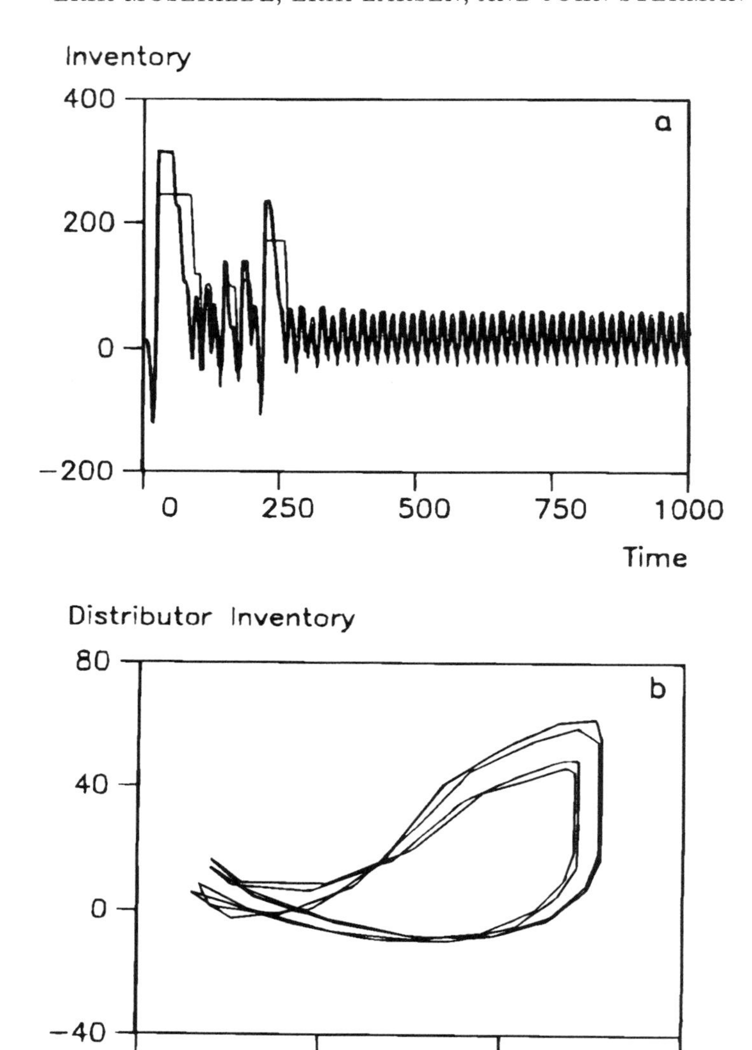

Figure 9. Example of a chaotic approach to a period-4 limit cycle. In this simulation $\theta = 0.25, \alpha_S = 0.425, \beta = 0.09$, and $Q = 17$. Part (b) shows a phase plot of the stationary period-4 solution. This figure was obtained by plotting simultaneous values of the distributor and factory inventories over a large number of oscillations. The sharp corners reflect the discrete-time approach applied in the model.

5. Mode-Distribution in Policy Space

The beer model contains 27 state variables. Compared with real managerial systems, the model is a vast simplification; compared with most systems investigated in nonlinear dynamics, however, the model is very complex. In certain regions of parameter space the distribution system has three positive Lyapunov exponents. Therefore one would expect the beer model to exhibit an unusually complicated variety of behaviors. Figure 10 shows the distribution of modes in the (α_S, β)-policy plane for $\theta = 0.25$ and $Q = 17$. Here, the stationary solutions of 200×200 simulations have been indicated by means of a grey tone code: light grey indicates stable behavior, dark grey indicates aperiodic behavior, and black indicates periodic behavior. We have not distinguished between quasiperiodic and chaotic behavior.

Inspection of Figure 10 shows that the policy plane contains several regions of unstable behavior, separated by fjords of stable behavior. Thus, in regions around $\beta = 0.50$ and $\beta = 0.70$ the model is stable for all values of α_S. While the narrow peninsula around $\beta = 0.72$ contains only small-amplitude periodic (and quasiperiodic) solutions, the other regions of unstable behavior are dominated by large-amplitude fluctuations. The occurrence of unstable behavior is most clearly seen in the lower-right corner, where α_S is large and β relatively small. Therefore, to stabilize the distribution chain it is necessary to employ an ordering policy in which both inventory discrepancies are adjusted relatively slowly and a significant fraction of the supply line is taken into consideration. It will shortly become clear, however, that β should not be taken to be too large. A large value of β increases the costs because the system then stabilizes in a state where the inventories are negative.

A particularly interesting feature of the mode distribution in Figure 10 is the fractal character of part of the boundary between policies leading to stable and unstable behaviors. To see this phenomenon a little more clearly, we have magnified the region $0.35 \leq \alpha_S \leq 0.45$, $0.02 \leq \beta \leq 0.12$ by a factor of 100 to produce the more detailed mode distribution depicted in Figure 11 (see the color plate in the center of the book). Again, a set of 200×200 simulations are involved. By inspection of Figure 11, we observe how fingers of stable behavior penetrate deeply into the region of unstable behavior. Crossing these fingers are fingers of periodic solutions for which the internally-generated oscillation locks on to the discrete-time period of the model. These latter fingers are often referred to as Arnol'd tongues [9, 11]. In certain re-

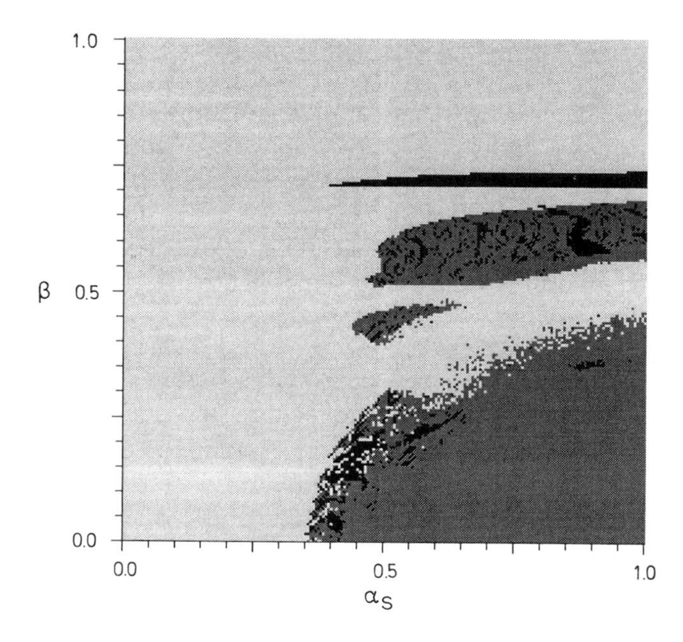

Figure 10. Distribution of modes in the (α_S, β) policy plane for $\theta = 0.25$ and $Q = 17$. The stationary solutions of 200×200 simulations have been indicated by means of a grey tone code: light grey indicates stable behavior, dark grey indicates aperiodic behavior, and black indicates periodic behavior.

gions of the policy plane, a stable solution is surrounded by aperiodic solutions on all sides. Here a small change in one of the policy parameters produces a qualitatively different solution. Further magnification reveals additional structure in the mode distribution. By contrast with the fractal character of part of the boundary between stable and unstable solutions, other parts of this boundary are quite regular. This is particularly true around the peninsula of small amplitude periodic solutions.

Figure 12 shows the variation in costs corresponding to the mode distribution in Figure 10. Here, costs are defined as the average weekly costs over the period from week 8500 to week 9000. The costs are indicated by a grey tone code with 7 intervals so that light grey represents solutions with costs of less than \$10/week, and black represents solutions with costs in excess of \$500/week. The details of the code are indicated in the figure. It follows approximately a logarithmic scale.

The variation in costs directly reflects the mode distribution with stable solutions producing significantly lower costs than unstable solutions. As expected, the costs are highest in the lower right corner of the

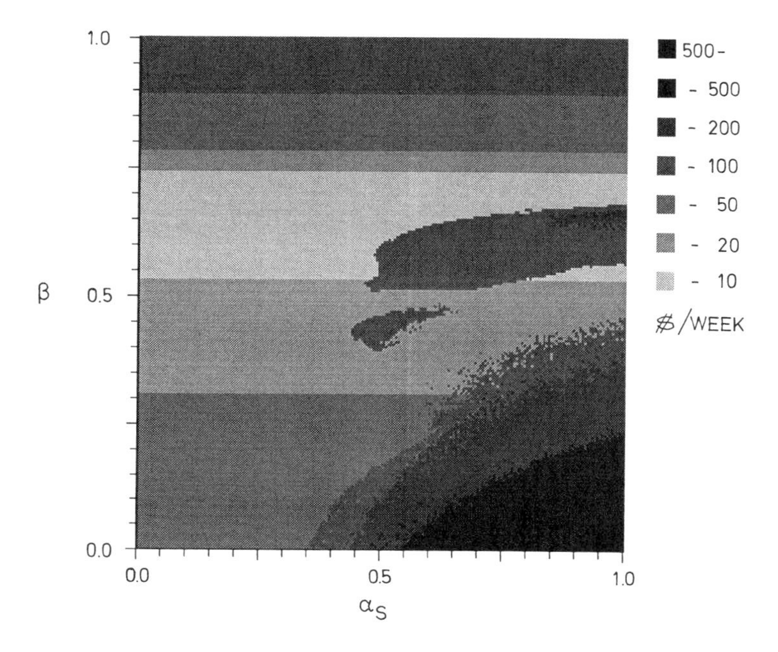

Figure 12. Variation in costs corresponding to the mode distribution in Figure 10. Stable solutions produce significantly lower costs than unstable solutions. Minimal costs are found for $\beta = 0.70$, where steady state operation is possible with costs of \$4/week. In the unstable regime close to the lower right corner, the costs exceed \$500/week.

(α_S, β) plane where large amplitude fluctuations occur. The minimal costs are approximately \$4/week. They are found in the stable regime around $\beta = 0.70$ where steady state operation with a few cases of beer in each inventory is possible. The peninsula of low-amplitude periodic solutions around $\beta = 0.72$ is not visible in the cost variation. This shows clearly that the amplitudes of these oscillations are very small.

It is interesting to note that the costs are independent of α_S for the stable solutions. This implies that the stationary state attained by the chain after all transients have died out is independent of the rate at which inventory discrepancies are eliminated. The parameter α_S is a significant determinant of the transient approach to this state, however, and α_S also influences the behavior of the unstable solutions. In the special case $\alpha_S = 0$, there is no check on either stocks or supply lines. Under these conditions the inventories will wander freely away from their desired values, and the costs can become very large. This is not visible in the figure because $\alpha_S = 0$ coinsides with one of the axes.

As previously noted, the minimal costs are attained for $\beta \approx 0.70$. As β is varied away from this value, the steady-state balance between the inventory and supply line of each sector shifts. This occurs because the ordering policy allows a low stock to be compensated for by a high supply line, and vice versa. If β is increased above 0.70, for instance, the stationary inventories become negative. The pressure to increase orders to bring the inventories back to their desired level is compensated for by a pressure to reduce orders to lower the supply lines. On the other hand, if β is reduced below 0.70 the stationary inventories become positive. Thus, for the stable solutions the variation of the costs with β reflects our original cost function. It is not likely, however, that a participant could determine the optimal value of β beforehand.

Figures 13 and 14 show the mode distribution in the (α_S, β) plane for $Q = 15$ and $Q = 12$, respectively. As previously defined, Q denotes the sum of the desired inventory and β times the desired supply line, with β being the fraction of the supply line accounted for in the ordering policy. As Q is reduced from 17 (in Figure 10), the regions of unstable behavior increase and start to merge. For $Q = 15$ (Figure 13) there is still a scattered band of stable solutions penetrating all the way

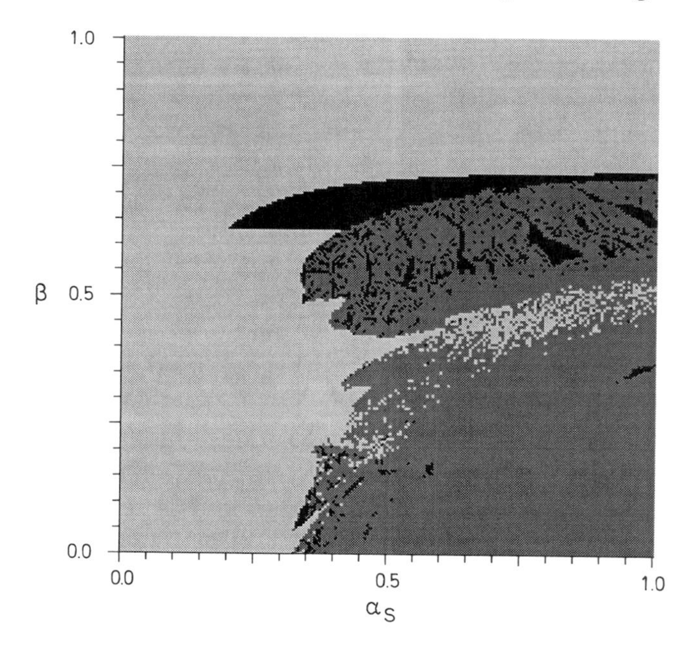

Figure 13. Mode distribution in the (α_S, β) plane for $Q = 15$. Compared to the mode distribution in Figure 10, the unstable regions have expanded.

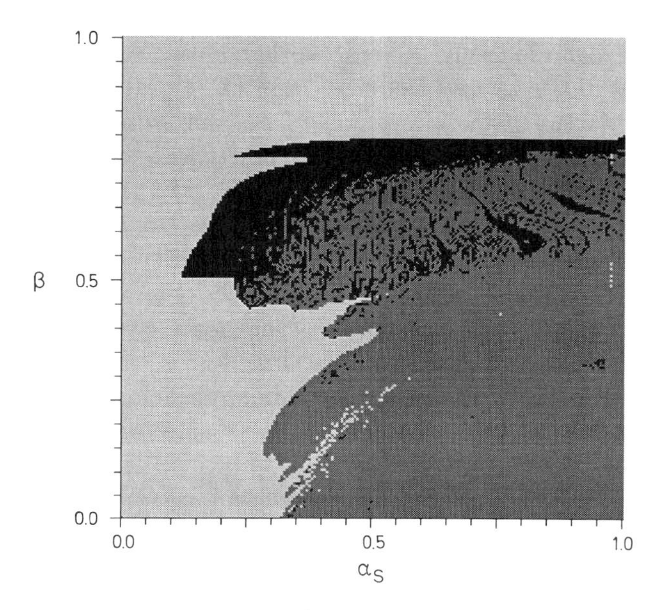

Figure 14. Mode distribution in the (α_S, β) plane for $Q = 12$. The unstable regions have now completely merged and cover more than half of the policy plane.

through the unstable regime. For $Q = 12$, however, the regions of unstable behavior have completely merged and now cover more than half of the policy plane. Therefore, to stabilize the distribution chain, one should operate with a relatively high desired inventory.

6. Discussion

The discovery of deterministic chaos and other highly nonlinear dynamic phenomena in physical and biological systems naturally motivates the search for similar behavior in human systems. Yet in that search we are faced with difficulties of the sort that do not plague the natural scientist, at least not to the same degree. Social systems are not easily isolated from the environment. The large temporal and spatial scales of most of these systems, the vast number of individual actors, considerations of cost, and ethical concerns often make controlled experiments virtually impossible.

The social scientist is left with three main options in the search for understanding the relevance of nonlinear dynamics in the domain of human systems. Formal models may be constructed and analyzed. Previous work in this direction has demonstrated the existence of chaotic modes in an array of different social and economic models. However, the relevance of these models to real-world problems may be questioned. Also, analysis of the models frequently indicates that the chaotic regime lies far outside the plausible region of parameter space. The second approach is to apply newly-developed statistical methods to analyze economic time series. However, aggregate data sufficient for strong empirical tests do not exist for most social systems. In addition, a variety of problems arise with respect to stationarity and external noise.

The third approach is to develop laboratory experiments with simulated social systems [34, 35]. Such experiments create microworlds in which subjects face institutional structures and reward systems similar to those of the real world. Adopting this approach, we have presented students of management and experienced managers with an apparently straightforward stock adjustment task. The applied ordering policies have been estimated in terms of decision heuristics supported by a variety of studies in behavioral science, and the estimated policies have been simulated numerically to determine the corresponding system behavior. In this work, 10 out of 44 parameter sets were found to produce chaotic and quasiperiodic behavior. In all cases, the policies were found to be highly suboptimal.

These findings are significant in several respects. Our demonstration that chaos can be produced by the decisionmaking behavior of real people strengthens the argument for the prevalence of complex dynamic phenomena in human systems. Thus, social scientists can ignore nonlinear dynamics only at their peril. Models of economic and social systems should portray the processes by which disequilibrium conditions are created and dissipated, and not assume that the economy is always in or near equilibrium—or that a stable equilibrium even exists! Models should be formulated so that they are robust in extreme situations. It is the nonlinearities required to ensure such robustness that control instabilities and complex dynamic phenomena.

This having been said, a number of questions concerning the consequences of chaos in social systems become pertinent. Real systems are constantly bombarded by external influences of many different types. It is obvious that such random blows severely reduce the predictability of these systems. But does the existence of stochastic shocks swamp the uncertainty caused by chaos? Much further investigation is required to answer this question for different systems. Nonetheless, there is a dramatic conceptual change in realizing that even without shocks, the development of a social system controlled entirely by its own locally rational logic can be completely unpredictable, and that the slightest change in policy can produce a different type of behavior.

Another interesting aspect of our investigation is the observation that the applied policies spread over most of the available policy space, and that none of the policies are close to being optimal. Consistent with both intuition and with formal analysis, low desired inventories, aggressive attempts to correct discrepancies between desired and actual stock, and inappropriate account of supply lines predispose the distribution chain to violent oscillations. Nontheless, a good many of the participants apply high α_S values, low Q values, and far too low β values in their ordering policies. This appears to be a consequence of misperceiving the delay structure of the chain, combined with an overemphasis on stock control in a narrow sense. This misadjusted emphasis may be created by the cost function defined for the game. Considering the simplicity of the beer distribution chain, it seems reasonable to suggest that similar misperceptions occur in the real world.

Acknowledgment

Jannet Sturis is acknowledged for her comments on a preliminary version of this paper. We would also like to thank Ellen Buchhave for her assistance in preparing the manuscript. The study was supported by The Social Science Research Council of Denmark.

References

1. Cvitanovic, P. *Universality in Chaos,* Adam Hilger Ltd., Bristol, UK (1984).

2. Berg, P., Y. Pomeau, and C. Vidal. *Order Within Chaos.* John Wiley and Sons, New York (1984).

3. Thompson, J. M. T. and H. B. Stewart. *Nonlinear Dynamics and Chaos,* John Wiley and Sons, New York (1986).

4. Lorenz, E. N. "Deterministic Nonperiodic Flow," *Journal of Atmospheric Science,* 20 (1963), 130–141.

5. Rössler, O. E. "An Equation for Continuous Chaos," *Physics Letters,* 57A (1976), 397–398.

6. Ueda, Y. "Randomly Transitional Phenomena in the System Governed by Duffing's Equation," *Journal of Statistical Physics,* 20 (1979), 181–196.

7. Olsen, L. F. and H. Degn. "Chaos in Enzyme Reaction," *Nature,* 267 (1977), 177–178.

8. Jensen, K. S., E. Mosekilde, and N.-H. Holstein-Rathlow: "Self-Sustained Oscillations and Chaotic Behavior in Kidney Pressure Regulation," in *Laws of Nature and Human Conduct,* I. Prigogine and M. Sanglier, eds. Brussels (1986), pp. 91–109.

9. Guevara, M. R., L. Glass, and A. Shrier: "Phase Locking, Period-Doubling Bifurcations, and Irregular Dynamics in Periodically Stimulated Cardiac Cells," *Science,* 214 (1981), 1350–1353.

10. Colding-Jørgensen, M. "A Model for the Firing Pattern of a Paced Nerve Cell," *Journal of Theoretical Biology,* 101 (1983), 541–568.

11. Feldberg, R., C. Knudsen, M. Hindsholm, and E. Mosekilde. "Non-Linear Dynamic Phenomena in Electron Transfer Devices," in *Computer-Based Management of Complex Systems,* P. M. Milling and E. O. K. Zahn, eds., Springer-Verlag, Berlin (1989).

12. Day, R. H. "Irregular Growth Cycles," *American Economic Review,* 72 (1982), 406–414.

13. Lorenz, H.-W. "Strange Attractors in a Multisector Business Cycle Model," *Journal of Economic Behavior and Organization,* 8 (1987), 397–411.

14. Lorenz, H.-W. "International Trade and the Possible Occurrence of Chaos," *Economic Letters,* 23 (1987), 135–138.

15. van der Ploeg, F. "Rational Expectations, Risk and Chaos in Financial Markets," *Economic Journal,* 96 (1985), 151–162.

16. Chiarella, C. "The Elements of Nonlinear Theory of Economic Dynamics," Ph.D. Thesis, University of New South Wales (1986).

17. Rasmussen, S., E. Mosekilde, and J. D. Sterman. "Bifurcations and Chaotic Behavior in a Simple Model of the Economic Long Wave," *System Dynamics Review,* 1 (1985), 92–110.

18. Rasmussen, D. R. and E. Mosekilde. "Bifurcations and Chaos in a Generic Management Model," *European Journal of Operational Research,* 35 (1988), 80–88.

19. Hald, B. G., C. N. Laugesen, C. Nielsen, E. Mosekilde, E. R. Larsen, and J. Engelbrecht. "Rössler Bands in Economic and Biological Systems," in *Computer-Based Management of Complex Systems,* P. M. Milling and E. O. K. Zahn, eds. Springer-Verlag, Berlin (1989).

20. Blatt, J. M. "Dynamic Economic Systems—A Post-Keynesian Approach," preprint, Armonk, NY (1983).

21. Brock, W. A. "Distinguishing Random and Deterministic Systems," *Journal of Economic Theory,* 40 (1986), 168–195.

22. Scheinkman, J. A. and B. LeBaron. "Nonlinear Dynamics and Stock Returns," Preprint, Department of Economics, University of Chicago (1986).

23. Brock, W. A. and C. L. Sayers. "Is the Business Cycle Characterized by Deterministic Chaos?," *Journal of Monetary Economics*, 22 (1988), 71–90.

24. Hogarth, R. *Judgement and Choice*, 2nd edition, John Wiley and Sons, New York (1987).

25. Forrester, J. W. *Industrial Dynamics*, MIT Press, Cambridge, MA (1961).

26. Jarmain, W. E. *Problems in Industrial Dynamics*, MIT Press, Cambridge, MA (1963).

27. Tversky, A. and D. Kahneman. "Judgement under Uncertainty: Heuristics and Biases," *Science*, 185 (1974), 1124–1131.

28. Simon, H. A. "Rational Decision Making in Business Organizations," *American Economic Review*, 69 (1979), 493–513.

29. Cyert, R. and J. March. *A Behavioral Theory of the Firm*, Prentice-Hall, Englewood Cliffs, NJ (1963).

30. Einhorn, H. J. and R. M. Hogarth. "Ambiguity and Uncertainty in Probabilistic Inference," *Psychological Review*, 92 (1985), 433–461.

31. Davis, H. L., S. J. Hoch, and E. K. Easton Ragsdale. "An Anchoring and Adjustment Model of Spousal Predictions," *Journal of Consumer Research*, 13 (1986), 25–37.

32. Wolf, A. "Quantifying Chaos with Lyapunov Exponents" in *Chaos*, A. V. Holden, ed., Manchester University Press, Manchester, UK (1986).

33. Rössler, O. E. "An Equation for Hyperchaos," *Physics Letters*, A 71 (1979), 155–157.

34. Smith, V. "Microeconomic Systems as an Experimental Science," *American Economic Review*, 72 (1982), 923–955.

35. Sterman, J. D. "Testing Behavioral Simulation Models by Direct Experiment," *Management Science*, 33 (1987), 1572–1592.

CHAPTER 10

Causality, Chaos, Explanation and Prediction in Economics and Finance

WILLIAM A. BROCK

Abstract

In this chapter we survey some new methods of time-series analysis that are good at detecting low-dimensional chaos as well as other nonlinearities that are useful for short-term prediction but might be missed by many more conventional statistical methods. The methods are extended to handle the problem of detecting nonlinear correlation between two different series above and beyond linear relationships. After development of these methods in Section Two, they are applied in Section Three. Section Three also discusses statistical issues raised by evaluating superior predictive performance of stock market technicians. Section Four discusses very speculative pathways by which fluctuations in economic activity arise.

1. Introduction

The chapter is organized as follows. Section One contains a brief introduction, while Section Two develops the analytical meat. The nonanalytical reader may prefer to skim this section and go on to Section Three, where we explain much of Section Two in plain English. Section Three also takes up the topic of designing tests for hidden structure in a time series where it matters where the series came from. What we are getting at is that many statistical methods could apply equally well to a series of insect population data as to a series of stock market returns. Finding predictable structure in a stock market series tends to require use of more delicate methods tailor made to this particular goal. We discuss how one might use well-known technical trading rules to build specification tests for the adequacy of popular parametric models used to fit stock market returns series. Section Four discusses some speculative sources of economic fluctuations, ranging from high to low frequency. The theme of all sections is the development and exposition of new methods for helping in the search for potential predictability in economic and financial time-series data. The last section consists of a short summary and conclusions.

2. *Searching for Nonlinear Predictive Ability in Economics*

Here we discuss some recent methods to detect nonlinear relationships in economic data. The methods are inspired by central limit theory in statistics, hypothesis testing theory in statistics, and measures of local spatial correlation used in chaos theory. Before going into the details, let's briefly touch upon the general issues of prediction and explanation in economics.

In economics, the notions of prediction and explanation are slippery ones. For example, "explanation" would seem to entail an understanding of what "causes" what. Does monetary tightness cause recessions? Does systematic expansion of the money supply cause inflation?

As a prime illustration of how tricky the idea of explanation is in economics, we have Milton Friedman's essay on positive economics and model building in economics, along with the controversy that it unleashed (*NP*-II, p. 422. Hereafter we will use *"NP"* to denote The New Palgrave). Friedman put forth the argument that the merit of a theory was in not the realism of its assumptions but in the quality of its predictions and in its simplicity.

To put it another way, he argued that it was silly to waste time arguing over whether people actually solve those complicated mathematical problems that pure theorists claim they act as if they do. Friedman argued that the optimization hypothesis is a rich generator of theories with tight, falsifiable predictions that can be tested on observables. Furthermore, the optimization hypothesis is a more reliable guide than rival theories such as Herbert Simon's bounded rationality (*NP*-I, p. 266) in the development of extensions of basic theory to explain anomalies in current theory. Friedman would argue that this is one source of value added, e.g., optimizing economic-theoretic models over "mechanical" models in income distribution theory. For an illustration of a "Friedmanian" (Gary Becker) debating a non-Friedmanian (Arthur Goldberger) in income distribution theory, see Arthur Goldberger's criticism of Gary Becker's work and Becker's rebuttal (Becker versus Goldberger (1989)).

Friedman's famous essay raises a related technical issue: Is it more productive to make strong, simplifying (although patently unrealistic) assumptions about the preferences, resources, and technology of the optimizing economic agents in the model in order to gain tractibility and derive all econometrics within that optimizing model context? Or, alternately, is it more productive to take a data-based approach to econometric inference and drop the insistence on explicitly modelling the optimizing agents?

The model-based route leads one to the rather onerous tasks of conducting estimation and inference exercises within a model structure, the unwanted byproduct being that the resulting theory of statistical inference is highly model bound, i.e., is highly sensitive to what some might think are economically irrelevant changes in model structure. This is especially the case when the model structure itself has been designed on the basis of tractibility considerations, e.g., single "stand-in" agents with recursive, additively separable preferences.

For the optimizing model-based approach to economic time series, see Thomas Sargent's (1981) article "Interpreting Economic Time Series." The Becker/Goldberger debate mentioned above is a good example of a dispute between the two points of views. These issues are unsettled in economics and the debate is likely to go on for a long time. One way to narrow the discussion is to focus on notions of prediction, causation, and explanation directed toward the goal of forecasting. One such notion is Weiner/Granger causality. We turn now to an exposition of this popular concept of causality in econometrics.

Weiner/Granger Causality

Following *The New Palgrave*-I (p. 380), the basic idea of Weiner/Granger causality in time-series analysis is the following: The time series $\{b(t)\}$ *causes* $\{a(t)\}$ if past b's help predict current and future a's, above and beyond the predictive content of past a's for current and future a's.

Note that this may have little to do with "philosophical" causality. One type of example to keep in mind arises in rational expectations economic models, where anticipated permanent increases in fiat money supply in the future can lead to an increase in the price level today. An econometrician using the Weiner/Granger concept of causality would conclude from data generated by such a model that rising price levels today "caused" the money supply increase in the future, and not vice versa. But in the context of the model just the opposite is true, and is what is in agreement with common sense and the quantity theory of money, a fact that has been well-established for long time scales.

This kind of spurious causality is very common in rational expectations models that generate dynamical equations that are forward looking rather than looking backward, as is the case with differential equation models in the natural sciences. This fact gives rise to a major difference between economic models and those of the natural sciences.

Economic models are populated by entities who know the dynamic laws of evolution of the model, act on that knowledge to optimize their own goals, and by so doing actually (re)create the laws of time evolution of the model. This self-consistent, forward-looking character of the dynamic laws of the model (e.g., a set of ordinary differential equations describing the evolution of prices and quantities), differentiates economic models from models used in the natural science (with the possible exception of those employed in biology). See the book by Brock and Malliaris (1989) for a development of this theme in more detail. The Santa Fe Institute Volume (Anderson et al. (1988)) is also well worth reading, not only for the ideas it contains, but also for the reactions of natural scientists when confronting this self-referential aspect of economic modelling. The implications of this self-referentiality, such as the "irrelevance" propositions on monetary policy, social security, and the national debt (Sargent (1987b)), are difficult for many natural scientists to grasp. Finding the frictions that neutralize these "irrelevance" propositions are part of the practical economist's art, and is all part of what makes economic forecasting so difficult. Now let's continue our development of causality tests.

We shall propose a new nonlinear causality test in this chapter. But to do this will require a digression into statistical methods that have been developed to test for chaos and other forms of nonlinearity in time series. The concept of average local spatial correlation plays a key role in these tests.

DEFINITION. *The "correlation integral" (Grassberger and Procaccia (1983)) generated by a time series of T numerical observations $\{x(t),\ t = 1, 2, \ldots, T\}$ is defined as follows:*

$$C(m, e, T) = \#\{(t, s) : ||x(t, m) - x(s, m)|| < e\}/T^2 \tag{1}$$

Here $x(t, m) = (x(t), \ldots, x(t-m+1))$, $x(s, m) = (x(s), \ldots, x(s-m+1))$, $||y|| = \max\{|y(i)|, i = 1, 2, \ldots, m\}$ for any m-dimensional vector y, $\#\{\cdot\}$ denotes the "number of elements," and T^2 denotes T squared.

When viewed as a function of e, the correlation integral $C(m, e, T)$ acts much like a cumulative distribution function. It measures the fraction of pairs of data strings embedded in m-dimensional space that differ in distance by an amount less than e. As e increases, the fraction of such pairs increases to a maximum of unity. By counting the number of additional pairs captured as e increases, one obtains a measure of spatial correlation. For example, the percentage rate of increase of new pairs captured by a one percent increase in e gives a measure of the "dimension," or number of active modes, at e. We shall see this in detail below. Hence, the quantity C can be used to generate measures of spatial correlation, dimension, and incremental structure beyond pure randomness. At the risk of repetition, we shall use the term "purely random" interchangeably with "independently and identically distributed (IID)." This is because no past information, including time, helps forecast any property of an IID stochastic process. This captures what we want "purely random" to mean—at least for the purposes of this discussion.

Brock and Dechert (1988) have found two basic properties of the correlation integral that are needed here. First, under modest regularity conditions, $C(m, e, T)$ has a limit $C(m, e)$ as $T \to \infty$. This limit is given by

$$\lim_{T\to\infty} C(m, e, ,T) = C(m, e) = \Pr\{||x(t, m) - x(s, m)|| < e\}, \quad (2)$$

where $x(t, m), x(s, m)$ are independent draws from the joint cumulative distribution function $\Pr\{X(r, m) < x(r, m)\}$, with $X(r, m) = (X(r), X(r+1), \ldots, X(r+m-1))$. Here X denotes a random variable and x a draw from the distribution of X.

One can see this result intuitively by writing $C(m, e, T)$ as a double average of indicator functions of events $\{||x(t, m) - x(s, m)|| < e\}$, averaging out (i.e., taking T to infinity) the m arguments of $x(s, m)$, and then averaging out the arguments of $x(t, m)$. Thus, we use the ergodic theorem for single averages twice. This is all made rigorous in Brock and Dechert (1988).

Secondly, if $\{x(t)\}$ is IID, then we have (letting log denote the natural logarithm):

$$\log[C(m, e)] = m \log[C(1, e)] \quad (3)$$

Hence, for purely random processes, letting $d(e, m)$ denote the elasticity of $C(e, m)$ with respect to e, i.e.,

$$d(e, m) = C'(e, m)e/C(e, m) = d[\log C]/d[\log e], \quad (4)$$

we have $d(e, m) = m$ for all embeddings of dimension m, where "$d[x]$" denotes the differential of x, and $C'(e, m)$ is the derivative of C with respect to e.

A Statistical Test for IID

The formula (3) leads naturally to a statistical test for randomness. That criterion is that $C(m, e) = [C(1, e)]^m$. In particular, when $\{x(t)\}$ is purely random, Brock, Dechert, and Scheinkman (1986) (hereafter, "BDS") have shown that

$$B(m, e, T) = \sqrt{T}\,[C(m, e, T) - C(1, e, T)^m], \tag{5}$$

converges in distribution to $N(0, V)$ as $T \to \infty$. Here $\sqrt{T}$, and $N(0, V)$ denote the square root of T and the normal distribution with mean 0 and variance V, respectively. Furthermore, the variance V can be consistently estimated from the data. Call the estimate $V(m, e, T)$. Hence, under the hypothesis that $\{x(t)\}$ is purely random, the statistic

$$W(m, e, T) = B(m, e, T)/\sqrt{V(m, e, T)}, \tag{6}$$

converges in distribution to $N(0, 1)$ as $T \to \infty$.

So we now have a randomness test that is easy to use: If you have a series that you want to test for randomness, just calculate W. If W is large in absolute value, you can reject the hypothesis of randomness at a significance level read off from a standardized normal table.

This randomness test is especially easy to use since W. D. Dechert (1987) has written menu-driven, user-friendly software for the IBM PC to calculate the statistic W, while D. Hsieh and B. LeBaron (1988a,b) (cf. Hsieh (1989)) have done a detailed performance evaluation of this BDS test.

The test easily detects many examples of low-dimensional, deterministic chaos. But there are examples of deterministic chaos that the BDS test cannot detect (the 10th iterate of the tent map will do). But we suspect that many other randomness tests will also fail to detect these alternatives to IID. See Hsieh and LeBaron (1988a, b) for the details. Now let's turn to some intuitive ideas about why dependence testing based upon measures of average local spatial correlation might be expected to work well.

The Intuition Behind Using Average Local Spatial Correlation Measures

The first problem concerns whether dependent series exist that fail to be detected by (6).

DEFINITION. *Let $\{x(t)\}$ be a strictly stationary, ergodic, scalar-valued stochastic process. Then we say that $\{x\}$ is C-IID if for all $e > 0$ and all positive integers m, we have*

$$C(e, m) = [C(e, 1)]^m$$

We pointed out before that IID implies C-IID. Unfortunately, however, Dechert (1988) has found examples showing that the converse does not hold.

Define the class $\mathcal{A}$ of strictly stationary, ergodic, stochastic processes that are not C-IID, i.e., $\mathcal{A}$ is the set of stationary, ergodic stochastic processes for which there is an $e > 0$ and an $m > 0$ such that $C(e, m)$ is not equal to $[C(e, 1)]^m$ which, in turn, equals the value of $C(e, m)$ when $\{x(t)\}$ is IID. It is easy to see that for processes of class $\mathcal{A}, W = C(e, m) - [C(e, 1)]^m$ rejects IID with probability 1 for some e, m. The class $\mathcal{A}$ can be given a natural interpretation. It is the set of processes for which there are expected gains from a near-neighbor forecasting scheme relative to IID. Consider the following identity, which holds as the sample length tends to infinity:

$$C(e, m+1) = \Pr\{|x(t+m) - x(s+m)| < e \text{ given } ||x_t^m - x_s^m|| < e\} C(e, m), \quad (7)$$

where $x_t^r = (x(r), \ldots, x(r+m-1))$, $r = s, t$.

By induction on m, we see that a sufficient condition for $C(e, m)$ to not equal $[C(e, 1)]^m$ is that for some $r < m+1$,

$$\Pr\{|x(t+r) - x(s+r)| < e \text{ given } ||x_t^r - x_s^r|| < e\} \neq \Pr\{|x(t+r) - x(s+r)| < e\} \quad (8)$$

Equation (8) captures the notion of the r-past helping to predict the future by using neighbors with "nearness e." Let's expand upon this point, making use of work by Savit and Green (1989).

Put $z(t) = (x(t), x(t-1))$. Consider the probability $\Pr\{|z(t) - z(s)| < e\}$, where we compute this quantity by drawing $z(t), z(s)$ at random from the distribution function of z. Set up the null hypothesis

$$H_0(1): \quad \Pr\{|x(t) - x(s)| < e \text{ given } |x(t-1) - x(s-1)| < e\} = \Pr\{|x(t) - x(s)| < e\}$$

This is equivalent to

$$H_0(2): \quad \Pr\{|z(t) - z(s)| < e\} = \Pr\{|x(t) - x(s)| < e\} \times \Pr\{|x(t-1) - x(s-1)| < e\}$$

From a sample of length T, as in (1), (2) above, estimate each "Pr" by the corresponding correlation integral, take the difference, multiply by the square root of T, and take the limit to get the distribution under the null of $\{x\}$ IID.

Near Neighbors Can Help Predict the Future

Intuitively, the foregoing development shows that our test would be expected to be very powerful against dependent alternatives for which a near neighbor $x(t-1)$ to $x(s-1)$ helps "predict" that $x(t)$ is close to $x(s)$. But this is exactly the case for a one-dimensional chaotic system like the tent map.

When the length of the history is extended from "1" to "L," one improves the ability of the test to detect higher-dimensional chaos and other nonlinearities. In multivariate generalizations, when the "width" of the history is extended by considering more series to help "predict" the future of one series using near neighbors, one improves the ability of the test to detect nonlinear relationships among the variables. Our test tends to work best when something akin to nearest neighbors is an effective prediction algorithm (Yakowitz, 1987, p. 235). We can put our heuristic argument another way.

Interpret the left-hand side (LHS) of $H_0(1)$ as the expected revenue from a prediction exercise in which you "predict" $x(t) = x(s)$ when $x(s-1)$ is near $x(t-1)$. You get a reward of 1 unit if the "prediction" $x(s)$ is within e of the actual value $x(t)$, and you get nothing otherwise. The conditional expected return is then just the LHS of $H_0(1)$. The frequency of opportunities to make such predictions is $\Pr\{|x(t-1) - x(s-1)| < e\}$. The unconditional expected prediction return (NPR), compared to that when $x(t)$ and $x(t-1)$ are independent, is just

$$NPR = [LHS(H_0(1)) - RHS(H_0(1))] \times \Pr\{|x(t-1) - x(s-1)| < e\}$$

The BDS-type tests just amount to replacing Pr's by their V-statistic estimates (correlation integrals) in NPR and multiplying by the square root of T. We believe the good performance of this type of test for dependence may be due to the fact that it uses V-statistics (closely related to U-statistics that are known to have good properties as estimators (cf. Serfling (1980)), and that it explores the whole space to check for opportunities to earn "net prediction rewards" relative to independence. We are going to build on this idea to develop a "Granger-like" nonlinear causality test below. But before we do this, let's look at some examples to motivate our approach.

Examples

Let r be a real number in the open interval $(-1, 1)$. Look at the process $\{a(s)\}$ generated as follows:

$$a(t) = ra(t-1) + b(t), \quad a(0) = a_0, \tag{9}$$

where $\{b(t)\}$ is any zero linearly-autocorrelated process and a_0 is an initial condition. Zero linear autocorrelation for a series $\{b(s)\}$ having mean zero means that $E[b(t)b(t-L)] = 0$ for all integers L. For example, one could put $b(t) = x(t) - \frac{1}{2}$, $x(t) = T(x(t-1))$, where $T(\cdot)$ is the tent map: $T(x) = 2x$ for x in $[0, \frac{1}{2}]$, $T(x) = 2 - 2x$ for x in $[\frac{1}{2}, 1]$. By construction, Eq. (9) looks like a well-specified linear autoregression of order 1, called an $AR(1)$, with $E[a(t-1)b(t)] = 0$. Here, since the driving "uncertainty" is actually deterministic as in a pseudorandom number generator, the "expectation" $E[a(t-1)b(t)]$ is calculated from a time series of length n by the time average

$$E^{\text{est}}[a(t-1)b(t)] = \frac{1}{n}\sum_{s=1}^{n} a(s-1)b(s), \tag{10}$$

where E^{est} denotes the estimator $\frac{1}{n}\Sigma$.

We are in a position to make our basic point. If $\{b\}$ is a zero-mean series that is zero linearly-autocorrelated and $\{a\}$ is generated as in (9), then linearly-based tests of incremental forecasting power for future a's of past b's, in addition to that forecasting power provided by past a's, will fail to detect the short-term predictability in $\{b\}$. A similar point can be made for any "innovation" process $\{b\}$ that is zero linearly-autocorrelated but displays nonlinear dependence. Here are three examples of model classes. The first class is called a nonlinear moving average process, the second is a nonlinear autoregressive process, while the third is termed an AutoRegressive Conditionally Heteroscedastic (ARCH) process.

To illustrate, let $\{n(s)\}$ be an IID zero-mean gaussian process and define

$$b_1(t) = n(t) + cn(t-1)n(t-2) \tag{11}$$
$$b_2(t) = n(t) + cn(t-1)b_2(t-2) \tag{12}$$
$$b_3(t) = h(t)n(t), \quad h(t) = ch(t-1) + [b_3(t-1)]^2 \tag{13}$$

where c is a nonzero constant. All three of these processes have zero covariances at all leads and lags L and all three can be used as "innovations" for (9) above. But linearly-based incremental predictive power tests will fail to alert the analyst to the fact that there is extra prediction potential to be had.

Another related kind of example that drives home this basic point is

$$x(t) = cy(t-L)x(t-M) + e(t), \tag{14}$$

where $\{e\}, \{y\}$ are mutually independent IID gaussian processes with mean zero and unit variance. Here all autocorrelations and cross correlations are zero. Therefore, past y's fail to help linearly predict future x's, yet past y's do help to nonlinearly predict future x's.

We turn now to using the correlation integral for the development of a nonlinear Granger causality test capable of detecting nonlinear predictive potential in the above examples.

Nonlinear Granger Causality

The maximum norm will be used in what follows, i.e. $|x| = \max\{|x(i)|, i = 1, 2, \ldots, n\}$ for all $x = (x(1), \ldots, x(n)) \in R^n$. We are finally ready to define the notion of "causality" to be used in this chapter. In order to focus on the main ideas, we consider only the case for two series. Let $\{x\}$, $\{y\}$ be two scalar-valued, stationary, ergodic time series. In order to obtain central limit theorems for various statistics, we will also require that temporal dependence die off "fast enough" as in Denker and Keller (1983).

DEFINITION. *Given lags L, M, the series $\{y\}$ fails to Granger cause the series $\{x\}$ if*

$$Pr\{|x(t) - x(s)| < e \text{ given } |X(t,L) - X(s,L)| < e, |Y(t,M) - Y(s,M)| < e\} = Pr\{|x(t) - x(s)| < e \text{ given } |X(t,L) - X(s,L)| < e\}, \tag{15}$$

where $X(t,L) = (x(t-1),\dots,x(t-L)), Y(t,M) = (y(t-1),\dots, y(t-M))$, *and "Pr" denotes probability. We will say that* $\{y\}$ *helps to predict* $\{x\}$ *if "equals" is replaced by "not equals" in (15).*

Obviously, the definitions involve e, m. Note that different lengths of past histories for each of the variables can be inserted into the definition. Also other lengths of future segments starting at $t+m$ and $s+m$ can be inserted instead of the $m=0$ length as above. Now let's motivate our scheme.

The conventional approach to Granger-type causality testing in economics is to assume a linear model and replace "Pr" in the above definition with "conditional expectation" and delete the variables with "s" arguments. One then operationalizes the concept by linearly regressing $x(t)$ on past x's and past y's and testing the null hypothesis that the coefficients on past y's are all zero. The idea is to test for the incremental ability of past y to help linearly predict future x. See Sargent (1981), (1987a) for the details. Applying the linear procedure on the examples listed above, you would conclude that for none of them does y help to predict x. But this is false from the formulae given above. The following approach is based on joint work with Ehung Baek of Iowa State University.

An Example of a Nonlinear Causality Test

We want to develop a test for nonlinear Granger causality that can be applied to estimated residuals of bivariate vector autoregressions. That is to say, we want the test statistic to have the same first-order asymptotic distribution whether you use estimated residuals or true residuals under the null hypothesis being tested. This property is important because the usual applications test for predictive relationships between the residuals of estimated *linear* models (Vector AutoRegressions (VAR's), for example) to see what extra predictive power there is to be had by going to nonlinear models. In this way we can test for the presence of incremental nonlinear predictive power of y for future x after the linear predictive power has been assessed. Consider the following idea, which we illustrate for the case $M = L = 1$. Write (15) in terms of joint distributions (using conditional probability formulas) and, to simplify notation, put

$$C(x,x',y) = \Pr\{|x(t)-x(s)| < e, |x(t-1)-x(s-1)| < e, |y(t-1)-y(s-1)| < e\}, \quad (16)$$

and write (15) in this new notation as

$$C(x, x', y)/C(x, y) = C(x, x')/C(x). \tag{17}$$

We set up a test of the null hypothesis that $\{x\}, \{y\}$ are IIDI (individually IID and mutually independent) based on replacing each $C(\cdot)$ in (17) above with its corresponding correlation integral $C(\cdot, T)$ computed from a sample of $\{x\}, \{y\}$ of length T. Under the null of IIDI (suppressing T in the C's), we have

$$T^{\frac{1}{2}}[C(x, x', y)C(x) - C(x, x')C(x, y)] \rightarrow N(0, V), \quad T \rightarrow \infty, \tag{18}$$

where the convergence is in distribution and the variance V is given by

$$V = 4[C(x)]^2 K(x)[K(x) - C(x)^2][K(y) - C(y)^2], \tag{19}$$

where

$$K(z) = \Pr\{|z(r) - z(s)| < e, |z(s) - z(t)| < e\}, \quad z = x, y \tag{20}$$

Here an exact expression for a consistent estimator of the variance can be found, as is done for the BDS test. The point to note about this test for nonlinear Granger causality is the following.

One would like to test (17) directly, assuming only one of Denker and Keller's (1983) mixing conditions sufficient to get convergence of the expression for the limiting variance (assuming sufficient regularity of $\{x\}, \{y\}$ so that the Denker/Keller conditions apply). However, the formula for the variance is a very messy infinite series. So our strategy is to formulate a test of IIDI that yields a simple formula for a variance like (19), which is easy to estimate consistently. This must be done in such a way that rejection of the null is not due to x helping to predict x' when y does not incrementally help to predict x. A test of IIDI based on

$$C(x, x', y)C(x) - C(x, x')C(x, y) = 0 \tag{21}$$

has this property. But a test of IIDI based on

$$C(x, x', y) - C(x)C(x')C(y), \tag{22}$$

does not.

Nonlinear causality tests based on the correlation integral as outlined above appear promising. For example, suppose you have residuals of some estimated linear model given by (14). Then based upon Monte Carlo work done in related situations (Baek and Brock (1988)), we are optimistic that nonlinearity tests of the form outlined above will do a good job of alerting the analyst to the extra predictive potential in residuals that are given by (14)—even though linear methods give no clue at all. Now let's take up a very closely related problem.

Econometricians are often faced with the following situation. Consider the regression problem

$$y(t) = f(I(t-1), a) + e(t), \tag{23}$$

where $I(t-1)$ is a set of regressors (the information set available at period $t-1$), a is a vector of parameters to be estimated, and $\{e(t)\}$ are errors whose conditional expectation $E[e(t)|I(t-1)] = 0$. Testing that you have fit the correct functional form to the data $\{y(t), I(t-1)\}$ often takes the form of testing the "orthogonality condition" that $E[e(t)|I(t-1)] = 0$. Applying a linear version of this methodology to the examples (11), (12), (13), you would miss the nonlinear predictive potential in these examples because the relative covariances are zero.

Furthermore, for many applications, especially in finance, it's of interest to know certain properties of the joint distribution of $\{e(t), I(t-1)\}$ beyond simple orthogonalility conditions. For example, it's useful to be able to predict higher-order conditional moments such as the variance. Even in macroeconomics you can do a more precise job of estimating parameters by taking advantage of the predictability of the conditional variance of errors. This is the intuitive content behind methods like generalized least squares. With this motivation in mind, consider testing the independence of $e(t)$ and $I(t-1)$ by the following criterion

$$T^{\frac{1}{2}}[C(e(t), I(t-1)) - C(e(t))C(I(t-1))] \to N(0, V), \tag{24}$$

where convergence in distribution as $T \to \infty$ takes place under the null hypothesis of independence of $\{e\}$ and $\{I\}$ (provided that we impose the stationarity and mixing conditions as in Denker and Keller (1983)). The formula for the variance V can be worked out under these assumptions. What's being captured here is a test of the following property

$$\Pr\{|e(t) - e(s)| < e \text{ given } |I(t-1) - I(s-1)| < e\} = \Pr\{|e(t) - e(s)| < e\} \tag{25}$$

In other words, if $I(s-1)$ near to $I(t-1)$ helps predict that $e(s)$ is near to $e(t)$, then there is usable structure in the relationship between I and e that can help predict y that's not captured in the regression model (23). Tests of the form (24) look for the absence of such useful incremental predictive structure. Monte Carlo work suggests that a test of the form (24) will detect a model like (9), where to a linear analyst the errors appear to be orthogonal to the set of regressors. The reason is that "near neighbors" is likely to detect dependencies like (9).

Monte Carlo work has shown that our type of test has the potential to detect dependence in linear models driven by nonlinearly dependent processes with zero linear autocorrelations like (12) and (13) as well as (11). However, the reasons for this good performance are still not clear to us. One possibility is that "near neighbors"-based tests for incremental predictive power that explore the whole space for potential "near-neighbor" predictability are good at catching any kind of intertemporal clustering in the data. And almost any kind of nonlinearity will at least occasionally cause such intertemporal clustering. A possibility is that all of our tests are based upon functions of U statistics, and these functions are zero at their population values when we assume the null hypothesis. It is well known that U-statistics are minimum-variance unbiased estimators of their population values in a specific context (Serfling (1980, Chapter 5)). All of our test statistics are functions of these minimum-variance unbiased estimators. If our functions were *linear* functions of these kinds of estimators, then, since a linear function of U-statistics of the same order is a U-statistic of the same order, we would have a minimum-variance unbiased estimator of a population parameter that is zero under the null hypothesis. This minimum variance property suggests good power against alternatives to the null. Now let's sum up.

It appears possible to develop useful tests of nonlinear Granger causality that can be applied, for example, to estimated residuals of vector autoregressions to search for omitted nonlinear structure that may have useful predictive content. Based on earlier discussions, we would expect this class of tests to be good at finding nonlinear structure that "near neighbors" is good at forecasting. A detailed discussion and performance evaluation of the testing methodology expounded here, as well as computer software written by W. D. Dechert and B. LeBaron to implement it, will be available in the forthcoming book (Brock, Hsieh, and LeBaron (1990)). Computer software for IBM compatibles has been developed by W. D. Dechert (1989).

3. Chartists and Prediction Evaluation in Stock Markets

In the preceding section we discussed general nonparametric methods for detecting patterns in time-series data. The section was rather technical. In this section I discuss some recent work on attempts to find and evaluate evidence of "extra" structure in stock market returns data. At the risk of much repetition, I'll try to explain the ideas in plain English. There are two key areas we want to discuss. First, we take up a comparison of the type of methods discussed here with "technical" methods used, for example, in many stock market newsletters. Then we take up methods for turning some of the ideas used by "technicians" into formal statistical inference.

Much of the work that I discuss here emerged from evidence acquired by applying some of the methods outlined in Section Two. Hsieh (1989), Frank and Stengos (1989), Scheinkman and LeBaron (1989), and LeBaron (1989) all contain evidence of "extra structure" in financial asset returns that seems difficult to account for using standard efficient markets models and conventional econometric models of the type that are commonly employed to fit asset returns data. A more conventional approach to accounting for "extra structure" in the context of liquidity-constrained, general-equilibrium, efficients-markets, asset-pricing models is given in Brock and LeBaron (1989), where it's shown that liquidity constraints on "small" firms can account for some of the difference between returns on small firms and returns on large firms.

Recall that the efficient markets hypothesis (EMH) imposes constraints on the degree of predictability of stock returns from publicly-available information such as past stock returns. If returns are generated completely by deterministic processes, then arbitrage would force the returns on all assets to be equal (Scheinkman and LeBaron (1989)). If returns were generated by deterministic chaoses buffeted by noise whose variance is small relative to the variance of the chaos, then arbitrage would force rates of return be be nearly equal. These conclusions are clearly in contradiction to the empirical facts: returns vary a lot across assets! Learning is a friction that could prevent arbitrage from eroding return differentials. But then the deterministic chaos would have to be hard to detect by traders using the nonparametric regression methods that will be discussed below. Hence, this argument suggests that the chances are small that a low-dimensional deterministic chaos generates asset returns. To begin our study of this point, let's look more closely at the logic of the efficient-markets argument.

The argument is simple. If future returns could be predicted from the past, then positions could be taken on the market to exploit these predictions. Prices today would move to the point where risk-adjusted profits on such positions would be zero. Therefore what predictability there is in stock market returns must reflect "nonprofitable" information, such as predictable movements in sources of systematic risk or predictable movements in liquidity scarcity (opportunity cost of funds), like on holidays or pay periods. Deeper arguments are presented in Brock and Malliaris (1989) and in Frank and Stengos (1989) and their references, especially the work of Sims. A large collection of facts purporting to challenge the EMH are reported in the "Anomalies" section of the *Journal of Economic Perspectives.*

Theory notwithstanding, there exists a large industry composed of people who claim to be able to predict future stock market returns using publically available data. Furthermore, these people claim supranormal profits from using their predictions, resulting in a thriving business in newsletters. Evaluation of these claims raises fundamental issues of statistical methodology.

What differentiates our work from most standard fundamentalist and chartist techniques is the use of the most recent econometric methods (the methods discussed above, as well as bootrapping techniques of Efron and Tsibirani (1986)) for testing temporal dependence. This is done in an attempt to do formal statistical inference in evaluating the likelihood that any evidence we find for superior predictability is due only to chance.

For example, we fit forecasting models to stock returns and other economic time series following standard practices in academic finance. We then study the time series of *forecast errors* of those models. This time series is constructed by taking the actual value of the time series being forecasted and subtracting from it the forecasted value given by the model. If the model is adequate, this series of forecast errors should be completely random. Methods like those presented in Section Two are used to test these forecast errors for IID. This is a stronger test than that usually employed in formal econometrics.

In formal econometrics, one popular technical criterion for model adequacy is that the series of forecast errors $\{e(t)\}$ be a martingale difference sequence, i.e., the conditional expectation of $e(t+1)$ on past information should be zero for each time period t (White (1984,p. 56)). This is not adequate for finance, where we want to predict conditional moments like the conditional variance, the conditional skewness, the conditional kurtosis, and possibly higher-order moments. So in finance it's important to be able to predict the conditional moment-generating function. If the sequence of forecast errors is IID, then the conditional moment-generating function is unpredictable.

There are practical reasons why we would like to be able to predict conditional moments. Consider the options market. Here if one could predict conditional variance better than the rest of the market, there may be opportunities to take profitable positions. If one can predict conditional skewness better than the rest of the market, one may be able to take profitable positions in puts and calls. If one can predict conditional kurtosis of returns better than the rest of the market, then, since many traders calibrate their decisions using the Black/Scholes option pricing model and this model assumes conditional normality (constant conditional kurtosis of 3), a prediction of an increase in conditional kurtosis beyond 3 could enable a trader to be more efficient at extracting profits from options positions. This is because the value of an option is, roughly speaking, an increasing function of thickness in the tails of returns on the underlying stock. For an interesting application of the forecastability of conditional variance, see Akgiray (1989). For stock returns, he shows that it's possible to forecast conditional variance one period ahead using GARCH models, even though there is little evidence for forecastability of the conditional mean one period ahead.

The sequence of forecast errors will turn out to be random under conventional statistical tests for many standard forecasting models. Our statistical tests are capable of detecting nonrandomness (extra forecastable structure) that many other tests cannot. See Hsieh (1989) for a published study reviewing some of our work and its application to exchange rates. Our statistical tests allow us to make precise statements about the level of burden of proof that our findings achieve. That is, our tests make clear what the probability is, under precise assumptions verifiable by other scientists, that any of our findings of "extra structure" are due soley to chance. A common fallacy in statistical work is finding "patterns" in data, the so-called patterns' appearance due only to chance. This was the message of the work by statistician Harry Roberts, which was discussed in the Haugen/Lakonishok book *The Incredible January Effect,* (Dow Jones-Irwin, 1988). A more recent (and more technical) discussion of data-snooping biases in statistical testing that may be especially relevant in evaluating claims to superior profits reported in financial newletters is found in Lo and MacKinlay (1989).

There will be instances where chartist techniques equipped with "other information" will compile forecasting records that look quite impressive. But with such a large number of practitioners, it's clear that some will compile outstanding records solely by chance. Also their records may be infected with Lo/MacKinlay data-snooping bias (Lo/MacKinlay (1989)). You will hear about the successes; you won't hear about the failures. This raises the question of measuring the burden of proof that we mentioned in the preceding paragraph.

Our statistical methods could be viewed as being similar to the methods of the chartists (technical analysts), in the sense that they can be used to look for inherent dynamics in stock prices that might be discovered by scientists locked in a room with stock price data as their only source of information. But our methods, like all statistical methods, can be used by chartists and fundamentalists alike. Our tests are nonparametric, so they have the potential of picking up any structure or pattern that's useful in prediction. But if you have an idea of what to look for, better parametric methods are available. Before going on, let me talk about the the predictability of conditional moments.

Prediction: Conditional Means, Conditional Variances, ...

Many people ask me about the predictability of conditional means, conditional variances, and higher moments like skewness, kurtosis, and so forth. These are technical notions. To explain them, consider the model

$$r(t) = F(I(t-1), u(t)),$$

where $r(t)$ is the stock return (the first difference of the logarithm of stock prices between time t and $t-1$), $I(t-1)$ is the information available at time $t-1$, and $\{u(t)\}$ is an IID stochastic process independent of $I(t-1)$ at each time t. The conditional mean is given by $m = E[r(t)|I(t-1)]$, while the conditional variance is $E\{[r(t)-m]^2|I(t-1)\}$. The kth conditional moment is, by definition, $E\{[r(t)-m]^k|I(t-1)\}$. In particular, the conditional skewness is the third conditional moment divided by the cube of the conditional standard deviation, while the conditional kurtosis is the fourth conditional moment divided by the fourth power of the conditional standard deviation minus 3. The conditional moment generating function is $M(s, I(t-1)) = E\{\exp[sr(t)]|I(t-1)\}$.

In plain English, the conditional mean is a measure of the likelihood that a stock price will increase or decrease as a function of the past information $I(t-1)$. So the term "conditional" here means conditioned on past information $I(t-1)$, which could be thought of as a set of regressors if one were to estimate the conditional mean from the actual data. In the standard efficient markets/random walk theories, the conditional mean is zero or, at least, constant. That is, past patterns (heads and shoulders, distances between successive peaks and troughs of chartist constructed curves from Horsey charts, etc.) predict nothing about the future movements in the stock price.

I hasten to add that this theory has been under attack in recent years. For example, in addition to higher returns in January, lower returns from Friday's close to Monday's close, and systematic return differences over the month, academic finance has found evidence for mean reversion in stock price data. Thus there may be linearly predictable 3–5 year swings in stock returns. Such evidence contradicts simple versions of the random walk theory. However, the mean reversion evidence is controversial. It is sensitive to the Great Depression years, and seems to disappear in large firms in the post-WWII period. Thus, it may just be due to chance. See Brock and LeBaron (1989) and especially Richardson (1989) for recent discussions of this evidence and for further references.

Above we were talking about conditional means. So let's turn now to a discussion about the variance, or volatility, of returns as well as the role played by higher moments such as the skewness.

Conditional variance is a measure of the volatility of the stock price as a function of past information. Past patterns of the stock price movements may have predictive power for forecasting future *volatility* of movement of the stock price, even though the past patterns have **no** predictive power for assessing whether the stock price itself will go up or down. Before considering this hypothesis, let's make a short digression to look at measures of central tendency and dispersion.

Skewness is a measure of symmetry. For example, it's zero for the bell-shaped normal curve, which is perfectly symmetric about its mean. Kurtosis is a measure of the peakedness, or fat-tailedness, of a distribution. Thus, it measures the likelihood of extreme values. For the normal distribution, the kurtosis is 3. Many people adjust things so that the kurtosis is zero for the normal. That way kurtosis is measured *relative* to the normal. Leptokurtic distributions have an adjusted kurtosis greater than zero (larger than the normal), while platykurtic distributions have an adjusted kurtosis less than zero (less than the normal). Leptokurtosis is very common for stock returns, which are usually measured by the difference of the logarithm of the stock's price in consecutive time periods. This means that it's very likely that stock returns will experience greater extremes than if they were normally distributed. This is especially true for high-frequency returns. Stock returns look more normally distributed at lower frequencies. Now let's return to our main theme.

How Our Methods Work

Repeating once more for clarity, recall that we use economic theory, the received wisdom in academic finance, and fundamental analysis, to guide us in choosing a forecasting model against which to compare our own methods. We then fit the model to the actual data, obtaining its forecast errors. Next we test the forecast errors for the presence of "extra structure" in the following manner: First, create a totally random series that has the same mean, variance, skewness, kurtosis, and so on of the original series of forecast errors. Second, calculate a measure (the Brock, Dechert, Scheinkman (BDS) statistic discussed in Section 2) of the difference between the original series of forecast errors and this constructed random comparison series. If the measure is zero, the series of forecast errors is random and there is no extra structure to be discovered and exploited. If the measure is not zero, we calculate the probability that the nonzero difference is due only to chance. If the probability that this difference arises solely from chance is less than an error tolerance level of, say, 5 percent, then we tentatively accept the conclusion that there is extra structure that the original forecasting model has missed. In this event, we use economic theory, fundamentalist analysis, and nonlinear techniques to revise the original forecasting model. Finally, we repeat the preceding steps until we obtain a series of forecast errors that satisfy the test for being IID.

Some Findings

Results of applications using the foregoing steps are reviewed in the papers cited below. Briefly, so far we have found that: (i) U.S. industrial production, U.S. pigiron production, U.S. unemployment and civilian employment behave nonlinearly in the sense that forecast errors from best-fit linear models (after trend removal) display evidence of extra structure (Brock and Sayers (1988)); (ii) Returns on the equal-weighted and value-weighted NYSE stock indices for both weekly and monthly data display extra structure in the forecast errors of fitted linear models, even after known regularities such as the January effect and the monthly effect and turn-of-the-week effects are taken out (Scheinkman and LeBaron (1989)); (iii) There has been a shift in the structure of stock returns between the period 1962–1974 and the period 1975–1985 (Brock (1988), LeBaron (1988)). This shift remains even after accounting for systematic movements in volatility over each period; (iv) Nominal Treasury bill returns have a low dimension even after removal of linear structure and trend (Brock (1988), Brock and Baek (1988)). I believe this last fact is due to movements in monetary policy.

More precisely, the series of T-bill forecast errors is generally quiet with infrequent bursts of volatility. This kind of series can give a low estimated dimension (because it is quiet most of the time and the dimension of a constant series is zero), even though there may be no forecastable structure in the series. Bursts of volatility may appear when traders are uncertain about future monetary policy shifts and such shifts are being considered by the Federal Reserve's Board of Governors. But this in turn feeds back into stock returns. To repeat: the structure of T-bill returns changes with the Fed's policy. For example, compare regimes of interest rate targeting with regimes of monetary base targeting. Finally, we have the result (v): Forecast errors from standard (in academic finance) forecasting models of the conditional variance (volatility) still contain extra, potentially forecastable, structure. This is true for both stock returns on the equal- and value-weighted indices, as well as for the exchange rates of the U.S. dollar against the Japanese yen, the British pound, and the German mark. This last result has been obtained by Blake LeBaron and José Scheinkman for stock returns (LeBaron (1989), Scheinkman and LeBaron (1989)) and David Hsieh for exchange rates (1989).

We have calculated the probability that the above results are due solely to chance and find that it is very low.

Comparison with Chartists

The reader is probably familiar with books such as Hurst's or Prechter's dealing with chartist schemes. I have seen chartist material where curves are fit to Horsey or TRENDLINE charts. Then distances between successive peaks and troughs for both upper- and lower-bound curves are related. This is done at every frequency ranging from day-to-day to decade-to-decade in an attempt to locate minor and major curves. I saw one chartist try to explain Black Monday as a trough in a "killer cycle" of several centuries. From a scientific point of view, the problems with this literature are manifold: (i) No one knows how to calculate the probability that the successful prediction of a given chartist is due solely to chance, partly because it's difficult to nail down precisely what rules the chartists use to predict turning points; (ii) It's not clear that two chartists, given, say, the same TRENDLINE chart, would make the same prediction even though they claim to be using the same technique. Scientific standards demand that two scientists using the same information and same technique come up with the same results. Furthermore, the method they use must be precisely definable in a manner that can be communicated to another scientist; (iii) Since a chartist technique is difficult to rigorously define (even for a particular chartist), when a chartist claims success for his methods it's difficult to measure how much of the success is due to the particular method used and how much is due to instinctive judgment based on other information. This other "outside" information (street smarts, for example) may feed into the identification process that the chartist uses to construct his curves. It's also impossible for chartists to calculate the probability that their successes are due solely to chance. But in order to have any possibility of having your findings accepted by the scientific community, it's necessary to be able to do this . Our methods satisfy each of these scientific standards.

An acid test of a chartist technique would be to feed a group of chartists artificial TRENDLINE charts based on random-walk processes of the type studied by academic finance, and then see if the chartists would identify "regularities" and "major and minor cycles with relationships between successive peaks and troughs." Of course, any such "regularities" would be spurious. Haugen and Lakonishok (1987) talk about this point, suggesting that people tend to see patterns in such purely random data even though there is absolutely nothing there that has any predictive value.

Nevertheless, there does appear to be predictable structure in stock returns. It's related to business-cycle movements, interest rate changes, tax policy, demographics, and so on. The real question is whether chartists are picking up more than this kind of forecastable structure, which is already known to the academic finance community. Statistical studies of the predictive power of chartists exist. Some of them are referenced in work discussed below. However, we believe that technical advances in statistical methodology allow us to improve upon these earlier studies. Let's now turn to a related test of the predictive power of charting.

Technical Analysis as a Model Specification Test

Brock, Lakonishok, and LeBaron (1990) have studied daily returns on the Dow-Jones Industrial Average index in an attempt to evaluate the prediction techniques used by technicians like Weinstein (1988). This article also contains references to earlier statistical studies of chartist predictive power. While it's extremely difficult to design statistical evaluation experiments to test technical trading strategies that respect the full richness of such strategies, one can take a simple trading strategy and study it. Here's a good example.

Form an m-day moving average of the stock's price. Buy when the stock's price cuts the moving average from below. Hold the buy position for h periods. Short when the stock's price cuts the moving average from above. Close out the position in h periods. Call this a "moving-average (m, h) $(MA(m, h))$ strategy." Before we begin, it's important to realize that an $MA(m, h)$ strategy is a very great simplification of the types of moving-average strategies put forth by, for instance, Weinstein (1988).

To begin with, Weinstein does not advocate acting on a buy/sell signal from a moving average unless it is "confirmed by volume." For example, he does not act on a buy signal unless the price cuts the moving average from below *when volume is rising.* Secondly, his holding and closing-out rules are more complicated than simply closing out the position in h periods. Finally, there is the issue of where to place your money when you are out of the market. Nevertheless, we can illustrate the philosophy of the BLL study with this $MA(m, h)$ strategy. The philosophy is not only to evaluate the statistical significance of apparent supranormal profits generated by such trading strategies, but also to use trading strategies to design specification tests of the standard econometric models in finance. Let's illustrate these points.

Consider the econometric model termed a Generalized AutoRegressive Conditionally Heteroscedastic in Mean (GARCH-M) model:

$$r(t) = p(t) - p(t-1) = a + bh(t) + e(t), \tag{1}$$

$$e(t) = h(t)z(t), \tag{2}$$

where $p(t)$ is the logarithm of the stock price at day $t, h(t)$ is a linear function of past h's and past squared e's, a, b are constants, and $\{z(t)\}$ is IID $N(0, 1)$, i.e., $\{z(t)\}$ is independently and identically distributed normally with mean zero and unit variance. The GARCH-M model is intended to capture the stylized facts that returns $\{r(t)\}$ have persistent volatility and, *ex ante,* the higher the conditional variance $h(t)$, the higher the expected returns. There are many variations on the basic idea of the model (1)–(2). See LeBaron (1989) for variations and references, especially to the work of Engle and Bollerslev. One particularly interesting variation due to LeBaron is the GARCH-EAR:

$$r(t) = a + f(h(t))r(t-1) + e(t) \tag{3}$$

$$f(h) = b + c\exp[-h/d] \tag{4}$$

Here $\{e(t)\}$ is the same as in (2), "exp" denotes the exponential function, d is a constant equal to the sample variance of the series being fitted, a, b, c are constants, and "EAR" stands for Exponential AutoRegression. LeBaron (1989) fits GARCH-EAR models because they capture an important fact that he's established, namely, that that stock returns are more locally autocorrelated following periods when the local variance is smaller. I will call this the "LeBaron Effect."

A third type of model has been motivated by recent work of Smith, Suchanek, and Williams (1988) (hereafter SSW) on experimental stock market bubbles. SSW find that on the first trial, experimental stock markets show a strong tendency to start out below the fundamental value, bubble up over the fundamental value on increasing volume, and crash back below fundamental value on decreasing volume. If the traders are brought back to the market for the same asset, on the second trial the bubble dampens. On the third trial the bubble disappears altogether, and most transaction prices hover around fundamental value. SSW's interpretation is that inexperienced traders generate behavioral uncertainty that disappears as experience is accumulated with asset markets. This relates to the literature on quasirational speculative bubbles and excess volatility discussed by West (1988). While Brock (1982) has shown that bubbles cannot exist in the standard rational expectations, general-equilibrium, asset-pricing model, reality may be different.

A process that might capture the idea of bubbles that grow and then burst is the following setup, written in continuous time for ease of exposition:

$$\begin{aligned} dP/P &= a(b, m(g), L)\,dt + b\,dZ + g\,dq \\ dq &= 1 \text{ with probability } L\,dt \end{aligned} \tag{5}$$

where dZ is IID $N(0, dt), dq$ is Poisson with arrival probability $L\,dt$ over $[t, t+dt], g$ is a random variable, and $m(g)$ denotes the probability measure of g. *Ex ante* expected returns conditional on information available at time t are given by

$$E\{dP/P)\} = a\,dt + LEg\,dt \tag{6}$$

For a model of an impending crash, one can put $a > a^*$ where a^* is the fundamental value with $L = 0$. Now let a depend upon L, b, and $m(g)$ in such a way that *ex ante* expected returns move so that it pays the investor to remain in the market even though there is a small chance

$L\,dt$ of a crash g with expected value $E[g] < 0$. In this way the value of a put on the stock may be greater than the value of a call, with both the put and the call being out of the money by the same amount. This is Bates's "crash premium" measure (Bates (1987)). We now have examples of three types of models that may be fit to stock market data in order to generate simulated data.

Return now to MA-based technical trading strategies. In the paper Brock, Lakonishok, and LeBaron (1990), profits from several $MA(m,h)$ strategies were estimated, where values of m and h were chosen based upon choices typically made by technical traders. This was done on daily Dow-Jones returns over different historical periods. There appear to be supranormal profits. This raises a puzzle: MA-based technical trading rules are the oldest and most publicly known of all trading rules. So why hasn't the market removed all supranormal profits attainable by these rules? My suspicion is that we are picking up some source of unmeasured risk that the market is valuing. For example, consider the model (5)–(6). Crash events are relatively infrequent in real markets, but are quite severe when they happen. Suppose MA-based rules tend to get one into the market at such risky times. Then most of the time they will yield apparently supranormal profits with only infrequent severe losses. Properly risk-adjusted profits, of course, could be normal. Before one can claim that evidence of supranormal profits using publicly available trading strategies has been found and, hence, conclude that the market is inefficient, one must exclude plausible alternative models in which the market is assumed to be efficient. Also one must eliminate possible sources of "data-snooping" statistical bias of the type that Lakonishok and Schmidt (1988) and Lo and MacKinlay (1989) have stressed. We have tried to control for data snooping by choosing an MA strategy and fixing its parameters at the same values throughout all subperiods. While this is not perfect control, it strengthens our confidence that data-snooping bias is not the cause of the apparent supranormal profits. One way to think of MA-based strategies is as an attempt to locate a trend in a stochastic process composed of temporal trends of random length and random sign. Let's now turn to other possible explanations and the use of technical trading strategies as specification tests of econometric models.

Consider the GARCH-M model (1)–(2). When investigators fit this kind of model to stock return data, they typically find the coefficient on the conditional variance in the expression for *ex ante* expected returns to be positive and statistically significant. They also find strong persistence in the conditional variance. This makes sense within the context of the efficient markets hypothesis. That is, we expect that returns should be high (low) when risk is high (low). Also, since variance displays momentum in fitted GARCH-M models, and MA-based technical strategies are designed to search out "trend" that is clouded by noise, we have a good candidate explanation for the apparent supranormal profits found by BLL.

In order to check this possibility, we estimated a GARCH-M model from our stock returns series, and then used the estimated model to generate R simulated stock return series. We then computed trading profits from our MA technical trading strategy for each of the R simulated series, forming a histogram of these technical trading profits. By choosing R large enough, it's possible to get quite a smooth histogram of trading profits under the null model, i.e. the fitted GARCH-M. Now we tick off a rejection region at a level of significance of, say, 5 percent. If the GARCH-M truly describes the data, actual technical trading profits will lie in the 5 percent rejection region 5 percent of the time, assuming our experiment is repeated many times. Therefore, finding technical trading profits in the 5 percent rejection region of the profit histogram under the null model constitutes strong evidence against the null model. This is exactly what we found. There are several observations to make about this finding.

First, we believe that this is an example of a specification-testing methodology that may be more useful in financial investigations than the conventional specification-testing methods employed in statistics and econometrics. This is because the specification test generates potentially valuable information, i.e., it's an avenue to possible trading profits when the null model is rejected by the data.

Second, there are many other plausible possibilities that may account for the apparent supranormal profits, such as versions of the standard general-equilibrium, asset-pricing model (Frank and Stengos (1989)), where risk moves over time with aggregate consumption. A model like (5)–(6), which is compatible with the EMH, may be constructed and may also account for the supranormal profits.

Third, there may be movements in the other moments, such as skewness and kurtosis, that the market values but which are overlooked and that may account for the results. We are presently investigating some of these alternative models.

Fourth, one could consider a diffusion/Poisson model like (5), drawing the drift from a distribution, the length of life of that drift then drawn from a lifetime distribution in such a way to simulate what one actually sees on stock charts. MA-based schemes would be of some potential use in locating the turning points of such randomly-lived trends. The risk of the Poisson part could be made large relative to the Brownian part by bunching the Poisson events rather than making them independent. If the MA schemes tended to expose one relatively more to Poisson bursts, then adjusting for this rather large relative risk might show the supranormal profits found by Brock, LeBaron, and Lakonishok (1989) to be normal, after all. This kind of process would be closer to the subordinated processes considered by Gallant, Hsieh, and Tauchen (1988).

Fifth, there is yet another risk that $MA(m,h)$ schemes might expose the investor to. In a market populated by bull and bear trends of random lengths, but where the average length of a bull trend is longer than the average length of a bear trend (because the economy tends to grow), an MA-based trading scheme (or many other trading schemes) exposes the investor to the risk that he will be out of the market in a bull trend. After all, Joe Granville sank out of sight when he missed the great bull market of the '80's. A buy-and-holder who simply plows back his dividends and plunks a fixed amount from his paycheck into an index portfolio each month will be hard to beat in a growing economy. He will tend not to buy too much stock when it is overpriced. Moreover, he will not be caught missing bull trends and shorting an upcoming bull trend. He also will not waste money churning his portfolio. Fidelity's *Investment Vision* (January/February 1990, p. 60) states that in a study of 120 newsletters, only one in four succeeded in matching the market averages, Joe Granville's track record being the worst of the lot!

We have not done such a rigorous test of the predictive power of a technical trading rule. We have only used a stylized rendition of a trading rule to create a model specification test that is tailor made for financial applications.

Nonparametric Evidence for the Predictability of Asset Returns

Most nonparametric evidence for incremental predictability of financial asset returns (above and beyond simple finance models such as random walks) is negative. Recent studies of this type include those by Diebold and Nason (1988), Meese and Rose (1989), Prescott and Stengos (1989), and Frank and Stengos (1989). Some of these studies use prediction algorithms that should be good at short-term predicting a dynamical process showing low-dimensional deterministic chaos. There is one study, Sentana and Wadhwani (1989), that looks at the stock market return in Japan and finds evidence of predictability that is difficult to explain away by standard risk adjustments. The bad news is that none of the studies (except that of Japan) are able to document predictive performance that beats simple standard models in finance, e.g., random walk models. It appears that parametric prediction methods having high power at detecting and forecasting structures "when you know what you are looking for" will have to be used in order to have much success at predicting financial asset returns. Also, it may be necessary to measure predictive gains in units like trading profits, which are more meaningful than usual measures such as reduction in mean-squared error. Furthermore, since the gains are likely to be small, more delicate statistical methods such as those used in the Brock, Lakonishok, and LeBaron (1990) study discussed above will have to be utilized to measure the incremental probability of success over chance.

We have talked a lot in the last two sections about searching for predictable structure, as well as about the search for more exotic structure like chaos and erratic dynamics. Now we turn to a discussion of possible pathways for "interesting" economic dynamics.

4. Possible Pathways for Erratic Economic Dynamics

Chaotic dynamics are a very special type of nonlinearity. See Brock (1986) for a formal definition and references. Perhaps this is why there is considerable controversy about the relevance of the findings of chaotic dynamics for economics—or even for the natural sciences. But there is little controversy about the presence of nonlinear dynamics in economics. Yet, for the most part, the bulk of econometric modelling ignores nonlinearity completely. Perhaps part of this lapse is due to the Wold Representation Theorem, which says that under modest regularity conditions, any second-order stationary time series $\{x(t)\}$ can be written as

$$x(t) = \sum_{i=0}^{\infty} a(i)e(t-i), \tag{W}$$

where $\{e\}$ is uncorrelated and $\{a\}$ is a square-summable sequence.

The expression (W) is a linear representation as far as covariance and spectral properties are concerned. Preistley's (1981) formal definition of *linear* is the same as (W) except that he requires the $\{e\}$ to be IID. The difference between being merely uncorrelated and IID is what gives his definition content. While the Wold representation might be one reason why the bulk of the economics profession clings to linear econometric methods, a more serious reason could be that economic time-series data are so noisy and nonstationary that only simple procedures like linear methods are likely to pick up any statistically significant structure.

Whatever the reason, many of the models used in economics (e.g. Sargent's books (1987a,b), the Real Business Cycle (RBC) models, King, Plosser and Rebelo (1989), Lucas (1987)) display two key properties: First, if you shut off the source of exogenous stochastic shocks, the model generates dynamics that are stable in the sense that the system trajectory converges to a steady state as time tends to infinity. And secondly, log linear stochastic approximations tend to be very good approximations to the nonlinear stochastic dynamics generated by these models. Hence, linear econometric methods are the methods of choice when confronting these models with data.

To proceed further we need a definition of erratic dynamics. Define dynamics to be *erratic* if when exogenous stochastic shocks are shut off, the resulting deterministic dynamics do not converge to a limit cycle or to a fixed point as time tends to infinity. Define the dynamical process to be *chaotic* if it displays sensitive dependence on initial conditions. A formal definition of chaos and some issues that arise in testing for it is given in Brock (1986).

In the Santa Fe Institute Volume (Anderson, et al. (1988)) Arthur, Boldrin, and I discuss pathways to chaos and to erratic dynamics in economics. See also Lorenz's recent book (1989). The basic theme set forth in these works is that economic models stress arbitrage, substitution, and intertemporal smoothing on the part of the microentities. These economic mechanisms tend to smooth out erratic and chaotic dynamics, making it difficult to get such dynamics with realistic model parameters. Therefore, empirically realistic frictions and autocatalytic, mutually-reinforcing external effects have to be found in order to make a case for empirically plausible pathways to erratic dynamics that would leave a trace in aggregative data. There's no doubt that examples of sensitive dependence upon initial conditions abound at the microlevel (e.g., Arthur in Anderson, et al.). The issue is whether there is a large enough mass of such microentities for sensitive dependence upon initial conditions to appear in aggregated statistics like real GNP.

The bulk of the empirical evidence suggests that low-dimensional (one to three) deterministic chaos is not likely in aggregated economic data (e.g., Brock and Sayers (1988), Frank and Stengos (1989), Lorenz (1989)) although there is still some controversy on the matter. Before proceeding, let's quickly say a few words about this controversy.

Part of the controversy surrounds the application of Brock's (1986) residual diagnostic, which says that if the data is chaotic the dimension of the residuals of a linear time-series model should be the same as the dimension of the raw data. This diagnostic was applied to macroeconomic data by Brock and Sayers (1988), leading to the conclusion that the evidence for low-dimensional deterministic chaos was weak. It was pointed out by Brock (1986) and Brock and Sayers (1988) that the residual diagnostic may reject deterministic chaos too many times when chaos is actually present. But for estimated models with a small number of parameters, the test should provide a useful caution to claims of having found low-dimensional deterministic chaos—especially in very low dimensions, say, between one and three. The residual diagnostic was applied to financial asset returns by Scheinkman and LeBaron (1989) and Frank and Stengos (1989). They obtained a correlation dimension estimate of about 6 that didn't change when recalculated for the residuals. As those authors pointed out, while this evidence is consistent with a chaos of dimension 6, other stochastic process models might also be consistent with the evidence.

Brock (1988) split the data sample of Scheinkman and LeBaron (1989), finding that the correlation dimension estimate jumped when moving from the period 1962–1974 to the period 1975–1983. Ramsey, Sayers, and Rothman (1988) found that the correlation dimension estimate was highly sensitive to removal of a few data points. Prescott and Stengos (1988) attempted to short-term forecast the Frank and Stengos gold returns series of dimension around 6, but were unable to forecast it even one period ahead using nonparametric kernel methods (which should work well for a low-dimensional chaos). LeBaron (1988) has not only shown that the correlation dimension estimate is unstable across time periods for the Scheinkman and LeBaron data, but also that it is possible to short-term forecast residuals of estimated models on 6–7 dimensional Mackey-Glass data generated to "look" like the stock return data. He was unable to short-term forecast the residuals of the same models fit to the actual data, which had an estimated dimension of 6–7. This suggests that if a chaos of dimension 6–7 is generating the data, it is more complicated than a 6–7-dimensional Mackey-Glass chaos and, furthermore, it's not stationary.

LeBaron (1989) found that stock returns are more autocorrelated following periods of low volatility than when following periods of high volatility. He also reports evidence that stock returns in the current period are negatively correlated with volatility in the next period. Simon Potter (1989) has fitted nonlinear models to macroeconomic data, discovering that his models do a good job of passing standard criteria for improvement over linear models.

To summarize, while investigators are discovering more structure in macroeconomic time-series and financial asset-returns data, there are still plenty of puzzles left to mystify investigators. A big mystery is where do economic and asset market fluctuations ultimately come from? Let us return to a consideration of the views of one major school of macroeconomics.

Most macroeconomic models lead to reduced form dynamics

$$x(t+1) = H(x(t), r(t+1)), \tag{1}$$

where $\{r\}$ is a low-order, "stable, stationary, and ergodic" Markov process, and

$$x(t+1) = H(x(t), a) \tag{2}$$

is a globally asymptotically stable dynamical system for a set of values of a that include the mean of $\{r\}$. Here x lies in m-dimensional space and r lies in n-dimensional space. Furthermore, log-linear approximations give good fits to the dynamics of (1)–(2). See Lucas (1987), King, Plosser, and Rebelo (1988) and Sargent (1987a,b).

A system like (1)–(2) doesn't possess sensitivity to initial conditions in the sense that if you start (1) with two initial conditions $x \neq y$ and let the same sequence of shocks $\{r(t)\}$ unfold, then the two paths generated by (1) will converge. Many components of $\{r(t)\}$ are unobservable by the econometrician, so it's difficult to design a statistical test for the presence of sensitive dependence upon initial conditions in the dynamics (1). Note also that the sources of statistical fluctuations are exogenous and left unexplained.

I want to outline here some possible pathways to "interesting" (i.e. sensitive dependence upon initial conditions and convergence to an attractor that is not a limit cycle or a fixed point) dynamics in economics. These pathways to endogenous fluctuations were not stressed or discussed in the aforementioned papers. The reader is warned that these are speculations, at best. They should be considered only as personal hunches. I will give examples of high-frequency, middle-frequency, and low-frequency pathways to interesting dynamics.

Information Congestion in the Stock Market

The first example concerns events on a fast or high-frequency time scale. Black Monday, October 19, 1987, still looms large in everyone's mind in view of the most recent crash after Black Monday. In my opinion, no satisfactory explanation has been found, and I will offer nothing here but a parable. Imagine a normally functioning stock market in which traders periodically update their information on fundamentals by calling upon specialists for each company. One reads about specialized security analysts who follow company fundamentals so carefully that the CEOs even call them to find out what's going on in their own firms! Imagine that communication links between traders and these company specialists are initially uncongested. Assume that traders form their bids-and-asks condition on both backward-looking information, such as past price and volume patterns, as well as on forward-looking information like projections of future net earnings payable to shareholders.

Now let a negative shock hit the market so that prices start to fall. Traders use the behavior of other traders, as well as checks on fundamentals, in an attempt to project the future course of prices. Let the negative shock be large enough and across enough stocks to create congestion in channels of communication to sources of fundamental information, e.g., telephone calls to specialized securities analysts. A trader now has no source of information other than the movement of price on the ticker tape and the observed behavior of other traders, most of whom are selling. The trader, especially if he is working for a client, is in the position of losing money conventionally (i.e., like the majority of traders) if he sells off now and takes a loss for his client. But he bears the risk of losing a lot more money unconventionally if he fails to join the selloff.

Dean LeBaron (1983) of Batterymarch has stressed that clients punish those managers who lose money unconventionally more than they punish those who lose money conventionally. Pension-fund and investment managers need to assure clients they are competent. This leads to reward functions based on comparison to others. In LeBaron's (1983) words, "The guiding principle in this environment seems to be that it is better to make a little money conventionally than to run even the smallest risk of losing a lot unconventionally." This creates an external effect from the activities of other managers upon the reward function of a money manager. If you join the sell off you can always say to your client, "I didn't lose any more on your portfolio than Goldman Sachs did." The reasoning works the same way for bull markets. If you fail to catch a bull market that the rest of the managers caught, you will be punished more by your clients than if you join the crowd and lose money trying to catch a bull that's turned into a bear.

What I am getting at is the following. A situation in which a selloff gets started can lead to *magnification* of the selloff force in the same direction when sources of fundamental information become congested during an emerging pulse of demand for information channels. In short, a positive feedback loop emerges. The reward schedules for fund managers are likely to amplify the strength of this feedback loop. It's possible that programmed trading might have accelerated this process, although Sanford Grossman (1987) has shown that the relationship between programmed trading and volatility is tenous, at best.

To sum up, a market move in one direction that is large enough and uniform across enough stocks can lead to such strong demand for fundamental information channels that congestion externalities arise. Traders cut off from sources of fundamental information must then rely on the ticker tape and the behavior of the traders around them. This can lead to autocatalytic positive-feedback behavior that can magnify an otherwise innocuous market move into a serious collapse. It's possible that trader unfamiliarity with the speed at which demands for trading are ignited by new technology like programmed trading could have overloaded the communication system, thus contributing to amplification of the collapse along the lines laid down above. If the collapse becomes large enough that the financial credibility of big traders is called into question, it can turn into a disaster unless the system is reliquified by the Federal Reserve. Fortunately, on Black Monday reliquification by the Fed did take place and the potential disaster was contained.

The literature contains very little work on the possibly autocatalytic effect of rewarding portfolio managers based upon a norm set by other portfolio managers or based upon a measure of relative performance, such as performance relative to the Standard and Poor's 500, in which the utility, or production, function is directly impacted by actions of others by pathways outside of the price system acting through budget or cost constraints. We have not seen any work on equilibrium asset-price dynamics in models of portfolio managers that are rewarded on relative performance measures like those actually used in practice. It would be interesting to see if the increased involvement of relative performance-rewarded managers magnifies stock market swings.

As an aside, there is an interesting positive feedback loop in volume that has been discussed in the literature. Admati and Pfliederer (1988) put forth a model involving two types of traders: informed and uninformed. The informed traders try to find some period when volume is relatively heavy in order to offload their position without a big effect on the price. This effect leads to volume bunching. Now let's turn to possible sources of fluctuations at a lower frequency, say two to four years.

Inventory Pulses

Inventories have long been associated with "cycles" in economic activity. Russell Cooper and John Haltiwanger (1989) argue that non-convexities in adjustment costs lead to production bunching in inventory replenishment. They offered this as an explanation for the stylized fact that production is more variable than sales, thus contradicting more conventional inventory models in which inventories are assumed to smooth production. They also found that output is more variable than consumption.

While production-bunching arguments are easy to muster at the microlevel, what's difficult is to draw any conclusions for the dynamics of macro aggregates. After all, there is a strong economic force at work to build up your inventories when others are not bidding up the price of the necessary inputs. That is, it's more profitable to build up your inventories out of phase with others. But what one can do, everyone can do. This force leads to a staggering of inventory buildups across firms, the aggregate effects being washed out. Cooper and Haltiwanger argue that in the case of strategic complements (your reaction function is upward sloping relative to your rivals' actions so that your marginal profits increase in proportion to your rivals' actions), inventory bunching by one member of a group of strategic complements will lead to bunching by all. If the group looms large enough, it will impact upon macroaggregates and we will see a bunching of the aggregate. In other words, if strategic complementarity is strong enough throughout the economy, we will tend to see the components of the economy run hot or cold together.

The trick is to come up with a convincing argument for something to play the role of these strategic complementarities. One can get them through "thick market externalities" (Murphy, Schleifer, and Vishny (1989)), positive external effects through the production process, or through "animal spirits" (Howitt and MaAfee (1989). But it remains to be seen whether estimates of strategic complementarities are strong enough to generate interesting endogenous dynamics in aggregates. A general model motivated by Ising models in physics is offered in Durlauf (1990). Durlauf shows that his model is capable of generating aggregative time series that show the strong persistence observed in actual aggregative output data.

Lags in Economic Behavior: Inventories

Lags have long played a key role in explanations suggested for endogenous economic fluctuations. Inventories play a key role in discussions of the propagation mechanisms of business cycles. The conventional wisdom (taken from *NP*–II, pp. 969-985) about inventories is as follows:

(1) While the level of inventory investment is small relative to the level of GNP, the average decline in inventory investment accounts for almost half of the average decline in GNP during periods of contractions. In expansions inventories change less as a fraction of the GNP changes relative to contractions. That is, inventories move asymmetrically over the cycle with drops during contractions larger than rises during expansions. It appears that a model like Durlauf's (1990) could generate this type of behavior. In Durlauf's model the interconnected businesses are more likely to pay start-up costs when neighboring businesses start up. In an expansion new businesses must decide whether to start up and acquire inventories. In a contraction they are already started. The relevant decision at the beginning of a contraction is to dispose of existing inventories and, possibly, to shut down the business if the contraction is perceived to be long enough. Since the start-up cost is not borne by incumbent businesses, asymmetric movement of inventories results.

(2) Retail inventories exhibit as much volatility as manufacturers' inventories.

(3) The variance of production usually exceeds the variance of sales. Inventories are more a propagating mechanism of the business cycle than a cause. To put it another way, inventory investment does not appear to be an important monetary transmission mechanism, but it is an important propagation mechanism and an important source of shocks (Ramey, 1989, p. 351).

Here are some more stylized facts from West's work (West (1988, NBER #2664, 1989, NBER #2992)):

(4) With the possible exception of Japan, the primary function of aggregate inventories is not to smooth aggregate output in the face of aggregate demand shocks. This result holds for

the G7 developed countries (Canada, France, West Germany, Italy, Japan, United Kingdom, United States).

(5) Inventory movements are procyclical and are very important at cyclical turning points of the business cycle.

(6) In a plausible context most of the variance of inventories is due to cost disturbances. Firms build up (deplete) inventories when costs are low (high). This is consistent with procyclicity of inventory stocks.

(7) Fluctuations in GNP appear to be roughly equally due to cost and demand.

(8) There is a small tendency to bunch production.

(9) Point estimates are quite noisy.

Summary: Empirical work suggests that inventories magnify exogenous demand shocks and such magnification is stronger for negative shocks (recessions) than for positive shocks (expansions).

The Beer Game

Let's now look at a hybrid study involving theory and experiment with the MIT Sloan School's beer inventory management game (Sterman (1989), Mosekilde, Larsen, and Sterman (1989), and Chapter 9 of this book). Subjects are asked to operate a four-stage production-distribution chain, consisting of a factory, a distributor, a wholesaler, and a retailer. Distributions of parameter estimates of econometric models are prepared from observations of the subjects' behavior. These estimated models are then simulated on a computer to see what fraction of the estimated models generate erratic (or even chaotic) behavior. While the majority are stable, a minority of around 40 percent of the models generate interesting dynamics, including cycles and even chaos. These studies represent an effective combination of experimental work and computer simulation to gain insight into the sources of endogenous dynamics in economics. Perhaps the same methodology could be applied to inventory models estimated from field data. In any event, a pathway to erratic dynamics through inventory management in a cascaded distribution system (chain of production) seems plausible. We still face the issue of whether such dynamics are intertemporally contiguous enough across sectors to leave a trace in aggregates.

The empirical work cited above suggests that there is enough procyclicity in inventories that aggregate implications are plausible.

Strategic complementarity across sectors would help the production-bunching effects show up in aggregate data.

Summary: We conclude that a very promising pathway to the emergence of erratic endogenous dynamics (or to sensitive dependence upon initial conditions) in macroaggregates is through lags in economic behavior like inventory accumulation. Recent theoretical work that's useful in this regard is that of Mackey (1989) and Benhabib and Rustichini (1989). A potentially fruitful methodology is to "estimate" Benhabib/Rustichini-type models on disaggregated data to get point estimates of model parameters and estimates of distribution parameters of point estimates. Then draw a model at random from this estimated distribution and simulate in the manner of Sterman (1989) in order to estimate the fraction of models that give chaotic dynamics. Then estimate aggregative linear models on aggregative data produced by the simulated Benhabib/Rustichini model. Finally, use the statistical methods of Brock and Sayers (1988) to see if the residuals of these models appear nonchaotic to the Brock-Denchert-Scheinkman (1986) test. In this way we would get insight, disciplined by field data, into whether lags in economic behavior are realistically capable of generating low-dimensional deterministic chaos, and whether the recent statistical methods for detecting low-dimensional chaos are biased against detecting it. Now let's consider sources of endogenous fluctuations at very low frequencies, say 20–30 years.

Very Low-Frequency Dynamics

We outline here a parable of long swings in economic activity at a frequency much lower than the business cycle frequency of three to five years.

A standard result in capital theory concerns an economy populated by a finite number of dynastic, long-lived "families" (countries) that can freely borrow and lend to each other. The proposition is that, in the long run, the "family" (country) that least discounts future payoffs ends up with all the economy's (world's) wealth. Applied to families, the result says that the dynastic family that discounts the welfare of future offsprings' consumption the least will eventually end up with all the wealth. This is a very strong mechanism for generating wealth and income inequality over the long haul. This inequality is likely to be magnified by birth rate and educational differences across income classes. Now insert a "rent-seeking" sector into this economy, where an individual family faces the following tradeoff: It can either work to produce something useful to sell to another family or it can invest in "lawyers" to file suits against the wealthy in an attempt to capture their assets.

To be specific, let's tell this story with two socioeconomic classes—capital (the low discounters) and labor (the high discounters). In models of long-lived "families," Robert Becker (1980) has shown that the lowest discounter ends up with all the wealth. This result is quite strong and hard to eliminate in standard capital-theoretic, intertemporal, competitive, general-equilibrium models, where the utility of the parents contains the utility of the children. Epstein and Hines (1983) discuss ways out of this dilemma. But the robustness of this result is what we need to launch our story. Let's continue.

As the wealth builds up in the capitalist class and inequality (both income and wealth inequality) grows, we can predict an increase in resources spent on redistributive activity through the political system (Magee, et al. (1989)). Endogenous policy theory suggests that a growth in attempts by labor to capture capitalist wealth through the political system (because as wealth inequality grows, it will increasingly pay labor more at the margin to try to get wealth through the political system than to work at producing useful goods) will lead to obfuscatory attempts by the capitalist party to protect its clients' wealth. So it appears that an increase in economic distortions caused by politicking is likely. The median voter effect in a two-party system is likely to lead to more action to tax the wealthy as the inequality grows. Presumably increases in distortions such as income taxes (with a lot of deductions for the wealthy, e.g., corporate hunting lodges) will start to choke off growth in the economy's income. As the standard of living begins to fall (especially relative to the rest of the world) even though income and wealth inequality have dropped, concern about this drop in living standard will rise.

For example, in the late 1980's a common slogan by restructuralists in New Zealand was, "thirty years ago New Zealand was first in the world in standard of living; now we are 30th." New Zealand was virtually a command economy until 1984 when the Labor (!) party took over and instituted a Thatcher-like restructuring—dramatic cuts in income taxes and the introduction of a comprehensive consumption tax.

As economic distortions like income taxes are lowered in an attempt to restart growth, wealth inequality will gradually start to build up again due to the relative differences in discounting discussed at the beginning of this parable. The process then repeats itself.

We caution the reader that we are only offering a parable of long waves here, not a seriously tested economic theory. Endogenous policy theory is not a well developed area of economics. See Magee, et al. (1989) for a survey of empirical and theoretical results from this literature. A main point is that the political system tends to redistribute income in an indirect and inefficient manner, leading to a lot of economic waste. Also, protective activity is countercyclical, i.e., when times are good people can make more money producing useful goods than investing in the political process to extract wealth from their fellows. The theory is not all that helpful in predicting what will happen when an economy like New Zealand's in 1983 is so overregulated and so many special interest groups are legislatively protected from competition that the overall standard of living is headed down. In any event, we tell this parable to stimulate thinking on how endogenous fluctuations in economic activity might appear at very low frequencies.

Summary and Conclusions

In Section Two of this article we surveyed some new methods for detecting potential forecasting power from hidden patterns in data that looks random to many of the standard statistical tests. We also showed how these methods could be extended to detect cases when one series helps predict another, even though to many methods the two series look independent.

Section Three discussed applications of the new methods to the detection of hidden patterns in asset-returns data. We also discussed whether there was any evidence that stock returns were predictable, the goal being to make supranormal profits from stock trading. A new approach to testing the adequacy of fitted parametric models to stock market returns was laid out. Evidence for supranormal profit potential is mixed, but there is strong evidence of extra structure in stock market returns that has not yet been captured by existing parametric models. Evidence of predictability by nonparametric methods is fairly weak. Parametric methods tailor made to financial applications may do better. Too many statistical methods used to model financial time series could equally well be applied to model insect population time series. Knowledge of where the series being modelled came from will have to be used to do better.

Finally, in Section Four three rather speculative sources of endogenous dynamics in economic time series were considered. They ranged from high frequency rapid changes in stock prices (crashes) to very low frequency dynamics in wealth inequality. The reader is warned that these are only my personal hunches.

Acknowledgements

Many of the ideas contained in this paper have evolved from interaction with colleagues at the Santa Fe Institute. The supportive atmosphere of Santa Fe Institute has been vital to my work on nonlinear science. The National Science Foundation under Grant #144-AHO1 and the Wisconsin Graduate School has given me essential financial support. None of the above are responsible for errors or shortcomings in this article.

References

This list of references contains not only those cited in the paper but also related references.

Admati, A., and Pfliederer, P., (1988), "A Theory of Intraday Patterns: Volume and Price Variability," *Review of Financial Studies,* 1, 3-40.

Akgiray, V., (1989), "Conditional Heteroscasdticity in Time Series of Stock Returns: Evidence and Forecasts," *Journal of Business,* 62, #1, 55-80.

Anderson, P., Arrow, K., Pines, D., (Eds.), *The Economy as an Evolving Complex System,* Addison Wesley, Redwood City, CA.

Ashley, R., and Patterson, D., (1989) "Linear Versus Nonlinear Macroeconomies: A Statistical Test," *International Economic Review,* 30, #3, 685-704.

Ashley, R., Patterson, D., and Hinich, M., (1986), "A Diagnostic Test for Nonlinearity and Serial Dependence in Time Series Fitting Errors," *Journal of Time Series Analysis,* 7, #3, 165-178.

Azariadis, C., and Guesnerie, R., (1986), "Sunspots and Cycles," *Journal of Economic Theory,* 40, 725-737.

Baek, E., (1987), "Contemporaneous Independence Test of Two IID Series," Department of Economics, University of Wisconsin, Madison.

Baek, E., (1988), PhD Thesis, Department of Economics, University of Wisconsin, Madison.

Baek, E., and Brock, W., (1988), "A Nonparametric Test for Temporal Dependence in a Vector of Time Series," Department of Economics, Iowa State University and the University of Wisconsin, Madison.

Bak, P., et al., (1988), "Self Organized Criticality," *Physical Review A,* July, 364-373.

Bates, D., (1987), "The Crash Premium," Department of Economics, Princeton University.

Barnett, W. and Chen., P., (1988), "The Aggregation-Theoretic Monetary Aggregates Are Chaotic and Have Strange Attractors," in Barnett, W., Berndt, E., and White, H., (eds.) *Dynamic Econometric Modelling,* Proceedings of the Third International Symposium in Economic Theory and Econometrics, Cambridge University Press, Cambridge.

Baumol, W. and J. Benhabib, (1989), "Chaos: Significance, Mechanism, and Economic Applications," *Journal of Economic Perspectives,* Winter, Vol. 3, #1, 77-105.

Becker, R., (1980), "On the Long Run Steady State in a Simple Dynamic Model with Heterogeneous Households," *Quarterly Journal of Economics,* 95, 375-382.

Benhabib, J., and Rustichini, A., (1989), "On Optimal Dynamics in Vintage Models," Department of Economics, New York University.

Black, F., (1986), "Noise," *Journal of Finance,* 41, July, 529-543.

Boldrin, M., and Woodford, M., (1990), "Equilibrium Models Displaying Endogenous Fluctuations and Chaos," *Journal of Monetary Economics* (forthcoming).

Bollerslev, T., (1986), "Generalized Autoregressive Conditional Heteroskedasticity," *Journal of Econometrics,* 31, 307-327.

Bollerslev, T., (1985), "A Conditionally Heteroskedastic Time Series Model for Security Prices and Rates of Return Data, *Review of Economics and Statistics* (to appear).

Bollerslev, T., (1988), "On the Correlation Structure for the Generalized Autoregressive Conditional Heteroskedastic Process," *Journal of Time Series Analysis,* Vol. 9, No. 2, 121-132.

Brillinger, D., et al, (1980), "Empirical Modelling of Population Time Series Data: The Case of Age and Density Dependent Vital Rates," *Lectures on Mathematics in the Life Sciences,* 13, 65-90.

Brock, W., and LeBaron, B., (1989), "Liquidity Constraints in Production Based Asset Pricing Models," National Bureau of Economic Research W.P. #3107.

Brock, W., Lakonishok, J., and LeBaron, B., (1990), "Technical Analysis as a Specification Test," University of Wisconsin and University of Illinois, Champaign/Urbana.

Brock, W., and Malliaris, A., (1989), *Differential Equations, Stability and Chaos in Dynamic Economics,* North Holland, Amsterdam.

Brock, W. (1982), "Asset Prices in A Production Economy," in J. J. McCall, (Ed.), *The Economics of Information and Uncertainty,* University of Chicago and NBER, Chicago.

Brock, W., and Dechert, W., (1988), "A General Class of Specification Tests: The Scalar Case," *Proceedings of the Business and Economic Statistics Section of the American Statistical Assocation,* 70-79.

Brock, W., (1986), "Distinguishing Random and Deterministic Systems," *Journal of Economic Theory,* Vol. 40, No. 1, October, 168-195.

Brock, W., and Sayers, C., (1988), "Is The Business Cycle Characterized by Deterministic Chaos?" *Journal of Monetary Economics,* July, 22, 71-90.

Brock, W., and Dechert, W., (1988), "Theorems on Distinguishing Deterministic and Random Systems," in Barnett, W., Berndt, E., and White, H., (op. cit.).

Brock, W., Dechert, W., Scheinkman, J., and LeBaron, B., (1988), "A Test for Independence Based Upon the Correlation Dimension," Working Paper, The Univ. of Wisconsin, Madison, The Univ. of Houston, and The Univ. of Chicago, Department of Economics.

Brock, W.A., Dechert, W. D., and Scheinkman, J., (1986), "A Test for Independence Based On the Correlation Dimension," Department of Economics, University of Wisconsin, Madison, University of Houston, and University of Chicago.

Brock, W. A., (1988), "Nonlinearity and Complex Dynamics in Finance and Economics," in Anderson, P., Arrow, K., Pines, D., eds., *The Economy as an Evolving Complex System,* Addison-Wesley, Redwood City, California.

Brock, W., Hsieh, D., and LeBaron, B., (1990), *A Test for Nonlinear Dynamics,* MIT Press, Cambridge, MA (forthcoming).

Brock, W., (1987a), "Nonlinearity in Finance and Economics," University of Wisconsin, Madison.

Brock, W., (1987b), "Notes on Nuisance Parameter Problems in BDS type Tests for IID," University of Wisconsin, Madison.

Brock, W. A., and Baek, E., (1988), "The Theory of Statistical Inference for Nonlinear Science: Gauge Functions, Complexity Measures, and Instability Measures," Deparment of Economics, The University of Wisconsin, Madison.

Cooper, R., and Haltiwanger, J., (1989), "Macroeconomic Implications of Production Bunching: Factor Demand Linkages," Department of Economics, University of Iowa, and University of Maryland.

David, P., (1988), "The Future of Path-Dependent Equilibrium Economics," Center for Economic Policy Research, Stanford University.

Dechert, W., (1988), "A Characterization of Independence for a Gaussian Process in Terms of the Correlation Integral," Department of Economics, University of Houston.

Diebold, F., and Nason J., (1988), "Nonparametric Exchange Rate Prediction?", Department of Economics, University of Pennsylvania and University of British Columbia.

Dechert, W., (1987), "A Program to Calculate BDS Statistics for the IBM PC," Department of Economics, University of Houston.

Denker, M., and Keller, G., (1983), "On U-Statistics and Von Mises Statistics for Weakly Dependent Processes," *Z. Wahrscheinlichkeitstheorie verw. Gebiete,* 64, 505-522.

Durlauf, S., (1990), "Locally Interacting Systems Coordinations Failure and the Behavior of Aggregative Activity," Department of Economics, Stanford University.

Eatwell, J, Milgate, M., Newman, P., (1987), *The New Palgrave,* Vols. I-IV, Macmillan Press, Ltd., London.

Engle, R., (1982), "Autoregressive Conditional Heteroscedasticity With Estimates of The Variance of U. K. Inflations," *Econometrica,* 50, 987-1007.

Engle, R., (1987), "Multivariate ARCH with Factor Structures-Cointegration in Variance," Department of Economics, University of California, San Diego.

Engle, R., Hendry, D., Trumble, D. (1985), "Small-sample properties of ARCH estimators and tests," *Canadian Journal of Economics,* 18, #1, February.

Efron, B., (1982), *The Jackknife, The Bootstrap, and Other Resampling Plans,* Philadelphia: Society for Industrial and Applied Mathematics.

Efron, B., and Tsibirani, R. (1986), "Bootstrap Methods for Standard Errors, Confidence Intervals, and Other Measures of Statistical Accuracy," *Statistical Science,* Vol. 1, #1, 54-77.

Epstein, L., and Hines., A., (1983), "The Rate of Time Preference and Dynamic Economic Analysis," *Journal of Political Economy,* 91, #4, August, 611-635.

Farmer, J., and Sidorowich, J., (1988), "Exploiting Chaos to Predict the Future and Reduce Noise," Theoretical Division and Center for Nonlinear Studies, Los Alamos National Laboratory, Los Alamos, NM.

Fama, D., and French, K., (1988), "Permanent and Temporary Components of Stock Prices," Center for Research on Security Prices Working Paper #178.

Fama, E., (1976), *Foundations of Finance,* Basic Books, New York.

Feigenbaum, M., (1983), "Universal Behavior in Nonlinear Systems," in Barenblatt, G., Iooss, G., and Joseph, D., (Eds.), (1983), *Nonlinear Dynamics and Turbulence,* Pitman, Boston.

Feinberg, M., (1980), "Chemical Oscillations, Multiple Equilibria, and Reaction Network Structure," in Stewart, W., Ray, W., and Conley, C., (Eds.), *Dynamics and Modelling of Reactive Systems,* Academic Press, New York.

Frank, M., and Stengos, T., (1988), "Chaotic Dynamics in Economic Time Series," *Journal of Economic Surveys,* 2, 103-133.

Frank, M., and Stengos, T., (1989), "Measuring the Strangeness of Gold and Silver Rates of Return," *Review of Economic Studies,* 56, No. 4, #188, October, 553-567.

Frank, M., and Stengos, T., (1989), "Nearest Neighbor Forecasts of Precious Metal Rates of Return," Department of Finance, University of British Columbia and the University of Guelph.

Froot, K., and Obstfeld, M., (1989), "Intrinsic Bubbles: The Case of Stock Prices," NBER #3091.

Gallant, A., Hsieh, D., and Tauchen, G., (1988), "On Fitting a Recalcitrant Series: The Pound/Dollar Exchange Rate, 1974-1983." Department of Economics, North Carolina State University, and Fuqua School of Business, Duke University.

Gallant, A., and White, H., (1988), *A Unified Theory of Estimation and Inference For Nonlinear Dynamic Models,* Basil Blackwell, Oxford.

Gallant, A. R., and Tauchen, G., (1986), "Seminonparametric Estimation of Conditionally Constrained Heterogeneous Processes: Asset Pricing Applications," Department of Economics, North Carolina State University and Duke University.

Geweke, J., (1989), "Inference and Forecasting for Chaotic Nonlinear Time Series." Institute of Statistics and Decision Sciences, Duke University.

Gleick, J., (1987), *Chaos,* Viking, New York, NY.

Goldberger, A., (1989), "Economic and Mechanical Models of Intergenerational Transmission," *The American Economic Review,* 79, #3, June, pp. 504–513. Also see Gary Becker's rebuttal in the same journal.

Granger, C., and Andersen, A., (1978), *An Introduction to Bilinear Time Series Models,* Vandenhoeck & Ruprecht, Göttingen, West Germany.

Granger, C., (1987), "Stochastic or Deterministic Non-linear Models? A Discussion of the Recent Literature in Economics," University of California, San Diego.

Grandmont, J., (1985), "On Endogenous Competitive Business Cycles," *Econometrica,* 53, pp. 995–1045.

Grandmont, J., ed., (1986), *Journal of Economic Theory Symposium on Nonlinear Economic Dynamics,* Vol 40, #1, October.

Grassberger, P., and Procaccia, I., (1983a), "Measuring the Strangeness of Strange Attractors," *Physica 9D,* 189-208.

Grassberger, P., and Procaccia, I., (1983b), "Estimation of the Kolmogorov Entropy From a Chaotic Signal," *Physical Review A,* 28, October, 2591-2593.

Grossman, S., (1987), "An Analysis of the Implications for Stock and Futures Price Volatility of Program Trading and Dynamic Hedging Strategies," Princeton University, Department of Economics.

Guckenheimer, J., (1978), "Comments on Catastrophe and Chaos," *Lectures on Mathematics in the Life Sciences,* 10, 1-47.

Haugen, R., and Lakonishok, J., (1988), *The Incredible January Effect,* Dow Jones-Irwin, Homewood, Illinois.

Hinich, M., and Patterson, D., (1985), "Evidence of Nonlinearity in Stock Returns," *Journal of Business and Economic Statistics,* January, 69-77.

Hinich, M., and Patterson, D., (1986), "A Bispectrum Based Test of The Stationary Martingale Model," University of Texas at Austin and Virginia Polytechnic Institute.

Hinich, M., (1982), "Testing for Gaussianity and Linearity of a Stationary Time Series," *Journal of Time Series Analysis,* 3, No. 3, 169-176.

Howitt, P., and McAfee, R., (1989), "Animal Spirits," Department of Economics, University of Western Ontario.

Hsieh, D., and LeBaron, B., (1988a,b), "Small Sample Properties of the BDS Statistic," Graduate School of Business, The University of Chicago.

Hsieh, D., (1989), "Testing for Nonlinear Dependence in Daily Foreign Exchange Rates," *Journal of Business,* 62, #3, 339-369.

King, R., Plosser, C., and Rebelo, S. (1988), "Production, Growth, and Business, Cycles: I. The Basic Neoclassical Model," *Journal of Monetary Economics,* 21, Nos. :2/3, 309-342.

LeBaron, B., (1987a), "Nonlinear Puzzles in Stock Returns," Department of Economics, The University of Chicago.

LeBaron, B., (1987b), "A Program to Calculate BDS statistics for the MacIntosh PC," Department of Economics, The University of Chicago and the University of Wisconsin, Madison.

LeBaron, B., (1988), "The Changing Structure of Stock Returns," Department of Economics, The University of Chicago.

LeBaron, B., (1988), "Stock Return Nonlinearities: Some Initial Tests and Findings," Ph.D Thesis, Department of Economics, The University of Chicago.

LeBaron, B., (1989), "Some Relations Between Volatility and Serial Correlations in Stock Market Returns," The University of Wisconsin, Madison.

LeBaron, D., (1983), "Reflections on Market Efficiency," *Financial Analysts Journal,* May/June, 16-23.

Lee, T., White, H., and Granger, C., (1989), "Testing for Neglected Nonlinearity in Time Series Models: A Comparison of Neural Network Methods and Alternative Tests," Department of Economics, The University of California, San Diego.

Levine, D., (1982), "Theory of Price Bubbles," UCLA Department of Economics Working Paper.

Lo, A., (1989), "Long Term Memory in Stock Market Prices," Sloan School of Management, MIT, Working Paper #3014-89-EFA.

Lo, A., and MacKinlay, C., (1989), "Data-Snooping Biases in Tests of Asset Pricing Models," Working Paper #3020-89-EFA, Sloan School of Management, MIT.

Long, J., and Plosser, C., (1983), "Real Business Cycles," *Journal of Political Economy,* 91, February, 39-69.

Litterman, R., and Weiss, L., (1985), "Money, Real Interest Rates, and Output: A Reinterpretation of Postwar U.S. Data," *Econometrica,* 53, January, 129-156.

Loeve, M., (1963), *Probability Theory,* Van Nostrand, Princeton.

Lo, A., and MacKinlay, C., (1987), "A Simple Specification Test of The Random Walk Hypothesis," Wharton School, University of Pennsylvania.

Lorenz, H. W., (1989), *Nonlinear Dynamical Economics and Chaotic Motion,* Springer Verlag Lecture Notes in Economics and Mathematical Systems, #334.

Lucas, R., (1978), "Asset Prices in An Exchange Economy," *Econometrica,* 46, 1426-46.

Lucas, R., (1987), *Models of Business Cycles,* Basil Blackwell, Oxford.

Mackey, M., (1989), "Commodity Price Fluctuations: Price Dependent Delays and Nonlinearities as Explanatory Factors," *Journal of Economic Theory,* 48, #2, August, 497-509.

Mackey, M., and Glass, L., (1988), *From Clocks to Chaos: The Rhythms of Life,* Princeton University Press, Princeton, NJ.

Madansky, A., (Ed.), (1978), "Symposium on Forecasting With Econometric Methods," *Journal of Business,* Vol. 51, No. 4, October, 547-600.

Magee, S., Brock, W., and Young, L., (1989), *Black Hole Tariffs and Endogenous Policy Theory,* Cambridge University Press, Cambridge.

Manski, C., (1989), "Regression," SSRI Working Paper #8917, University of Wisconsin, Madison.

Meese, R., and Rose, A., (1989), "An Empirical Assessment of Nonlinearities in Models of Exchange Rate Determination," School of Business, University of California, Berkeley.

Mosekilde, E., Larsen, E., Sterman, J., (1989), "Coping with Complexity: Deterministic Chaos in Human Decision Making Behavior," Physics Laboratory III, The Technical University of Denmark, Lyngby, Denmark.

Neftci, S., and Policano, A., (1984), "Can Chartists Outperform the Market: Market Efficiency Tests for 'Technical Analysis,'" *The Journal of Futures Markets,* 4, #4, 465-478.

Nelson, C., and Plosser, C., (1982), "Trends and Random Walks in Macroeconomic Time Series," *Journal of Monetary Economics,* 10, 139-162.

Prescott, D., and Stengos, T., (1988), "Do Asset Markets Overlook Exploitable Nonlinearities? The Case of Gold," Department of Economics, University of Guelph.

Potter, S., (1990), Ph.D Thesis, The University of Wisconsin, Madison.

Powell, J., Stock, J., and Stoker, T., (1989), "Semiparametric Estimation of Weighted Average Derivatives," *Econometrica,* (forthcoming).

Priestley, M., (1981), *Spectral Analysis And Time Series,* Vols. I, II, Academic Press, New York.

Prigogine, I., (1980), *From Being to Becoming,* W. H. Freeman, New York.

Prigogine, I., and Stengers, I., (1984), *Order Out of Chaos,* Bantam Books, New York.

Ramey, V., (1989), "Inventories as Factors of Production and Economic Fluctuations," *American Economic Review,* 4, #4, 338–354.

Ramsey, J., Sayers, C., and Rothman, P., (1988), "The Statistical Properties of Dimension Calculations Using Small Data Sets: Some Economic Applications," Department of Economics, New York University and the University of Houston, *International Economic Review,* forthcoming.

Richardson, M., (1988), "Temporary Components of Stock Prices: A Skeptic's View," Graduate School of Business, Stanford University.

Robinson, P., (1989), "Hypthesis Testing in Semiparamatric and Nonparametric Models for Econometric Time Series," *Review of Economic Studies,* 56, (4), #188, 511-534.

Sargent, T., (1981), "Interpreting Economic Time Series," *Journal of Political Economy,* 89, #2, 213-248.

Sargent, T., (1987a), *Macroeconomic Theory,* Academic Press, New York.

Sargent, T., (1987b), *Dynamic Macroeconomic Theory,* Harvard University Press, Cambridge, MA.

Savit, R., and Green, M., (1989), "Time Series and Dependent Variables," Physics Department, The University of Michigan, Ann Arbor.

Scheinkman, J., and LeBaron, B., (1989), "Nonlinear Dynamics and Stock Returns," *Journal of Business,* 62, #3, 311-337.

Scheinkman, J., and LeBaron, B., (1987b), "Nonlinear Dynamics and GNP Data," Forthcoming in Barnett, W., Geweke, J., and Shell, K., (Ed.), *Economic Complexity: Chaos, Sunspots, Bubbles, and Nonlinearity,* Proceedings of The Fourth International Symposium in Econometric Theory and Econometrics, Cambridge University Press, Cambridge.

Serfling, R., (1980), *Approximation Theorems of Mathematical Statistics,* Wiley, New York.

Smith, V., (1986), "Experimental Methods in The Theory of Exchange," *Science,* Vol. 234, No. 4773, October 10, 167-173.

Stein, D., (ed.), (1989), *Lectures in the Sciences of Complexity,* Santa Fe Institute Studies in The Sciences of Complexity, Addison Wesley, Redwood City, Calif.

Sterman, J., (1989), "Deterministic Chaos in An Experimental Economic System," *Journal of Economic Behavior and Organization,* 12, 1-28.

Suess, Dr., (1961), *The Sneetches, and Other Stories,* Random House, New York.

Swinney, H., (1985), "Observations of Complex Dynamics and Chaos," in Cohen, E., Ed., *Fundamental Problems in Statistical Mechanics: VI,* North Holland, Amsterdam.

Tong, H., (1983),*Threshold Models in Nonlinear Time Series Analysis,* Lecture Notes in Statistics, #21, Springer Verlag, New York.

Weinstein, S., (1988), *Secrets for Profiting in Bull and Bear Markets,* Dow Jones-Irwin, Homewood, Illinois.

West, K., (1988), "Bubbles, Fads, and Stock Price Volatility Tests: A Partial Evaluation," *Journal of Finance,* 43, July, 639-55.

West, K., (1988), "Evidence From Seven Countries on Whether Inventories Smooth Aggregate Output," National Bureau of Economic Research Working Paper #2664.

West, K., (1989), "The Sources of Fluctuations in Aggregate Inventories and GNP," National Bureau of Economic Research Working Paper #2992.

White, H., and Domowitz, I., (1984), "Nonlinear Regression with Dependent Observations," *Econometrica,* 52, #1, 143-161.

White, H., (1984), *Asymptotic Theory for Econometricians,* Academic Press, New York.

White, H., (1987), "Specification Testing in Dynamic Models," in T. Bewley, ed., *Advances in Econometrics,* New York: Cambridge University Press.

Woodford, M., (1987a) "Three Questions About Sunspot Equilibria as an Explanation of Economic Fluctuations," *American Economic Review,* May, 93-98.

Woodford, M., (1987b), "Equilibrium Models of Endogenous Fluctuations," Siena Lectures, Siena, Italy and Department of Economics and the Graduate School of Business, The University of Chicago.

Yakowitz, S., (1987), "Nearest-Neighbor Methods for Time Series Analysis," *Journal of Time Series Analysis,* 8, #2, 235-247.

CHAPTER 11

Chaos, Gödel, and Truth

John L. Casti

Abstract

This chapter is, for the most part, an introductory, informal exposition of the interrelations between attractors of dynamical systems, chaotic dynamics, algorithms, Gödel's Theorem, Chaitin's Theorem, and the notion of truth versus proof. Examination of the interconnections among these topical themes leads to a strong case in support of the conjecture that Chaitin's Theorem $\Rightarrow$ chaos $\Rightarrow$ Gödel's Theorem. These themes are then tied together by relating their overall message to the basic issue of the ability of modern science to predict and/or explain natural and human phenomena.

1. Chaos and Praxis

In 1963 theoretical meteorologist Edward Lorenz published his now classic paper "Deterministic Nonperiodic Flow," marking the birth of the modern theory of chaotic dynamical processes. While this paper languished in the meteorology literature, where it was seldom read and even less well understood, in the mid-1970s physicists and mathematicians like Mitchell Feigenbaum and James Yorke were fleshing out Lorenz's ideas with mathematical and computational results leading to a more general framework for the kinds of structures described in Lorenz's seminal paper. All this developmental work notwithstanding, it's still probably more correct to say that the birth time of chaos *as a recognizable intellectual endeavor* is of much more recent vintage, the midwife being James Gleick's enormously successful 1987 book *Chaos.*

In a famous remark on the fleeting nature of fame, Andy Warhol claimed that in today's helter-skelter world each person would have his or her fifteen minutes as a celebrity. With the presence of Gleick's volume on the bestseller lists for months on end, as well as the presence of a critical mass of scientists and mathematicians in the area, chaos was by 1987 ready to claim its allotted quarter of an hour. The result has been nothing short of breathtaking. Over the past couple of years it's been literally impossible to make contact with the media in any form—newspaper, television, radio, books, magazines—without soon encountering those weird and wonderful pictures of Mandelbrot sets,

Lorenz attractors, and all the other finery of the dynamical system theorist turned graphic artist. And this is just in the general press. If anything, the pace within the halls of academia and the labs of science has been even more frantic, as a quick glance at a randomly-selected journal from a randomly-selected academic discipline will quickly attest. Professors from aardvarkology to zymurgology are suddenly uncovering chaotic structure where none was even suspected before. Indeed, chaos and its many offshoots, now loosely labeled "complex systems," is probably the biggest growth industry in academia, rivalled perhaps only by the cognitive sciences.

Oddly enough amidst this plethora of applications of chaotic concepts, one of the few areas that's been left almost untouched is mathematics itself. Of course, there is a virtual army of workers in dynamical system theory and computational mathematics who have been cranking out theorems and programs at a rate that makes one's head swim. But what I have in mind is not this mathematical and computational development of the concepts and principles of chaotic motion itself. Rather, my interest in this chapter is with the implications of these foundational concepts to other areas *within* mathematics. In particular, my arguments here are directed toward showing that there are some very interesting connections between the notion of a strange attractor, the foundation upon which the chaos revolution rests, and the ideas of Gödel, Turing, and, more recently, Chaitin involving truth and proof.

My basic claim is that there is a direct chain of connection linking the existence of strange attractors, Chaitin's results on algorithmic complexity, and Gödel's Incompleteness Theorem, and that a sufficiently perceptive logician or computer scientist might just as easily ended up being the father of chaos as the meteorologist Edward Lorenz. A far-from-trivial side benefit of this "logical route" to chaos is that it dispels one of the most commonly-held misperceptions about the subject, at least in the general public's eye. That is the notion that the existence of chaotic processes means that the real world is even more unpredictable than we had previously believed. Stated in more logical terms, this "chaotic uncertainty" translates to the belief that the gap between truth and proof has been widened to a forever unbridgeable, Grand-Canyon-sized chasm. My claim here will be just the opposite: The existence of strange attractors is an absolutely necessary condition for the *narrowing* of that gap.

Since this chapter is intended to be basically an exposition of the interconnections between dynamical systems, Gödelian logic, and computation, I will follow a leisurely path in my presentation leading up to the foregoing conjectures. Most of the chapter will be devoted to an informal presentation of work on dynamics, logic, computation, and complexity, showing how they are all different sides of the same three-sided coin. The paper will then conclude with the arguments for why I believe that not only does order reside in chaos, so does truth.

2. A Strangeness in the Attraction

Stripped to its bare essentials, a *dynamical system* is a gadget composed of three elements: (1) a *manifold of states* M on which the system "lives," (2) a *vector field* $f: M \to M$, which is simply a rule stating, in effect, that "if you're currently at the point $m \in M$, move to the point $f(m)$ during the next time period," and (3) a *time set* T characterizing the moments at which changes of state take place. It's the combination of the structure of the manifold M, the nature of the vector field f, and the form of the time set T that gives rise to the individual character of any particular dynamical system. Here are a few important examples:

- $M = R^n \qquad f \in C^\infty(M, M), \qquad T = R$
- $M = R^n \qquad f \in C^1(M, M), \qquad T = \mathbb{Z}^+$
- $M = \overset{p}{\mathsf{X}}(\mathbb{Z}/k)^\infty, \quad f: M \to M, \qquad T = \mathbb{Z}^+$

The first case leads to what's termed a *smooth dynamical system,* while the second is a *discrete dynamical system,* and the last is a *p-dimensional cellular automaton,* whose "cells" can assume any of k different values at each instant of time.

The most important thing we can know about a dynamical system is its *attractor set.* This is the set of points in M to which the system trajectory moves as the time $t \to \infty$. Just as the motion of a piece of flotsam on the ocean is governed completely by a nearby whirlpool, even before the flotsam is eventually sucked up into the maelstrom, so it is with dynamical systems, too. The overall motion of the system trajectory is dictated by the character of the attractor set. Of course, it's perfectly possible for the trajectories from different starting points in M to move to different attractors. But to each point of M there corresponds only a single attractor, with the overall motion from that point governed completely by that attractor.

Classical dynamical systems have only two types of attractors: *fixed points* and *limit cycles.* Speaking informally, a fixed point of the vector field f is simply a point of M carrying the message "stay right here." In short, at a fixed point the system trajectory stops dead in its tracks. A limit cycle, on the other hand, is a closed curve in M. Once the system trajectory enters this kind of attractor, it just keeps "going around in circles" forever. Both of these types of classical attractors can be either stable or unstable, in the sense that if a point on the attractor is perturbed a little bit to a point off the attractor, the trajectory from the perturbed point either moves back to the attractor if the attractor is stable or is repelled from the attractor if it is unstable. So that's the story for classical attractors: fixed points and limit cycles, stable or unstable.

All the excitement surrounding chaos has been generated by the discovery and development of a third type of behavior, what we now call a *strange attractor.* Such an object involves a combination of unstable periodic orbits, along with a new ingredient: *aperiodic orbits.* Roughly speaking, a strange attractor is a subset of M composed of a countable number of unstable periodic orbits, together with an uncountable number of aperiodic orbits. A good image of what such an attractor looks like is to think of a bowful of spaghetti. Figure 1 shows each of the main types of attractors. Before leaving this topic, for completeness let me mention a fourth type of attractor: *quasiperiodic motion.* This is a type of behavior that's basically a combination of limit cycles that *almost* intersect. The typical example of this sort of attractor is a solenoid-like curve winding around the surface of a torus. While such attractors are of great interest in many areas, their principle role is to serve as a sort of bridge between a limit cycle and fully developed chaos. Thus, quasiperiodic attractors don't play much of a role in the arguments of this paper, and I'll leave it to the interested reader to consult the literature cited in the Notes and References section for more information.

The most convenient type of dynamical system to focus our subsequent discussion upon is a 1-dimensional cellular automaton. For simplicity, we consider such a system for which the state manifold is $M = (\mathbb{Z}/2)^{\infty}$. Thus, at each moment in time $t = 0, 1, \ldots$, the system state is simply a countably infinite sequence, each of whose entries is either 0 or 1. The vector field f is a prescription telling what the value of each element of the sequence will be at the next tick of the clock. The system's trajectory is then given by a succession of such sequences. Extensive computer experimentation and analytic work on 1-dimensional cellular automata has shown that the attractor set of such a system can be one of the four characteristic types shown in Figure 2, giving rise to the standard categorization of cellular automata by the not very imaginative labels Types A, B, C, and D.

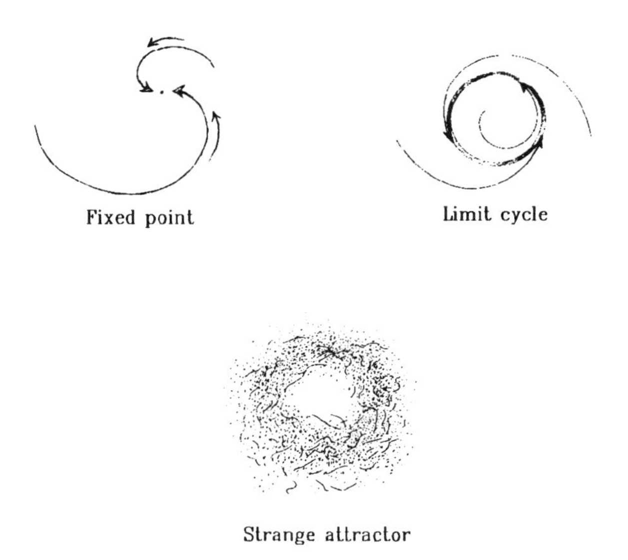

Figure 1. Classical and Strange Attractors

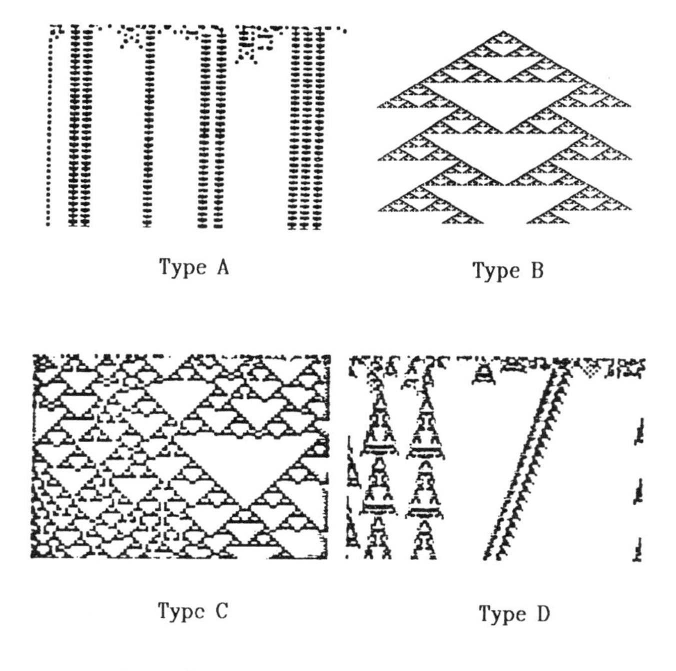

Figure 2. Cellular Automata Attractors

In Figure 2, the top line of each diagram represents the initial state of the system at time $t = 0$, where each location on the line is either black (if the cell has the value 1) or white (if the cell has the value 0). The successive states on the trajectory at times $t = 1, 2, \ldots$ are then represented by the successive lines as we move down the diagram from top to bottom. Thus, with the Type A system shown in the figure, we see that after a short transient period of 8 or 9 steps, the system settles into a fixed state that persists thereafter, i.e., it moves to a fixed point. With this picture in mind, we find the following matchups between the attractor types discussed earlier and the four types of cellular automata.

Fixed point	$\Longleftrightarrow$	Type A
Limit cycle	$\Longleftrightarrow$	Type B
Strange attractor	$\Longleftrightarrow$	Type C
Quasiperiodic orbit	$\Longleftrightarrow$	Type D

I should point out that these matchups do not follow the pattern generally given in the cellular automata literature. In particular, for reasons I find a little difficult to understand, cellular automata theorists, while carefully distinguishing Type C and Type D behavior for their own analyses, lump both types into the same category of "strange attractor" when trying to make contact with the traditional dynamical systems literature. Fortunately, this point is of no consequence for the arguments I want to make here. Nevertheless, it seems to me that it would be desirable for researchers to make a more concerted effort to match the four types of classical and modern attractors with the four types of cellular automata behavior, if for no other reason than for standardization.

Happily, these elementary ideas about attractors are all we need to know about dynamical systems, at least for this chapter. But since my overall goal is to make contact with ideas of truth and proof, let me move on now to a consideration of the vehicle by which mathematicians grapple with these eternal notions: formal systems.

3. Speaking Formally

In his famous epigram on the nature of mathematics, Bertrand Russell claimed that "pure mathematics is the subject in which we do not know what we are talking about, or whether what we are saying is true." This pithy remark summarizes the content of Hilbert's Program for the formalization of mathematics: The development of a purely syntactic framework for establishing the truth or falsity of all mathematical assertions. Hilbert believed that the way to eliminate the possibility of paradoxes and inconsistencies arising in mathematics was to create a purely formal, essentially "meaningless," framework within which to speak about the truth of mathematical statements. Such a framework is now termed a *formal system,* and it constitutes the jumping-off point for our investigation of the gap between what can be proved and what is actually true in the universe of mathematics.

The "meaningless statements" of a formal system are finite sequences of abstract *symbols,* usually called *symbol strings.* A finite number of these strings are taken as the *axioms* of the system. To complete the system, there are a finite number of *transformation rules* telling us how a given string of symbols can be converted into another such string. The general idea is to start from one of the axioms and apply a finite sequence of transformations, thereby converting the axiom into a finite sequence of new strings in which each string in the sequence is either an axiom or is derived from its predecessors by application of the transformation rules. The terminal string in such a sequence is called a *theorem* of the system. The totality of all theorems constitutes the *provable strings* or *statements* of the system. But note carefully that these so-called "statements" don't actually say anything; they are just strings of abstract symbols. Let's see how this setup works with a simple example.

Suppose the symbols of our system are the three more or less culturally-free objects ★ (star), ✠ (maltese cross), and ☁ (cloud). Let the two-element string ✠☁ be the sole axiom of the system, and take the transformation rules to be:

Rule I: x☁ ⟶ x☁★
Rule II: ✠x ⟶ ✠xx
Rule III: ☁☁☁ ⟶ ★
Rule IV: ★★ ⟶ —

In these rules, "x" denotes an arbitrary finite string of stars, crosses, and clouds, while ⟶ means "is replaced by." The interpretation of Rule IV is that anytime two stars appear they can be dropped from the string. Now let's see how these rules can be used to derive a theorem in this simple formal system.

Starting with the single axiom ✠☁, we can deduce that the string ✠★☁ is a theorem by applying the transformation rules in the following sequence:

⟶ ✠☁	⟶ ✠☁☁	⟶ ✠☁☁☁☁	⟶ ✠★☁
(Axiom)	(Rule II)	(Rule II)	(Rule III)

Such a sequence of steps, starting from an axiom and ending at a statement like ✠★☁, is termed a *proof sequence* for the theorem represented by the last string in the sequence. Observe that when applying Rule III at the final step, we could have replaced the last three ☁s from the preceding string rather than the first three, thereby ending up with the theorem ✠☁★ instead of ✠★☁.

You will probably have also noted that all the intermediate strings obtained in moving from the axiom to the theorem begin with a ✠. It's easy to see from the action of the transformation rules for this system that every string will have this property. This is a *metamathematical* property of the system, since it's a statement *about* the system rather than one made *in* the system itself. The distinction between what the system can say from the inside (its strings) and what we can say about the system from the outside (properties of the strings) is of the utmost importance for Gödel's results.

But what does all this abstract nonsense have to do with mathematics? What do maltese crosses, clouds, and stars have to do with things like the sum of the first n positive integers or the angles of a triangle? Indeed, what do these symbols have to do with *anything?* The answer to this eminently sensible query lies in one word: *interpretation.* Depending on the kind of mathematical structure under consideration, we have to make up a dictionary to translate (i.e., interpret) the abstract symbols and rules of the formal system into the objects of the mathematical structure in question. By this dictionary-construction step, we attach semantic meaning to the abstract, purely syntactic structure of the symbols and strings of the formal system. Thereafter all the theorems of the formal system can be interpreted as true statements about the associated mathematical objects. The diagram on the next page illustrates this crucial distinction between the purely syntactic world of formal systems and the meaningful world of mathematics.

Let me again emphasize that there are two entirely different worlds being mixed in this setup: the purely syntactic world of the formal

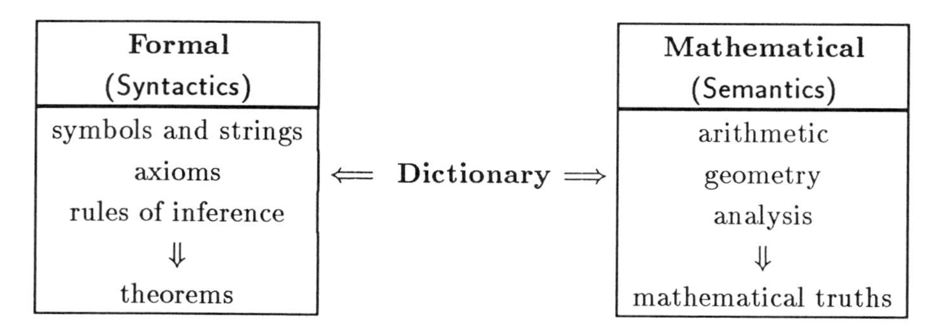

The Worlds of Formal Systems and Mathematics

system and the meaningful world of mathematical objects and their properties. And in each of these worlds there is a notion of truth: theorems in the formal system, correct mathematical statements such as "$2+5=7$" or "the sum of the angles of a triangle equals 180 degrees" in the realm of mathematics. The connection between the two worlds lies in the interpretation of the elements of the formal system in terms of the objects and operations of an associated mathematical structure. Once this dictionary has been written and the associated intepretation established, then we can hope along with Hilbert that there will be a perfect, one-to-one correspondence between the true facts of the mathematical structure and the theorems of the formal system. Speaking loosely, we seek a formal system in which every mathematical truth translates into a theorem, and conversely. Such a system is termed *complete.* Moreover, if the mathematical structure is to avoid contradiction, a mathematical statement and its negation should never both translate to theorems, i.e., both be provable in the formal system. Such a system in which no contradictory statements can be proved is termed *consistent.*

By the time Hilbert proposed his Program, it was already known that the problem of the consistency of mathematics as a whole was reducible to the determination of the consistency of arithmetic. So the problem became to give a "theory of arithmetic," i.e., a formal system that was (i) finitely describable, (ii) complete, (iii) consistent, and (iv) sufficiently powerful to represent all the statements that can be made about the natural numbers. By the term "finitely describable" what Hilbert had in mind is not only that the number of axioms and rules of the system be finite in number, but also that every provable statement in the system, i.e., every theorem, should be provable in a finite number of steps. This condition seems reasonable enough, since you don't really have a theory at all unless you can tell other people about it. And you certainly can't tell them about it if there are an infinite number of axioms, rules, and/or steps in a proof sequence.

A central question that arises in connection with any such *formalization* of arithmetic is to ask if there is a finite procedure by which we can "decide" every arithmetical statement. So, for instance, if we make the statement: "The sum of two prime numbers greater than 2 is always an even number" (Goldbach's Conjecture), we want a finite procedure, essentially a computer program that halts after a finite number of steps, telling us whether that statement is true or false, i.e., provable or not in some formal system powerful enough to encompass ordinary arithmetic. An example of such a decision procedure for the ✠-★-⌘-formal system considered earlier is given by the far-from-obvious criterion: "A string is a theorem if and only if it begins with a ✠ and the number of ⌘s in the string is not divisible by 3."

The question of the existence of a procedure to decide all statements about arithmetic, i.e., about the natural numbers, is Hilbert's famous *Entscheidungsproblem* (Decision Problem). Hilbert was convinced that such a formalization of arithmetic was possible, and his manifesto presented at the 1928 International Congress of Mathematicians in Bologna, Italy challenged the world's mathematical community to find/create it. It's somehow comforting to know how dramatically and definitively wrong even a man as great as Hilbert can be!

In 1931, less than three years after Hilbert's Bolognese call to arms, Kurt Gödel published the following metamathematical fact, perhaps the most famous, broad-sweeping mathematical (and philosophical) result of this century:

Gödel's Incompleteness Theorem—Informal Version

Arithmetic is not completely formalizable

Remember that for a given mathematical structure like arithmetic, there are an infinite number of ways we can choose a finite set of axioms and rules of a formal system in an attempt to mirror syntactically the mathematical truths of the structure. What Gödel's result says is that **none** of these choices will work; there does not and cannot exist a formal system satisfying all the requirements of Hilbert's Program. In short, there are no rules for generating **all** the truths about the natural numbers .

Gödel's result is shown graphically in Figure 3 for a given formal system **M**, where the entire square represents all possible arithmetical

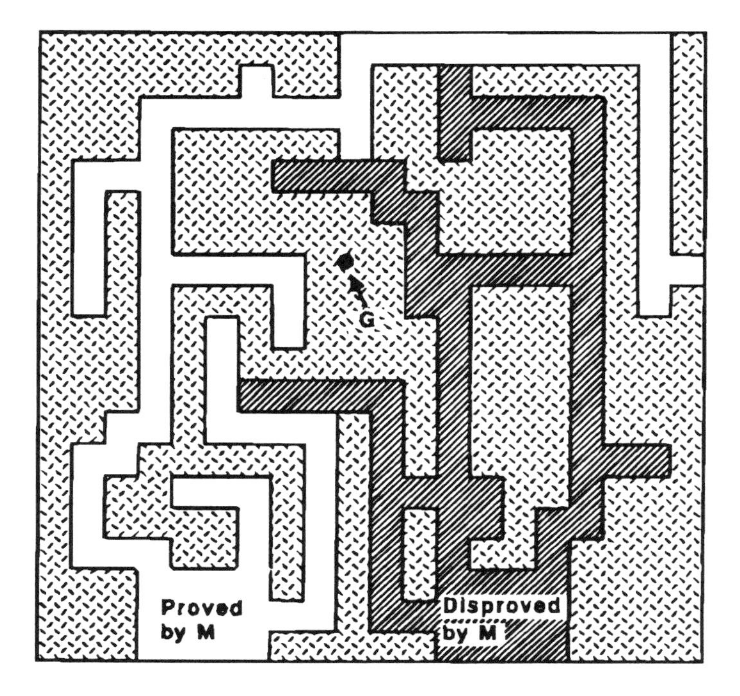

Figure 3. Gödel's Theorem in Logic Space

statements. As we prove an arithmetical statement true using the rules of the formal system **M**, we color that statement white; if we prove the statement false, we color it black. Gödel's Incompleteness Theorem (which henceforth I'll label more compactly as just Gödel's Theorem) says that there will always exist statements like **G** that are eternally doomed to a life in the shadow world of gray; it's impossible to eliminate the gray and color the entire square in black and white. And this result holds for *every* possible formal system **M**, provided only that the system is consistent. That is, for every consistent formal system **M** there is at least one statement **G** that can be seen to be true but is yet unprovable in **M**. So there's no washing away all the gray! We call a statement like **G** *undecidable* in **M**, since it can be neither proved nor disproved within the framework of that formal system. And if we add that undecidable statement **G** as an axiom, thereby creating a new formal system **M**′, the new system will have its own *Gödel sentence* **G**′—true, but unprovable.

By his Theorem Gödel snuffed out once and for all Hilbert's flickering hope of providing a complete and total axiomatization of arithmetic, hence of mathematics. Since Gödel's Theorem represents one of the pinnacles of human intellectual achievement, not to mention forming the basis for a whole host of related developments in mathematics, philosophy, computer science, linguistics, and psychology, let me spend a few pages looking at how one could ever prove such a profound, mind-boggling result.

4. The Undecidable

In arriving at his proof of the incompleteness of arithmetic, Gödel's first crucial insight was to recognize that every formalization of a branch of mathematics is itself a mathematical object in its own right. So if we create a formal system intended to capture the truths of arithmetic, that formal system can be studied not just as a set of mindless rules for manipulating symbols, but also as an object possessing mathematical as well as syntactic properties. In particular, since Gödel was interested in the truths about numbers, he showed how it would be possible to represent any formal system purporting to encompass arithmetic within arithmetic itself. In short, Gödel saw a way to mirror all statements about relationships between the natural numbers by using these very same numbers themselves.

To understand this "mirroring" operation a little more clearly, consider the familiar situation at a post office or airline ticket counter where customers are given a number when they enter in order to indicate their position in the queue. Suppose Clint and Brigitte each want to return home for the holidays. To arrange their flights they go to the local Pan Am office, where upon entering Clint receives the service number 4, while Brigitte comes in a bit later and gets number 7. By this service-assignment scheme, the real-world fact that Clint will be served before Brigitte is "mirrored" in the purely arithmetical truth that 4 is less than 7. In this way a truth of the real world has been faithfully translated, or mirrored, by a truth of number theory. Gödel used a tricky variant of this kind of numbering scheme to code all possible statements about arithmetic in the language of arithmetic itself, thereby using arithmetic as both an interpreted mathematical object and as an uninterpreted formal system with which to talk about itself. It's revealing to see how this *Gödel numbering* scheme actually works.

For sake of exposition, let's consider a somewhat streamlined version of the Russell-Whitehead language of logical calculus due to Ernest Nagel and James R. Newman. In this language there are elementary signs and variables. To follow Gödel's scheme, suppose there are the

Sign	Gödel Number	Meaning
~	1	not
$\vee$	2	or
$\supset$	3	if ... then
$\exists$	4	there exists
$=$	5	equals
0	6	zero
s	7	the immediate successor of
(	8	punctuation
)	9	punctuation
′	10	punctuation

Table 1. Gödel Numbering of the Elementary Signs

ten logical signs shown in Table 1 along with their Gödel number, an integer between 1 and 10.

In addition to the elementary signs, the language of the *Principia* contains three types of variables: numerical, sentential, and predicate. Numerical variables are symbols like x, y and z for which we can substitute numbers or numerical expressions. Sentential variables, usually denoted $p, q, \ldots$, can be replaced by formulas (sentences). Finally, for predicate variables $P, Q, \ldots$ we can substitute properties such as "prime," "less than," and "odd." If we have only 10 elementary signs, then in Gödel's numbering system numerical variables are coded by prime numbers greater than 10, sentential variables by the squares of prime numbers greater than 10, and predicate variables by the cubes of prime numbers greater than 10, all the prime numbers taken in numerical order.

To see how this numbering process works in practice, consider the logical formula $(\exists x)(x = \mathbf{s}y)$, which translated into English reads: "There exists a number x which is the immediate successor of the number y." Since x and y are numerical variables, the coding rules dictate that we make the assignment $x \rightarrow 11$, $y \rightarrow 13$, since 11 and 13 are the first two prime numbers larger than 10. The other symbols in the formula can be coded using the correspondence in Table 1. Carrying out this coding yields the sequence of numbers $(8, 4, 11, 9, 8, 11, 5, 7, 13, 9)$, corresponding to reading the logical expression symbol-by-symbol and substituting the appropriate number according to the coding rule.

While this sequence of 10 numbers pins the logical formula down unambiguously, for a variety of reasons it's more convenient to be able to represent the formula by a single number. Gödel's procedure for doing this is to take the first 10 prime numbers (since there are 10 symbols in the formula) and multiply them together, each prime number being raised to a power equal to the Gödel number of the corresponding element in the formula. So, since the first ten prime numbers in order are 2, 3, 5, 7, 11, 13, 17, 19, 23, and 29, the final Gödel number for the above formula is

$$(\exists x)(x = \mathrm{s}y) \longrightarrow 2^8 \times 3^4 \times 5^{11} \times 7^9 \times 11^8 \times 13^{11} \times 17^5 \times 19^7 \times 23^{13} \times 29^9$$

Using this kind of numbering scheme, Gödel was able to attach a unique number to each and every statement and sequence of statements about arithmetic that could be expressed in the logical language of *Principia Mathematica.*

By this coding procedure every possible proposition about the natural numbers can itself be expressed as a number, thereby opening up the possibility of using number theory to examine its own truths. The overall process can be envisioned by appealing to the metaphor of a locomotive shunting boxcars back and forth in a freight yard. This idea, due to Douglas Hofstadter, is shown in Figure 4. In the upper part of the figure we see the boxcars with their uninterpreted numbers painted on the sides of the cars, while looking down from the bird's-eye view we see the interpreted symbols inside each car. The shuffling of the cars according to the rules for manipulating logical symbols and formulas is mirrored by a corresponding transformation of natural numbers, i.e., statements of arithmetic—and vice-versa.

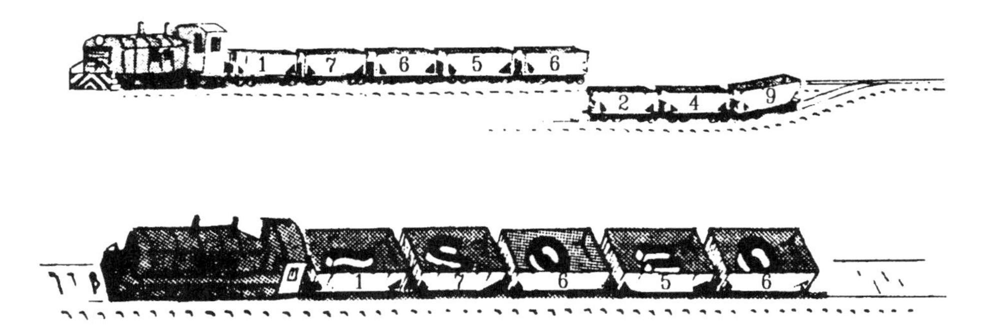

Figure 4. Freight Train View of Gödel Numbering

Deep insight and profound results necessarily involve seeing the connection between several ideas at once. In the proof of Gödel's Theorem there are two crucial notions that Gödel had to juggle simultaneously, Gödel numbering being the first. Now for Big Idea #2.

Logical paradoxes of the sort that worried Hilbert are all based on the notion of self-reference. The granddaddy of all such conundrums is the so-called Epimenides Paradox, one version of which is

This sentence is false.

What Gödel wanted to do was find a way to express such paradoxical self-referential statements within the framework of arithmetic. However, a statement like the Epimenides Paradox involves the notion of truth, something that logician Alfred Tarski had already shown could not be captured within the confines of a formal system. Enter Gödel's Big Idea #2.

Instead of dealing with the eternally slippery notion of truth, Gödel's insight was to replace "truth" by something that **is** formalizable: the notion of *provability*. Thus, he translated the Epimenides Paradox above into the Gödel sentence:

This statement is not provable.

This sentence, of course, is a self-referential claim about a particular "statement," the statement mentioned in the sentence. However, by his numbering scheme Gödel was able to mirror this assertion by a corresponding self-referential, metamathematical statement in the language of arithmetic itself. Let's follow through the logical consequences of this mirroring.

If the statement is provable, then it's true; hence, what it says must be true and it is *not* provable. Thus, the statement and its negation are both provable, implying an inconsistency. On the other hand, if the statement is not provable then its assertion is true. In this case the statement is true but unprovable, implying the formal system is incomplete.

Gödel was able to show that for *any* consistent formalization of arithmetic such a Gödel sentence must exist; consequently, the formalization must be incomplete. The bottom line then turns out to be that in **every** formal system powerful enough to contain all statements about the natural numbers, there exists a statement that is unprovable by the rules of the system. Nevertheless, that statement represents a true assertion about the natural numbers, one that we can see is true by "jootsing," to use Douglas Hofstadter's colorful term for "jumping out of the system." Almost as an aside, Gödel also showed how to construct an arithmetical statement **A**, which translates into the metamathematical claim "arithmetic is consistent." He then demonstrated that the statement **A** is not provable, implying that the consistency of arithmetic cannot be established by arguments that can be made using the formal system of arithmetic itself. Putting all these notions together, we come to:

Gödel's Theorem—Formal Logic Version

For every consistent formalization of arithmetic, there exist arithmetic truths unprovable within that formal system

Since the sequence of steps leading up to Gödel's startling conclusions are both logically tricky and intricately intertwined, I have summarized the principal landmarks along the road in Table 2.

Gödel's Theorem has many profound implications, both for science and for philosophy. Before entering into a detailed consideration of some of these developments, it's worth pausing for a moment to summarize what Gödel's conclusions have to say about the limits of human reasoning. When all the mathematical smoke clears away, Gödel's message is that mankind will never know the final secret of the universe by rational thought alone; it's impossible for human beings to ever formulate a complete set of rules for describing **all** the properties of the natural numbers. There will always be arithmetic truths that escape our ability to fence them in by means of rational analysis. As logician and science-fiction author Rudy Rucker has expressed it, Gödel's Theorem leaves scientists in a position similar to that of Joseph K. in Kafka's novel *The Trial.* We scurry about, running up and down endless corridors, buttonholing people, going in and out of offices and, in general, conducting investigations. But we will never achieve ultimate success; there is no final verdict in the courtroom of science giving us absolute truth. However, Rucker notes, "to understand the labyrinthine nature of the castle [courtroom] is, somehow, to be free of it." And there's no understanding of the castle of science that digs deeper into its foundations than the understanding given by Gödel's Theorem.

Let's now shift our attention from formal systems to computer systems as the last preliminary in the journey from chaos to Gödel to truth.

Gödel Numbering: Development of a coding scheme to translate every logical formula in *Principia Mathematica* into a "mirror-image" statement about the natural numbers.

⇓

Epimenides Paradox: Replace the notion of "truth" by that of "provability," thereby translating the Epimenides Paradox into the assertion "This statement is unprovable."

⇓

Gödel Sentence: Show that the sentence "This statement is unprovable" has an arithmetical counterpart, its Gödel sentence **G**, in every conceivable formalization of arithmetic.

⇓

Incompleteness: Prove that the Gödel sentence **G** must be true if the formal system is consistent.

⇓

No Escape Clause: Prove that even if additional axioms are added to form a new system in which **G** is provable, the new system with the additional axioms will have its own unprovable Gödel sentence.

⇓

Consistency: Construct an arithmetical statement asserting that "arithmetic is consistent." Prove that this arithmetical statement is not provable, thus showing that arithmetic *as a formal system* is too weak to prove its own consistency.

Table 2. The Main Steps in Gödel's Proof

5. Turing Around

In 1935 Alan Turing was an undergraduate student at Cambridge, sitting in on a course of lectures on mathematical logic. During the course Turing was introduced to Hilbert's *Entscheidungsproblem,* which as we know involves determining whether or not there exists an effective procedure for deciding if a given proposition follows from the axioms of a formal system. The central difficulty Turing had in trying to come to grips with this problem was that there was no clear-cut notion of what was to count as an "effective procedure." Despite the fact that humans had been calculating for thousands of years, in 1935 there was still no good answer to the question: "What is a computation?" Turing set out to overcome this difficulty and solve the *Entscheidungsproblem.* To do so he had to invent a theoretical gadget that ended up serving as the keystone in the arch of the modern theory of computation.

Just as Gödel had to replace the intuitive notion of truth by the formalizable concept of provability, Turing had to find a replacement for the intuitive idea of an "effective process." What he came up with is what we now call an *algorithm,* an idea he modeled on the steps a human being actually goes through when carrying out a computation. In essence, Turing saw an algorithm as a rote process or set of rules that tells one how to proceed under any given set of circumstances. Let's look at an example.

Consider the well-known Euclidean algorithm for finding the largest number dividing two given whole numbers a and b. Assume that a is larger than b and let "rem $\{\frac{x}{y}\}$" denote the remainder after dividing x by y. Then the Euclidean algorithm consists of calculating the sequence of integers $\{r_1, r_2, \ldots\}$ by the rules

$$
\begin{aligned}
r_1 &= \text{rem}\left\{\frac{a}{b}\right\} \\
r_2 &= \text{rem}\left\{\frac{b}{r_1}\right\} \\
r_3 &= \text{rem}\left\{\frac{r_1}{r_2}\right\} \\
&\vdots
\end{aligned}
$$

where the process continues until we obtain a quantity r_n such that the rem $\{\frac{r_{n-1}}{r_n}\} = 0$. The process halts with the number r_n, which indeed turns out to be the largest integer dividing both a and b. To illustrate the process concretely, suppose we let $a = 14$ and $b = 6$. Then following the steps of the Euclidean algorithm, we obtain

$$
\begin{aligned}
r_1 &= \text{rem}\left\{\frac{14}{6}\right\} = 2 \\
r_2 &= \text{rem}\left\{\frac{6}{2}\right\} = 0
\end{aligned}
$$

Thus, we conclude that 2 is the greatest common divisor of 14 and 6.

What's important for us here is that the steps of the Euclidean algorithm are rigidly prescribed and unvarying. One and only one operation is specified at each step, and there is no interpretation of the intermediate results or any skipping of steps—just a boring, basically mechanical repetition of the operations of division and keeping the remainder. This blind following of a set of rules is the distilled essence of what constitutes an algorithm. To reflect the mechanical nature of what's involved in carrying out the steps of an algorithm, Turing invented a hypothetical kind of computer now called a *Turing machine.* He then used the properties of this "machine" in order to formalize his attack on the *Entscheidungsproblem.* Here's how.

A Turing machine consists of two components: (i) an infinitely long tape ruled off into squares that can each contain one of a finite set of symbols, and (ii) a scanning head that can be in a finite number of internal states. The head can read the squares on the tape and write one of the symbols onto each square. The behavior of the Turing machine is controlled by an algorithm, or what we now call a *program.*

The program is composed of a finite number of instructions, each of which is selected from the following set of possibilities: change/retain the internal state of the head, print a new/keep the old symbol on the current square, move left/right one square; stop. Which of these nine simple possibilities the head takes at any step of the process is determined by the current state of the head and what it reads on the square. But rather than continuing to speak in these abstract terms, let's just run through an example to get the gist of how such a device operates.

Imagine we have a Turing machine with 3 internal states A, B, and C, and that the tape symbols that the head can read/write are "0" and "1." Let's suppose we want to use this machine to carry out the addition of two whole numbers. For definiteness, we agree to represent an integer n by a string of n consecutive 1's on the tape. I claim that the program shown in Table 3 serves to add any two whole numbers using this 3-state Turing machine.

	Symbol Read	
State	1	0
A	1, R, A	1, R, B
B	1, R, B	0, L, C
C	0, STOP	STOP

Table 3. A Turing Machine Program for Addition

The reader should interpret the entries in Table 3 in the following way: The first item is the symbol the head should print, the second element is the direction the head should move, R(ight) or L(eft), and the final element is the internal state the head should move into. Note that the machine stops as soon as the head goes into state C. Let's see how it works for the specific case of adding 2 and 5.

Since our interest is in using the machine to add 2 and 5, we place two 1's and five 1's on the input tape, separating them by a 0 to indicate that they are two distinct numbers. Thus the machine begins by reading the input tape

...	0	0	0	**1**	1	0	1	1	1	1	1	0	0	0	0	...

Assume the head starts in state A, reading the first nonzero symbol on the left. (Note: Here, and in the other examples of this chapter, the position of the reading head will be denoted in boldface.) Since this symbol is a 1, the program tells the machine to print a 1 on the square and move to the right, retaining its internal state A. The head is still in state A and the current symbol read is again a 1, so the machine repeats the previous step and moves one square further to the right. Now, for a change, the head reads a 0. The program tells the machine to print a 1, move to the right, and switch to state B. I'll leave it to the reader to complete the remaining steps of the program, verifying that when the machine finally halts the tape ends up looking just like the input tape above, except with the 0 separating 2 and 5 having been eliminated, i.e., the tape will have seven 1's in a row, as required. Adventurous readers might like to consider the action of the 6-state Turing machine having three tape symbols 0, 1, and 2, whose program is given in Table 4. The answer can be found in the Notes and References. Assume the input tape is

...	0	0	0	**1**	2	2	1	0	0	0	0	0	0	0	0	...

Before looking at the revolutionary implications of Turing's idea, let me pause here to emphasize that Turing machines are definitely not "machines" in the usual sense of being material devices. Rather they are "paper computers," completely specified by their programs once we have agreed on the number of states and the set of symbols that can be written onto their tapes. Thus, when we use the term "machine" in what follows, the reader should read "program" or "algorithm,"

	Symbol Read		
State	0	1	2
A	Print YES, STOP	0, R, B	0, R, C
B	0, L, D	1, R, B	2, R, B
C	0, L, E	1, R, C	2, R, C
D	STOP	0, L, F	0, Print NO, STOP
E	STOP	0, Print NO, STOP	0, L, F
F	0, R, A	1, L, F	2, L, F

Table 4. A Turing Machine Program for ? ?

i.e., software, and put all notions of hardware out of sight and out of mind. This abuse of the term "machine" should have been clear from Turing's idea of an *infinite* storage tape, but it's important to make the distinction as hard and fast as possible: Turing machine equals program. Period.

Modern computing devices, even home computers like the one I'm using to write this chapter, look vastly more complicated and powerful in their calculational power than a Turing machine with its handful of internal states and very circumscribed repertoire of scanning-head actions. Nevertheless, this turns out just not to be the case, and a large part of Turing's genius was to recognize that *any* algorithm, i.e., program, executable on *any* computing machine, idealized or otherwise, can also be carried out on a particular version of his machine, termed a *universal Turing machine* (UTM). So except for the speed of the computation, which definitely *is* hardware dependent, there's no computation that my machine (or anyone else's) can do that can't be done with a UTM.

To specify his UTM, Turing realized that the program for any computer could be coded by a series of 0's and 1's. Consequently, the program itself can be regarded as another kind of input data and written onto the input tape along with the data it is to operate on. With this key insight at hand, Turing constructed a program that could simulate the action of any other program P when given P as part of its input, i.e., he created a UTM. The operation of a UTM is simplicity itself.

Suppose we have a particular Turing machine with program P. Since a Turing machine is completely determined by its program, all we need do is feed the program P into the UTM along with the input data. Thereafter the UTM will simulate the action of P on the data. In short, there will be no recognizable difference between running the program P on the original machine or having the UTM pretend it *is* the Turing machine P.

Turing's work finally put the idea of a computation on a solid scientific footing, enabling us to pass from the vague, intuitive idea of an "effective process" to the precise, well-defined notion of an algorithm. In fact, Turing's work, along with that of the American logician Alonzo Church, forms the basis for what has come to be called the

• Turing-Church Thesis •

Every effective process is implementable by running a suitable program on a UTM

The universal Turing machine also gives us a tool for identifying just what kinds of quantities are actually computable. By definition, a number is *computable* if and only if it can be obtained as the output of a program P run on a UTM. But isn't every number computable? The surprising fact is that almost every real number is *not* computable. We'll return to this point later. For now, let's have some fun with Turing machines and look at an example of just one such uncomputable quantity.

Suppose you're given an n-state Turing machine and an input tape filled entirely with 0's. The challenge is to write a program for this machine displaying the following characteristics: (i) the program must eventually halt, and (ii) the program should print as many 1's as possible on the tape before it stops. Obviously, the number of 1's that can be printed is a function only of n, the number of internal states available to the machine. Equally clear is the fact that if $n = 1$, the maximum number of 1's that can be printed is only one, a result following immediately from the requirement that the program cannot run forever. If $n = 2$, it can be shown that the maximum number of 1's that can printed before the machine halts is four. Such programs that print a maximal number of 1's before halting are called *n-state Busy Beavers.* Table 5 gives the program for a 3-state Busy Beaver, while Figure 5 shows how this program can print six 1's on the tape before stopping.

Now for our uncomputable function. Define the quantity $BB(n)$ to be the number of 1's written by an n-state Busy Beaver program. Thus, the Busy Beaver function $BB(n)$ is the maximal number of 1's that any

	Symbol Read	
State	0	1
A	1, R, B	1, L, C
B	1, L, A	1, R, B
C	1, L, B	1, STOP

Table 5. A 3-State Busy Beaver

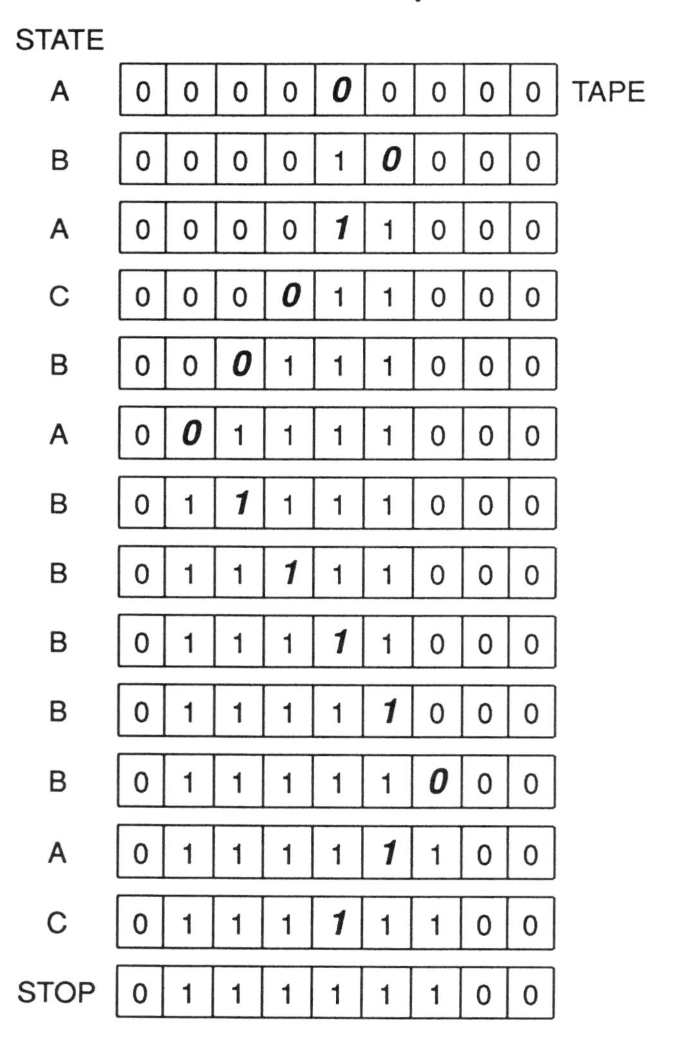

Figure 5. The Action of a 3-State Busy Beaver

halting program can write on the tape of an n-state Turing machine. We have already seen that $BB(1) = 1$, $BB(2) = 4$, and $BB(3) = 6$. From these results for small values of n, you might think that the function $BB(n)$ doesn't have any particularly interesting properties as n gets larger. But, just as you can't judge a book by its cover (or title), you also can't judge a function from its behavior for just a few values of its argument. In fact, it's been shown that

$$BB(12) \geq 6 \times 4096^{4096^{4096^{\cdot^{\cdot^{\cdot^{4096^{4}}}}}}}$$

where the number 4096 appears 166 times in the dotted region! So in trying to calculate the value of $BB(12)$, we reach the point where it becomes impossible to distinguish between the finite and the infinite. It turns out that for large enough values of n, the quantity $BB(n)$ exceeds that of *any* computable function for that same argument n. In other words, the Busy Beaver function $BB(n)$ is uncomputable. So for a concrete example of an effectively uncomputable number, just take a Turing machine with a large number of states n. Then ask for the value of the Busy Beaver function for that value of n. The answer is, for all intents and purposes, an uncomputable number. Now let's get back to Turing's resolution of the *Entscheidungsproblem.*

Comparing the workings of a Turing machine and the operations we went through earlier using the transformation rules of a formal system, it doesn't look as if there's any real difference between the two. And so it is: Given any formal system F, there is a Turing machine M such that the possible theorems of F coincide with the possible outputs of M. Conversely, given any Turing machine, we can find a formal system such that the possible outputs of the machine are exactly the possible theorems of the formal system. Using this "isomorphism" between machines and formal systems, Turing was able to translate Hilbert's *Entscheidungsproblem* involving theorems in a formal system into a closely-related problem expressible in the language of machines. This new problem has the property that a negative solution implies the same for the *Entscheidungsproblem.*

When phrased in the language of logic and computing, Turing's machine-theoretic question is known as the

• Halting Problem •

Is there a general algorithm for determining if a program will halt?

Put more precisely, we ask for a program that will accept another program P and input data set I, outputting a YES if the program P eventually halts when processing the data I, a NO if it doesn't. Of course, for some combinations of P and I such an algorithm certainly does exist (e.g., the Euclidean algorithm discussed above always halts). But the Halting Problem asks a much stronger question: Does there exist a **single** algorithm (program) that will give the correct answer in *all* cases?

The connection between formal systems and machines makes clear the very close relationship between the Halting Problem and Hilbert's original *Entscheidungsproblem.* Turing settled the matter once and for all in the negative with his 1936 result that given a program P and an input data set I, there is no way in general to say if P will ever finish processing the data I. As noted, a negative solution to the *Entscheidungsproblem* follows immediately as a corollary. Since we'll discuss the proof of this result within the context of another problem in a later section, let's close by noting that the equivalence between Turing machines and formal systems implies that there must be a machine-theoretic version of Gödel's Theorem. And, indeed, this is the case.

Gödel's Theorem—Turing Machine Version
No computer program can ever generate all the truths of arithmetic

Cursory as they are, the foregoing considerations of dynamical systems, formal systems, and computing systems strongly suggest that they are in fact just three sides of the same coin. The diagram below summarizes the whole matter. The discussion on Turing machines

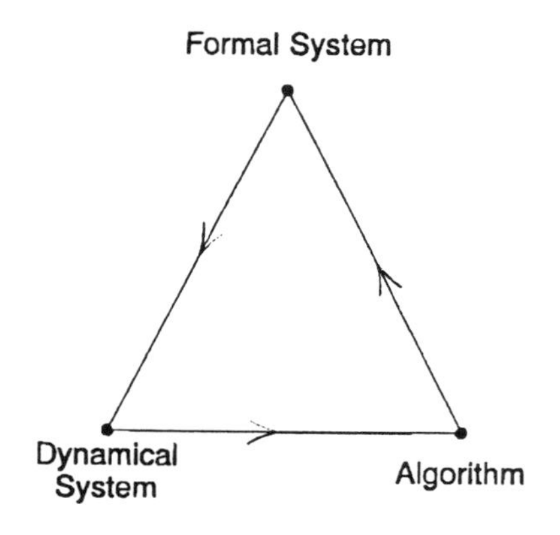

has already filled in the "Algorithm-Formal System" leg of this triangle, while the "Dynamical System-Algorithm" leg follows in an almost self-evident fashion by considering the way one would determine numerically the trajectory of a dynamical system. Thus, only the "Formal System-Dynamical System" remains to complete the triangle of implications. The appropriate dictionary is shown in Table 6. With these matters settled, let me finally turn to the main point of the paper: How this constellation of "systems" relates to chaos and truth.

Formal System	Dynamical System
symbol alphabet	state manifold
symbol string	state
grammar	constraints
axioms	initial condition
rules of inference	vector field
proof sequence	trajectory
theorem	attractor

Table 6. Formal Systems versus Dynamical Systems

6. The Importance of Being Arbitrary

The replacement of the Earth by the Sun as the center of heavenly motions is widely (and rightly) seen as one of the great scientific paradigm shifts of all time. But what is often misunderstood is the reason why this Copernican "revolution" eventually carried the day with the scientific community. The commonly-held view is that Copernicus's heliocentric model vanquished the competition, especially the geocentric view of Ptolemy, because it gave better predictions of the positions of the celestial bodies. In actual fact, the predictions of the Copernican model were a little *worse* than those obtained using the complicated series of epicycles and other curves constituting the Ptolemaic scheme, at least to within the accuracy available using the measuring instruments of the time. No, the real selling point of the Copernican model was that it was much *simpler* than the competition, yet still gave a reasonably good account of the observational evidence.

The Copernican revolution is a good case study in how to wield Occam's Razor to slit the throat of the competition: When in doubt, take the simplest theory that accounts for the facts. The problem is that it's not always easy to agree on what's "simple." The notion of simplicity, like "truth," "beauty," and "effective process" is an intuitive one, calling for a more objective characterization, i.e., formalization, before we can ever hope to agree about the relative complexities of different theories.

In 1964 Ray Solomonoff, a researcher at the Zator Corporation, published a pioneering article in which he presented a scheme to measure the complexity of a scientific theory objectively. He based his idea on the premise that a theory for a particular phenomenon must encapsulate somehow the empirical data available telling about that phenomenon. Thus, Solomonoff proposed to identify a theory with a computer program that reproduces the empirical data. He then argued that the complexity of such a theory could be taken to be the "length" of the shortest such program, measured perhaps by the number of keystrokes needed to type the program or some similarly unambiguous scheme.

By this definition of the complexity of a theory, Solomonoff was anticipating an observation made later by Gregory Chaitin and by the mathematician and philosopher René Thom, who noted that the point of a scientific theory is to reduce the arbitrariness in the data. If a program (read: theory) reproducing the data has the same length as the data itself, then the theory is basically useless since it contributes nothing toward reducing the arbitrariness in the data. Such a putative theory doesn't in any way "compress" the data, and we could just as well account for the observations by writing them out explicitly. But we don't need a theory to do that. So if a set of observations can only be reproduced by programs of the same length as the observations themselves, then we're justified in calling the observations "random" in the sense that they can be neither predicted nor explained.

At about the same time Solomonoff was developing these ideas about the complexity of scientific theories, Gregory Chaitin was enrolled in a computer-programming course being given at Columbia University for bright high-school students. At each lecture the professor would assign the class an exercise requiring writing a program to solve it. The students then competed among themselves to see who could write the shortest program solving the assigned problem. While this spirit of competition undoubtedly added some spice to what were otherwise probably pretty dull programming exercises, Chaitin reports that no one in the class could even begin to think of how to actually prove that the weekly winner's program was really the shortest possible.

Even after the course ended Chaitin continued pondering this shortest-program puzzle, eventually seeing how to relate it to a different question: How can we measure the complexity of a number? Is there any way that we can objectively claim π is more complex than, say, $\sqrt{2}$ or 759? Chaitin's answer to this question ultimately led him to one of the most surprising and startling mathematical results of recent times.

In 1965, now an undergraduate at the City University of New York, Chaitin arrived independently at the same bright idea as Solomonoff: Define the complexity of a number to be the length of the shortest program for a UTM that will cause the machine to print out the number. Since this concept of complexity is closely related to Gödel's results, let me dig a little deeper into the details.

Since the data of any experiment can be encoded as a set of numbers that can be expressed as a string of binary integers, there is no real difference between Solomonoff's idea about the complexity of a scientific theory and Chaitin's (and, independently, the Russian mathematician Andrei Kolmogorov's) regarding the complexity of a number. In a similar fashion, since we saw above that it's possible to code every program for a UTM in the same way that we code its input data, there is also no real difference in talking about complexity for numbers or for programs. In fact, this is the very essence of Chaitin's definition: the complexity of a number equals the complexity of the shortest program that generates that number.

Using these ideas, we can transfer the notion of a random, or incompressible, data string to numbers, coming up with Chaitin's definition of a random number. A number is *random* if there is no program for calculating the number whose length is shorter than the length of the number itself. Or, expressed another way, a number is random if it is maximally complex. Here, of course, we take the length of a number (program) to be the number of bits in the number's (program's) binary expression.

But do random numbers really exist? The surprising fact is that almost all numbers are random! To see why, let's compute the fraction of numbers of length n having complexity less than, say, $n-5$. There are at most $1+2+\ldots+2^{n-5} = 2^{n-4}-1$ programs of length no greater than $n-5$. Consequently, there are at most this many numbers of length n having complexity less than or equal to $n-5$. But there are a total of 2^n numbers of length n. Thus, the proportion of these numbers having complexity no greater than $n-5$ is at most $(2^{n-4}-1)/2^n \leq \frac{1}{16}$. So we see that less than one number in sixteen can be described by a program whose length is at least 5 less than the length of the number. Similarly, less than one number in 500 has a length 10 or more greater than its shortest program, i.e., its complexity. Using this kind of argument and letting $n \to \infty$, it's fairly easy to prove that the set of real numbers with complexity less than their length forms an infinitesimally small subset of the set of all real numbers. In short, almost every number is of maximal complexity, i.e., random. Now let's get back to the problem of shortest programs.

The starting point for Chaitin's remarkable results is the seemingly innocent query: "What is the smallest number that cannot be expressed in words?" This statement seems to pick out a definite number. Let's call it $\mathcal{U}$ for "unnameable." But thinking about things for a moment, we see that there appears to be something fishy about this labeling. On the one hand, we seem to have just described the number $\mathcal{U}$ in words. But $\mathcal{U}$ is supposed to be the first natural number that *cannot* be described in words! This paradox, first suggested to Bertrand Russell by a certain Mr. Berry, a Cambridge University librarian, plays the same role in Chaitin's thinking about the complexity of numbers and programs as the Epimenides Paradox played in Gödel's thinking about the limitations of formal systems.

Recall that to bypass the issue of formalizing truth Gödel had to substitute a related notion, provability, and talk about a statement being unprovable within a given formal system. Similarly, the Berry Paradox contains its own unformalizable notion, the concept of denotation between the terms in its statement and numbers. Part of Chaitin's insight was to see that the way around this obstacle was to shift attention to the phrase, "the smallest number not computable by a program of complexity n." This phrase *can* be formalized, specifying a certain computer program for searching out such a number. What Chaitin discovered was that no program of complexity n can ever produce a number having complexity greater than n. Therefore, the program of complexity n can never halt by outputting the number specified by Chaitin's phrase.

More generally, this result shows that even though there clearly exist numbers of all levels of complexity, it's impossible to prove this fact. That is, given any computer program, there always exist numbers having complexity greater than that program can recognize, i.e., generate. In Chaitin's words, "A ten-pound theory can no more generate a twenty-pound theorem than a one-hundred pound pregnant woman can birth a two-hundred-pound child." Speaking somewhat informally, Chaitin's Theorem says that no program can calculate a number more complex than itself. Here's an outline of the proof of this very fundamental result linking computing, complexity, and information.

Suppose we have a binary string that we suspect of having complexity greater than some fixed level N and we want to prove it. Assume such proofs exist. Then we can use a program of length $\log N + K$ to search for these proofs, since it takes only $\log N$ symbols to represent a number having magnitude N. The quantity K is of fixed size, representing the overhead in the program for things like reading in the number N, communicating with the printer, and so on. With this program we can search through all proofs of length 1, length 2, and so on until we come to the one that proves that the complexity of a specific number is greater than N. When such a proof is found, the program of length $\log N + K$ will have generated a string of complexity greater than N. But there will always be some number N such that N is much larger than $\log N + K$, since K is fixed. Thus, on the one hand we have computed a string of complexity N with a program having a length much shorter than N, while on the other hand we have proved that the string has complexity greater than N, which by definition can only be computed with a program of length greater than N—a contradiction. So we conclude that there is no such proof.

The implication of this result is that, for sufficiently large numbers N (those bigger than $\log N + K$), it cannot be proved that a particular string has complexity greater than N. Or equivalently, there exists a level N such that no number whose binary string is of length greater than N can be proved to be random. Nevertheless, we know that almost every number is maximally complex, hence random. We just can't prove that any *given* number is random. The problem is that in a random sequence each digit carries positive information since it cannot be predicted from its predecessors. Thus, an infinite random sequence contains more information than all our finite human systems of logic put together. Hence, verifying the randomness of such a sequence lies beyond the powers of constructive proof. Or, looking at the problem in another way, in order to write down an "arbitrarily long" string we need to give a general rule for the entries of the string. But then this rule is shorter than suitably large sections of the string, so the string can't be random after all! As one might suspect by now, this result is deeply intertwined with the other decision problems considered earlier.

To make this connection, consider a formal system whose axioms can be expressed in a binary string of length N. Chaitin's Theorem then says that there is a program of size N that does not halt, but we cannot prove that fact within this axiomatic system. On the other hand, this system can allow us to determine exactly which programs of size less than N halt and which do not. So Chaitin's result gives us another solution of the Halting Problem, since it says that there always exist programs for which we cannot determine in advance whether or not they will stop. And, of course, Chaitin's Theorem offers another perspective on Gödel:

> **Gödel's Theorem—Complexity Version**
> *There exist numbers having complexity greater than any theory of mathematics can prove*

Consequently, if we have some theory of mathematics, i.e., a formal system, there always exists a number t such that our theory cannot prove that there are numbers having complexity greater than t. Nevertheless, by "jootsing" we can clearly see that such strings exist. To generate one, simply toss a coin a bit more than t times, writing down a "1" when a head turns up and a "0" for tails.

It's thought-provoking to consider the degree to which Chaitin's result imposes limitations on our knowledge about the world. Rudy Rucker has made the following estimate: Suppose $\mathcal{K}$ represents our best present-day knowledge, while $\mathcal{M}$ denotes a UTM whose reasoning powers equal those of the smartest and cleverest of human beings. Rucker estimates the number t in Chaitin's Theorem as

$$t = \text{complexity } \mathcal{K} + \text{complexity } \mathcal{M} + 1 \text{ billion}$$

where the last term is thrown in to account for the overhead in the program of the machine $\mathcal{M}$. Plugging some plausible numbers for $\mathcal{K}$ and $\mathcal{M}$ into this expression, Rucker concludes that t is less than 3 billion. The bottom line then is that if any worldly phenomenon generates observational data having complexity greater than around 3 billion, no such machine $\mathcal{M}$ (read: human) will be able to prove that there is some short program (i.e., theory) explaining that phenomenon. Thus, recalling René Thom's idea of scientific theories as arbitrariness-reducing tools, Chaitin's work says that our scientific theories are basically powerless to say anything about phenomena whose complexity is much greater than about 3 billion. This does *not* mean that there is no simple explanation for these phenomena. Rather, it says that we will never understand this "simple" explanation—it's too complex for us! Complexity 3 billion represents the outer limits to the powers of human reasoning; beyond that we enter the "twilight zone," where reason and systematic analysis give way to intuition, insight, feelings, hunches, and just plain dumb luck.

7. The Tenth Problem

If asked to name the top ten theorems of all time, just about every mathematician I know would reserve a place somewhere on the list for the Pythagorean Theorem relating the lengths of the sides of a right triangle. If a and b are the lengths of the short sides of such a triangle, c being the length of the hypotenuse, the Pythagorean Theorem says that the equation $a^2 + b^2 = c^2$ linking the quantities a, b, and c always holds. This is an example of an integer-coefficient polynomial equation in three variables a, b and c. Of special mathematical interest are the so-called *Diophantine equations,* polynomial equations with integer coefficients for which we demand integer solutions. Thus the term "Diophantine" refers more to the character of the set of solutions we seek than it does to the equation itself.

The number of solutions of a given polynomial equation may vary from finite to infinite, depending upon whether or not we regard it as a Diophantine equation. For example, the Pythagorean equation has an infinite number of both real and integer solutions. On the other hand, the equation $a^2 + b^2 = 4$ has only the four integer solutions $a = \pm 2$, $b = 0$ or $a = 0$, $b = \pm 2$, but an infinite number of real solutions (e.g., a any real number between $+2$ and -2, $b = \sqrt{4 - a^2}$). So, regarded as a Diophantine equation, this equation has a finite solution set. But thought of as a general polynomial equation, it has an infinite number of solutions. Our concern with Diophantine equations here comes from the surprising connection between the nature of the set of solutions to Diophantine equations and the Halting Problem for a UTM.

In Hilbert's famous lecture to the 1900 International Mathematical Congress in Paris in which he outlined a set of problems for the coming century, the tenth problem on this list involved Diophantine equations. What Hilbert asked for was a general algorithm enabling us to decide whether or not an arbitrary Diophantine equation has any solution. Note carefully that Hilbert did not ask for a procedure to decide if the solution set was infinite, but only for an algorithm to determine if there is *any* solution.

It turns out that there exists an algorithm for listing the set of solutions to any Diophantine equation. So in principle all we have to do to decide if the solution set is empty is to run this program, stopping the listing procedure if no solution turns up. The difficulty is that it might take a very long time (like forever!) to decide whether or not a solution will appear. For instance, the first integer solution of the simple-looking Diophantine equation $x^2 - 991y^2 - 1 = 0$ is

$x = 3795164009068119306380148960 80$, $y = 120557357903313594474 42538767$. This example shows that to solve Hilbert's Tenth Problem we can't rely upon a brute force search for the first solution—a solution might not exist or it might be so large that we'll get tired of looking. In either case, a direct search gives no guarantee of ever coming up with the correct answer about a particular equation's solvability. We need to do something a bit more clever.

In our discussion of Turing machines we introduced the idea of a computable number as one whose digits can be successively calculated by some UTM program. This idea can be extended to sets of integers by saying that a set is computable if, given any integer in the set as input, the program prints a 1 and halts. But if the given number is not in the set, the program prints a 0 and stops. It turns out that this notion is a bit too strong for many purposes, and it's convenient to introduce a weaker version usually called *listability.* We say that a set of integers is *listable* if there is a program that, given any integer as input, prints a 1 and stops if the integer is in the set. But if the integer is not in the set, the program may print a 0 and halt or it may not stop at all. So the difference between a set being computable and being listable is that if the set is listable, the program may or may not halt. But the program always stops when the set is computable. Obviously, computable sets are listable—but not conversely. This distinction forms the basis for an attack mounted on Hilbert's Tenth Problem by the well-known mathematical logician Martin Davis. Let's look at his strategy.

Davis's idea was to prove that for every listable set of integers S, there is a corresponding polynomial $P_S(k, y_1, \ldots, y_n)$ with integer coefficients such that a positive integer k^* belongs to the set S if and only if the solution set of the Diophantine equation $P_S(k^*, y_1, \ldots, y_n) = 0$ is not empty, i.e., the equation has at least one solution in integers. Thus, we see that the solvability or unsolvability of the Diophantine equation $P_S(k, y_1, \ldots, y_n) = 0$ serves as a decision procedure for membership in the listable set S. Here we subscript the polynomial with a small S to indicate that there may be a different polynomial for each listable set S. Davis showed that Hilbert's Tenth Problem can be resolved negatively if such a polynomial can be found for every listable set of integers. But what's the connection between a particular Diophantine equation having a solution and the existence of a general algorithm for the solvability of such equations? The reasons are worth examining in detail, since they begin to reveal for us the interconnection between the Tenth Problem and the Halting Problem.

The logical chain of reasoning underwriting Davis's approach to the Tenth Problem is composed of the following links:

A. Suppose there were a Diophantine decision algorithm of the type that Hilbert wanted, and let S be some listable but not computable set of integers.

B. Then by the assumed existence of the algorithm, there is a Turing machine program (call it $\mathcal{D}$ for Diophantine) that, given the integer k^* as input, halts with output 1 if the Diophantine equation $P_S(k^*, y_1, \dots, y_n) = 0$ has a solution, and halts with output 0 if there is no solution.

C. But the relationship between S and P_S implies that the existence of such a program $\mathcal{D}$ would mean that S is computable, since $\mathcal{D}$ definitely stops with a 0 or a 1 as output.

D. But this contradicts the assumption that the set S is not computable. Hence, no such program $\mathcal{D}$ can exist. That is, there is no algorithm of the sort sought by Hilbert and the Tenth Problem is settled negatively.

Unfortunately, Davis was unable to prove the existence of such a polynomial P_S for every listable set S. However, later work by Davis, Julia Robinson, and Hilary Putnam showed that if there were even one Diophantine equation whose solutions grew at an exponentially-increasing rate in just the right way, then Davis's polynomial P_S would have to exist. In 1970 Yuri Matyasevich, a 22-year-old mathematician at the Steklov Mathematical Institute in Leningrad, found just such a Diophantine equation. Amusingly, Matyasevich made crucial use of the famous Fibonacci sequence of numbers in constructing the long-sought equation, a sequence originally introduced by Leonardo of Pisa in 1202 to explain the explosive growth of a rabbit population in the wild. Evidently, the well-known procreation habits of rabbits gives rise to just the kind of rapid growth Matyasevich needed to resolve negatively yet another of Hilbert's conjectures.

An interesting corollary of Matyasevich's proof is that for any listable set of natural numbers S, there exists some polynomial $P_S(y_1, y_2, \dots, y_n)$ with integer coefficients such that as the variables $y_1, \dots, y_n$ range over the nonnegative integers, the positive values of the polynomial are exactly the set S. To illustrate this curious result, such a polynomial involving 26 variables whose positive values are the set of prime numbers is given in the chapter Notes and References.

By now the reader should be highly sensitized to the connection between negative solutions to decision problems and Gödel's Theorem.

So before continuing our pursuit of the connection between Diophantine equations and the Halting Problem, let's pause to give Gödel his due—again.

Gödel's Theorem—Diophantine Equation Version

There exists a Diophantine equation having no solution—but no theory of mathematics can prove the equation's unsolvability

Buried deep within the Theoretical Physics Division of the IBM Research Laboratories in Yorktown Heights, New York is a broomcloset-sized office, whose spartan furnishings consist of a bare desk, three empty bookshelves, a spotless blackboard, a Monet landscape on the wall, and a computer terminal. After having spent over twenty years as an IBM salesman, systems engineer, and programmer, Gregory Chaitin now calls this office home. And in 1987 it was from these stark surroundings that Chaitin hurled forth a lightning bolt so electrifying that the editors of the *Los Angeles Times* wrote in their June 18, 1988 editorial that "Chaitin's article makes the world shake just a little." And who'd be a better judge of what does and doesn't make the world shake than an Angeleno? But what kind of mathematical result could possibly send the general press into such a state of rapture? As it turns out, nothing less than a proof that the very structure of arithmetic itself is random. As what must surely stand as being about as close to the final word on mathematical truth, proof, and certainty as we'll ever get, let's see how Chaitin managed to extend Gödel's results to come up with such an astounding conclusion.

Suppose we have a UTM and consider the set of all possible programs that can be run on this machine. As we already know, every such program can be labeled by a unique string of 0's and 1's, so it's possible to "name" each program by its own unique positive integer. Consequently, it makes sense to consider listing the programs one after the other and talk about the kth program on the list, where k can be any positive integer. Now consider the question: "If we pick a program from the list at random, what is the likelihood that it will halt when run on the UTM?" It turns out that this question is intimately tied up with the solvability of Diophantine equations, leading ultimately to Chaitin's remarkable result.

The work of Davis, Robinson, and Putnam on Hilbert's Tenth Problem showed that the solvability of decision problems can be expressed as assertions about the solvability of certain Diophantine equations. In particular, this work showed that there exists a Diophantine equation $P(k, y_1, \ldots, y_n) = 0$ that has a solution if and only if the kth computer program halts when run on a UTM.

The key step in Chaitin's route to ultimate randomness was to consider not whether a Diophantine equation has *some* solution or not, but the sharper question of whether the equation has an infinite or a finite number of solutions. The reason for asking this more detailed question is that the answers to the original query are not logically independent for different values of k. In other words, if we know whether or not a solution exists for a particular value of k, this information can be used to infer the answer to the question for some other values of k. But if we ask whether or not there are an infinite number of solutions, the answers are logically independent for each value of k; knowledge of the finiteness or not of the solution set for one value of k gives no information at all about the answer to the same question for another value.

Following this reformulation of the basic question, Chaitin's next step was a real *tour de force.* He proceeded to construct explicitly a Diophantine equation involving an integer parameter k, together with over 17,000 additional variables. Let's call this Diophantine equation $\chi(k, y_1, y_2, \ldots, y_{17,000+}) = 0$, using the Greek symbol χ in Chaitin's honor. From this equation we can form a binary number consisting of an infinite string of 0's and 1's in the following manner: Let k run through the values $k = 1, 2, \ldots$, setting the kth entry in our string to 1 if the equation $\chi = 0$ has an infinite number of solutions for that value of k, and setting it to 0 if the equation has a finite number of solutions (including no solution). Chaitin labeled the number created in this way by the last letter in the Greek alphabet Ω. And for good reason, too, as the properties of Ω show that it's about as good an approximation to "The End" as the human mind will ever make.

Chaitin showed that the quantity Ω is an uncomputable number. Furthermore, he proved that any N-bit formal system (read: program of length/complexity N) can yield at most a finite number N of the binary digits of Ω. Consequently Ω is random, since there is no program shorter than Ω itself for producing all of its digits, and the digits of Ω are logically and statistically independent. Finally, if we put a decimal point in front of Ω, it represents some decimal number between 0 and 1. And, in fact, when viewed this way Ω can be interpreted as the probability that the UTM will halt if we present it with a randomly-selected program. Indeed, Chaitin's equation was constructed precisely so that Ω would turn out to be this halting probability.

So while Turing considered the question of whether a given program would halt, Chaitin's extension produces the probability that a randomly chosen program will stop. As an aside, it's worth noting that the two extremes Ω equals zero or one cannot occur, since the first case would mean that no program ever halts, while the second would say that every program will halt. The trivial, but admissible, program "STOP" deals with the first case, while I'll leave it to the reader to construct an equally primitive program to deal with the second.

But the real bombshell, the one that shook up the *Los Angeles Times*'s editorial staff, is that the structure and properties of Ω show that arithmetic is fundamentally random. To see why, take the integer k to be finite, but "sufficiently large." For example, let k be greater than the Busy Beaver function value $BB(12)$ considered earlier. For values of k larger than this, there is no way to determine whether the kth entry of Ω is 0 or 1. And there are an infinite number of such undecidable entries. Yet each such entry corresponds to a simple, definite arithmetical fact: for that value of k, either Chaitin's Diophantine equation $\chi = 0$ has a finite or an infinite number of solutions. But as far as human reasoning goes, which of the two possibilities is actually the case may as well be decided by flipping a coin—it is completely undecidable, hence effectively random. So Chaitin's work shows that there are an infinite number of arithmetic questions with definite answers that cannot be found using any formal axiomatic procedures. The answers to these questions are uncomputable and are not reducible to other mathematical facts. Extending Einstein's famous aphorism about God, dice, and the universe, Chaitin describes the situation by saying that, "God not only plays dice in quantum mechanics, but even with the whole numbers." It's fitting to conclude this section with our final tribute to Gödel:

Gödel's Theorem—Dice-Throwing Version

There exists an uncomputable number Ω whose digits correspond to an infinite number of effectively random arithmetic facts

8. Truth and Proof

If there's any message for mankind at all in the results of Gödel, Turing, and Chaitin it's that there is a forever unbridgeable gap between what's true and what can be proved. So where does chaos and strange attractors fit into the overall scheme of things? My contention is that the existence of chaotic dynamical processes forms a natural link between Chaitin's complexity results and Gödel's Incompleteness Theorem, and that the existence of a rich variety of real-world truths that we can know for sure depends in an essential way upon the existence of such attractors. Here's why.

We have already seen that the theorems of a formal system, the output of a UTM, and the attractor set of a dynamical process (e.g., a 1-dimensional cellular automaton) are completely equivalent; given one, it can be faithfully translated into either of the others. But the idea of a provable real-world truth coincides with the decoding of a theorem in a formal system. Therefore, let's use $\mathcal{T}$ to represent the universe of true statements, while letting $\mathcal{P}$ denote the set of theorems provable in some formal system. Of course, Gödel's Theorem just states that $\mathcal{P} \subset \mathcal{T}$.

From the discussion on complexity, it's clear that there exist computable numbers of arbitrary complexity. But these computable quantites correspond to the attractor of some cellular automaton. Thus, since there are an infinite number of such computable strings, there must exist cellular automata whose attractor set is infinite. But fixed points and limit cycles are both finite attractors. Hence, there must exist something "larger." But Chaitin's Theorem tells us that the attractor set must be smaller than the whole state manifold M, since it asserts that there are strings that can never be computed. In short, Chaitin's Theorem implies the existence of some kind of cellular automaton attractor beyond Types A and B, i.e., a strange attractor.

Now let's assume that such strange attractors exist. Since they do not fill up the entire manifold M, there must exist states that cannot be attained from any given initial state. But from the equivalence of formal systems and cellular automata, this is just another way of saying that $\mathcal{P} \subset \mathcal{T}$, i.e., Gödel's Theorem. Putting these two sets of arguments together, we find the implications

$$\begin{array}{c}\text{Chatin's Theorem}\\ \Downarrow \\ \text{strange attractors (i.e., chaos)}\\ \Downarrow \\ \text{Gödel's Theorem}\end{array}$$

As the *pièce de résistance* of our tour, we come to the perhaps surprising fact that chaos implies truth, in the sense that a world without chaos would be very impoverished in the number of mathematical theorems that could be proved. And this, in turn, would imply that whatever real-world truths might exist, the overwhelming majority of them could not be formally proved. Of course, from this standpoint we might already be living in such a world. But at least the existence of strange attractors gives us hope that the gap between proof and truth can continue to be narrowed—even if it can never be closed.

Notes and References

Perhaps not surprisingly, compared to physics, biology, and the other sciences, there have been relatively few attempts by research mathematicians to explain the meaning, concerns, and the methods of mathematics to the general reader. Two outstanding efforts in this direction that I can recommend unreservedly are the volumes

Davis, P. and R. Hersh. *The Mathematical Experience.* Boston: Birkhäuser, 1980.

Stewart, I. *The Problems of Mathematics.* New York: Oxford University Press, 1987.

To the best of my knowledge, the first account of Gödel's Theorem written expressly for the general reader, and still one of the best, is the short volume

Nagel, E. and J. R. Newman. *Gödel's Proof.* New York: New York University Press, 1958.

This work notwithstanding, Gödel's results remained more or less buried in academic obscurity until the appearance of the Pulitzer Prize-winning account of "The Theorem" given in

Hofstadter, D. *Gödel, Escher, Bach: An Eternal Golden Braid.* New York: Basic Books, 1979.

§ Chaos and Praxis

Two excellent books for source material on chaos are the collections

Hao, Bai-Lin. *Chaos.* Singapore: World Scientific, 1984.

Cvitanović, P. *Universality in Chaos.* Bristol, UK: Adam Hilger, 1984.

James Gleick's runaway bestseller that brought the chaotic word to the masses is

Gleick, J. *Chaos.* Viking: New York, 1987.

§ A Strangeness in the Attraction

The basic elements of dynamical systems can be found in treatments at almost every level of mathematical development. Volumes that I've found particularly helpful are

Hirsch, M. and S. Smale. *Differential Equations, Dynamical Systems, and Linear Algebra.* New York: Academic Press, 1974.

Irwin, M. *Smooth Dynamical Systems.* New York: Academic Press, 1980.

Guckenheimer, J. and P. Holmes. *Nonlinear Oscillations, Dynamical Systems, and Bifurcation of Vector Fields.* New York: Springer, 1983.

Following the pioneering work on cellular automata by von Neumann, Ulam and others, there was a long hiatus before the field really took off again. This time lag was due to several factors, not the least of which was the lack of ready availability of the cheap computing power needed to really dig into the behavior of these dynamical devils. The modern era of cellular automata research can probably be traced to the following foundational paper by Stephen Wolfram:

Wolfram, S. "Statistical Mechanics of Cellular Automata." *Reviews of Modern Physics,* 55 (1983), 601–644.

Much additional information and many subsequent developments can be found in the collections

Theory and Application of Cellular Automata, S. Wolfram, ed. Singapore: World Scientific, 1986.

Cellular Automata, D. Farmer, T. Toffoli, and S. Wolfram, eds. Amsterdam: North-Holland, 1984.

§ Speaking Formally

A simple, easy-to-understand introduction to formal systems is given in Hofstadter's treatise cited earlier, as well as in

Levine, H. and H. Rheingold. *The Cognitive Connection.* New York: Prentice Hall Press, 1987.

For a more technical account emphasizing the connections between formal systems and languages, see the book

Moll, R., M. Arbib, and A. Kfoury. *An Introduction to Formal Language Theory.* New York: Springer, 1988.

The star-maltese cross-cloud system introduced in the text is based on the MIU-system originally presented in Hofstadter's treatise on Gödel. For a proof of the decision procedure given for this system, as well as for a discussion of some other results related to it, see the article

Swanson, L. and R. McEliece, "A Simple Decision Procedure for Hofstadter's *MIU*-System." *Mathematical Intelligencer,* 10, No. 2 (1988), 48–49.

§ The Undecidable

An English translation of Gödel's pioneering paper, as well an enlightening biographical account of his life, can be found in the first volume of Gödel's collected works.

Kurt Gödel: Collected Works, Volume 1, S. Feferman, et al., eds. New York: Oxford University Press, 1986.

An assessment of Gödel's Theorem from both a philosophical as well as mathematical point of view is contained in the collection of reprints

Gödel's Theorem in Focus, S. Shanker, ed. London: Croom and Helm, 1988.

People often wonder whether or not long-standing, seemingly intractable mathematical questions like Goldbach's Conjecture (every even number is the sum of two primes) are undecidable in the same way that Cantor's Continuum Hypothesis turned out to be undecidable. Musings of this sort give rise to the consideration of whether or not Gödel's results really matter to mathematics, in the sense that there are important mathematical questions that are truly undecidable. With the recent work of Chaitin and others, the comforting belief that there are no such problems seems a lot less comforting than it used to. For a discussion of some other "real" mathematical queries that are genuinely undecidable, see

Kolata, G. "Does Gödel's Theorem Matter to Mathematics?" *Science,* 218, (19 November 1982), 779–780.

Many details of Gödel's personality, views on life and philosophy, as well as an assessment of both his mathematical and philosophical work, are found in the following book written by the well-known mathematical logician Hao Wang, who was a long-time acquaintance of Gödel's:

Wang, H. *Reflections on Kurt Gödel.* Cambridge, MA: MIT Press, 1987.

Additional information about Gödel's life is given in

Dawson, J. "Kurt Gödel in Sharper Focus." *Mathematical Intelligencer,* 6, No. 4 (1984), 9–17.

Kreisel, G. "Kurt Gödel: 1906–1978." *Biographical Memoirs of Fellows of the Royal Society,* 26 (1980), 148–224.

The text discussion of "mirroring" and Gödel numbering follows that given in the Nagel and Newman book noted above. Hofstadter's switching-yard metaphor for Gödel numbering and transformations in a formal system can be found in the expository paper

Hofstadter, D. "Analogies and Metaphors to Explain Gödel's Theorem." *College Mathematics Journal,* 13 (March 1982), 98–114.

The discussion by Rudy Rucker likening Gödel's Theorem to the plight of Joseph K. in his frustrated wanderings through Kafka's *The Trial,* may be found in the very enlightening, but slightly technical, book

Rucker, R. *Infinity and the Mind.* Boston: Birkhäuser, 1982.

In connection with Gödel as a person, this book is especially recommended for its account of several meetings that Rucker had with Gödel in the years shortly before Gödel's death in January 1978.

§ Turing Around

An excellent introductory account of the circle of problems surrounding computation, formal systems, Turing machines, the Halting Problem, Gödel's Theorem, complexity, and Hilbert's Tenth Problem is available in the article

Davis, M. "What Is a Computation?" in *Mathematics Today: Twelve Informal Essays,* L. A. Steen, ed. New York: Springer, 1978, pp. 241–267.

For a general-readership development of the idea and workings of a Turing machine, see

Hoffman, P. *Archimedes' Revenge.* New York: Norton, 1988.

Rucker, R. *Mind Tools.* Boston: Houghton-Mifflin, 1987.

More technical accounts of Turing machines and their connections with not only decision problems but also languages, the reader is directed to the textbook

Davis, M. and E. Weyuker. *Computability, Complexity, and Languages.* Orlando, FL: Academic Press, 1983.

A very stimulating collection of essays reviewing current knowledge about Turing machines and their many implications and ramifications in other areas is presented in

The Universal Turing Machine, R. Herken, ed. Oxford: Oxford University Press, 1988.

The Turing machine program given in Table 4 enables the machine to decide if the string of 1's and 2's given on the input tape read the same forward and backward, i.e., if the string constitutes a *palindrome.*

For more details on the construction and operation of a UTM, the reader should see either of the Rucker books *Infinity and the Mind* or *Mind Tools* referenced above.

The Turing-Church Thesis lies at the heart of the currently fashionable artificial intelligence debate, which revolves about the question of whether or not a computer can think like a human being. *If* human thought processes can be shown to all be "effective," and *if* the Turing-Church Thesis is correct, then it necessarily follows that there is no barrier, in principle at least, between the "thought processes" of machines and those of humans. But both of these "ifs" are very big ifs indeed, and no one has yet been able to give a knockdown argument resolving either half of this conundrum. For an account of the current state of play, as well as for an extensive bibliography on the whole issue, see Chapter Five of

Casti, J. *Paradigms Lost: Images of Man in the Mirror of Science.* New York: Morrow, 1989.

A much more technical, philosophically-oriented approach to the implications of the Turing-Church Thesis for both psychology and the philosophy of mathematics is presented in

Webb, J. *Mechanism, Mentalism, and Metamathematics.* Dordrecht, Holland: Reidel, 1980.

In this same connection, see also

Arbib, M. *Brains, Machines, and Mathematics,* 2nd Edition. New York: Springer, 1987.

Penrose, R. *The Emperor's New Mind.* Oxford: Oxford University Press, 1989.

The Busy Beaver Game was dreamed up by Tibor Rado of Ohio State University in the early 1960s. A compact, introductory discussion of what's currently known about this problem and about the Busy Beaver function can be found in the articles

Dewdney, A. " Busy Beavers" in *The Armchair Universe.* New York: Freeman, 1988, pp. 160–171.

Brady, A. "The Busy Beaver Game and the Meaning of Life," in *The Universal Turing Machine,* R. Herken, ed. Oxford: Oxford University Press, 1988, pp. 259–277.

In 1973 Bruno Weimann discovered that the four-state Busy Beaver can write 13 ones on the tape before halting. Thus, $BB(4) = 13$. So far no one knows the value $BB(5)$, although in 1984 George Uhing showed that $BB(5) \geq 1,915$. The program for this remarkable result is

	Symbol Read	
State	0	1
A	1, R, B	1, L, C
B	0, L, A	0, L, D
C	1, L, A	1, L, STOP
D	1, L, B	1, R, E
E	0. R, D	0, R, B

Uhing's 5-State Turing Machine Program for the Busy Beaver Game

Here is a slick proof due to Ian Stewart showing the unsolvability of the Halting Problem: Suppose such a Halting Algorithm exists and let d be the input data. Consider the following UTM program:

1 Check to see if d is the code for a UTM program P. If not, go back to the start and repeat.
2. If d is the code for a program P, double the input string to get $d \cdot d$.
3. Use the assumed Halting Algorithm for the UTM with input data $d \cdot d$. If it stops, go back to the beginning of this step and repeat.
4. Otherwise, halt.

Call the above program H. Now since H is a program it has its own code h. Thus, we can ask, "Does H halt for input h?" It surely gets past step 1, since by definition h is the code for the program H. And H gets past step 3, as well, if and only if the UTM doesn't halt with input $h \cdot h$. Thus we conclude that H halts with input data h if and only if the UTM does not halt with input data $h \cdot h$. But the UTM simulates a program P by starting with the input data $P \cdot d$, and then behaving just like P operating on input data d. Therefore, we see that P halts with input data d if and only if the UTM halts with input data $P \cdot d$. So if we put $P = H$ and $d = h$, then we find that H halts with input data h if and only if the UTM halts with input data $h \cdot h$—a direct contradiction of the result obtained a moment ago. Thus we conclude that there is no such Halting Algorithm. This proof, along with much, much more about the state of modern mathematics, can be found in Stewart's book on the problems of modern mathematics noted earlier under General References.

§ The Importance of Being Arbitrary

The citation for Solomonoff's original paper on the complexity of scientific theories is

Solomonoff, R. "A Formal Theory of Inductive Inference." *Information and Control,* 7 (1964), 224–254.

The complete story of Chaitin's independent discovery of algorithmic complexity and its connection with randomness is contained in his collection of papers

Chaitin, G. *Information, Randomness, and Incompleteness,* 2nd Edition. Singapore: World Scientific, 1990.

Quite independently of both Chaitin and Solomonoff, the famous Russian mathematician Andrei Kolmogorov also hit upon the idea of defining the randomness of a number by the length of the shortest computer program required to calculate it. His ideas were presented in the classic paper

Kolmogorov, A. "Three Approaches to the Quantitative Definition of Information." *Problems in Information Transmission,* 1 (1965), 3–11.

The original formulation of Berry's Paradox involved a statement like: "The smallest number that cannot be expressed in fewer than thirteen words." Since the preceding phrase contains twelve words, the paradox follows for exactly the same reasons as given for the more general phrase used in the text. A fairly complete account of the Berry Paradox and its relationship to complexity and Gödelian logic is available in the Rucker book *Infinity and the Mind* already noted. This volume also contains the background assumptions underpinning Rucker's claim that 3 billion is an upper limit to the complexity of phenomena that the human mind will ever be able to rationally encompass and comprehend. An account that differs somewhat from Rucker's, showing that the estimate should probably be more like 17 or 18 billion is found in

Casti, J. *Searching for Certainty.* New York: Morrow, 1991.

§ The Tenth Problem

A very easy-to-understand, illuminating discussion of Hilbert's Tenth Problem is given in Chapter Six of

Devlin, K. *Mathematics: The New Golden Age.* London: Penguin, 1988.

A somewhat more technical account is presented in the Davis article "What is a Computation?" noted above, as well as in the volume

Salomaa, A. *Computation and Automata.* Cambridge: Cambridge University Press, 1985.

Each of the foregoing sources also gives a good account of Matyasevich's resolution of the problem.

An odd byproduct of the work on the Tenth Problem is the following polynomial equation in the 26 letters of the alphabet, whose positive values are the set of prime numbers:

$$\begin{aligned} P(a,b,\dots,z) = (k+2)\{1 &- [wz+h+j-q]^2 \\ &- [(gk+2g+k+1)(h+j)+h-z]^2 \\ &- [2n+p+q+z-e]^2 \\ &- [16(k+1)^3(k+2)(n+1)^2+1-f^2]^2 \\ &- [e^3(e+2)(a+1)^2+1-o^2]^2 \\ &- [(a^2-1)y^2+1-x^2]^2 - [16r^2y^4(a^2-1)+1-u^2]^2 \\ &- [((a+u^2(u^2-a))^2-1)(n+4dy)^2+1-(x+cu)^2]^2 \\ &- [n+l+v-y]^2 - [(a^2-1)l^2+1-m^2]^2 \\ &- [ai+k+1-l-i]^2 \\ &- [p+l(a-n-1)+b(2an+2a-n^2-2n-2)-m]^2 \\ &- [q+y(a-p-1)+s(2ap+2a-p^2-2p-2)-x]^2 \\ &- [z+pl(a-p)+t(2ap-p^2-1)-pm]^2\} \end{aligned}$$

As the letters a through z run through all the integers, the polynomial P takes on positive and negative integer values. The positive values are exactly the set of prime numbers; the negative values may or may not be the negatives of primes. Incidentally, the reader will note that the expression for P is given in terms of two factors, seeming to contradict the definition of a prime number as one that has no factors other than itself and 1. The apparent contradiction is resolved by noting that the formula produces only positive values when the factor $(k+2)$ is a prime and the second factor equals 1. This polynomial for primes was first published by James Jones, Daihachiro Sato, Hideo Wada, and Douglas Wiens in 1977.

An introductory account of Chaitin's fabulous Diophantine equation straight from the horse's mouth, so to speak, is found in

Chaitin, G. "Randomness in Arithmetic." *Scientific American*, 259 (July 1988), 80–85.

Creation of Chaitin's "monster" equation followed the flowchart below, involving the creation of a sequence of machine-language and LISP programs:

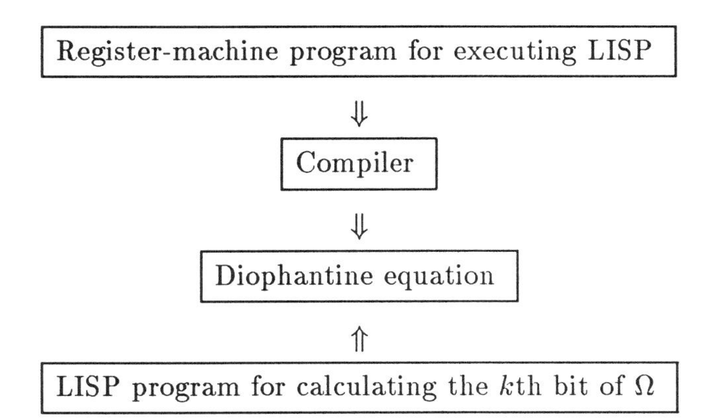

The technical details surrounding this monumental intellectual and programming effort are given in the book

Chaitin, G. *Algorithmic Information Theory.* Cambridge: Cambridge University Press, 1987.

INDEX

D

E